Introduction to Languages, Machines and Logic

Springer
London
Berlin
Heidelberg
New York
Barcelona
Hong Kong
Milan
Paris
Singapore
Tokyo

Alan P. Parkes

Introduction to Languages, Machines and Logic

Computable Languages, Abstract Machines and Formal Logic

Springer

Alan P. Parkes, BSc, PhD
Distributed Multimedia Research Group, Computing Department,
Lancaster University, Lancaster, LA1 4YR

British Library Cataloguing in Publication Data
Parkes, Alan
    Introduction to languages, machines and logic: computable languages, abstract
    machines and formal logic
    1. Machine theory
    I. Title
    511.3

    ISBN 1852334649

Library of Congress Cataloging-in-Publication Data
Parkes, Alan
    Introduction to languages, machines and logic: computable languages, abstract
    machines and formal logic/Alan P. Parkes.
        p. cm.
    Includes index.
    ISBN 1-85233-464-9 (alk. paper)
    1. Formal languages   2. Machine theory.   I. Title.

    QA267.3.P37 2001
    511.3–dc21   2001054282

ISBN 1-85233-464-9 Springer-Verlag London Berlin Heidelberg
a member of BertelsmannSpringer Science+Business Media GmbH
http://www.springer.co.uk

© Springer-Verlag London Limited 2002
Printed in Great Britain
2nd printing 2003

Typesetting: Gray Publishing, Tunbridge Wells, Kent
Printed and bound at the Athenæum Press Ltd., Gateshead, Tyne and Wear
34/3830-54321   Printed on acid-free paper   SPIN 10935873

Contents

Part 2 Machines and Computation

Part 3 Computation and Logic

Chapter 1
Introduction

1.1 Overview

This chapter briefly describes:

- what this book is about
- what this book tries to do
- what this book tries not to do
- a useful feature of the book: the exercises.

1.2 What This Book Is About

This book is about three key topics of computer science, namely *computable languages*, *abstract machines*, and *logic*.

Computable languages are related to what are usually known as "formal languages". I avoid using the latter phrase here because later on in the book I distinguish between *formal* languages and *computable* languages. In fact, computable languages are a special type of formal languages that can be processed, in ways considered in this book, by computers, or rather *abstract machines* that *represent* computers.

Abstract machines are formal computing devices that we use to investigate properties of *real* computing devices. The term that is sometimes used to describe abstract machines is *automata*, but that sounds too much like real machines, in particular the type of machines we call *robots*.

The logic part of the book considers using different types of formal logic to represent things and reason about them. The logics we consider all play a very important role in computing. They are *Boolean* logic, *propositional* logic, and *first order predicate* logic (FOPL).

This book assumes that you are a layperson, in terms of computer *science*. If you are a computer science undergraduate, as you might be if you are reading this book, you may by now have written many programs. So, in this introduction we will draw on your experience of programming to illustrate some of the issues related to formal languages that are introduced in this book.

The programs that you write are written in a formal language. They are expressed in text which is presented as input to another program, a *compiler* or perhaps an *interpreter*. To the compiler, your program text represents a *code*, which the compiler knows exactly how to handle, as the rules by which that code is constructed are completely specified inside the compiler. The type of language in which we write our programs (the *programming language*), has such a well-defined syntax (rules for forming acceptable programs) that a machine can decide whether or not the program you enter has the right even to be considered as a proper program. This is only part of the task of a compiler, but it is the part in which we are most interested.

For whoever wrote the compiler, and whoever uses it, it is very important that the compiler does its job properly in terms of deciding that your program is syntactically valid. The compiler writer would also like to be sure that the compiler will always *reject* syntactically *in*valid programs as well as accepting those that are syntactically valid. Finally, the compiler writer would like to know that his or her compiler is not wasteful in terms of precious resources: if the compiler is more complex than it needs to be, if it carries out many tasks that are actually unnecessary, and so on. In this book we see that the solutions to such problems depend on the type of language being considered.

Now, because your program is written in a formal language that is such that another program can decide if your program *is* a program, the programming language is a computable language. A *computable* language then, is a *formal* language that is such that a computer can understand its syntax. Note that we have not discussed what your program will actually *do*, when it is run: the compiler does not really understand that at all. The compiler is just as happy to compile a program that does not do what you intended as it is to compile one that does (as you will know, if you have ever done any programming).

The book is in three parts. Part 1: Languages and Machines is concerned with the relationship between different types of formal languages and different types of theoretical machines (abstract machines) that can serve to process those languages, in ways we consider later. So, the book is not really *directly* concerned with *programming* languages. However, much of the material in the first part of the book is highly relevant to programming languages, especially in terms of providing a useful theoretical background for compiler writing, and even programming language design. For that reason, some of the examples used in the book are from programming languages (particularly Pascal in fact, but you need not be familiar with the language to appreciate the examples).

In Part 1, we study four different types of formal languages, and see that each of these types of formal language is associated with a particular type of abstract machine. The four types of languages we consider were actually defined by the American linguist Noam Chomsky, and the classification of formal languages he defined has come to be called the *Chomsky hierarchy*. It *is* a hierarchy, since it defines a general type of language (type 0), then a restricted version of that general type (type 1), then a restricted version of *that* type (type 2), and then finally the most restricted type of all (type 3).

The types of language considered are very simple to define, and even the most complex of abstract machines is also quite simple. What is more, all of the types of abstract machine we consider in this book can be represented in diagrammatic form. The most complex abstract machine we look at is called the Turing machine (TM), after Alan Turing, the mathematician who designed it. The TM is, in some sense, the most powerful computational device possible, as you will see in Part 2 of the book.

The chapters of Part 1 are shown in Figure 1.1.

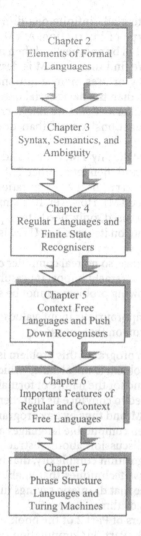

Figure 1.1. The chapters of Part 1: Languages and Machines

By the end of Part 1 of the book you should have an intuitive appreciation of computable languages. Part 2: Machines and Computation investigates computation in a wider sense. We see that we can discuss the types of computation carried out by *real* computers in terms of our abstract machines. We see that a machine, called a finite state transducer, which appears to share some properties with real computers, is not really a suitable general model of real computers. Part of the reason for this is that this machine cannot do multiplication or division. We see that the TM is capable of multiplication and division, and any other tasks that real computers can do. We even see that the TM can run programs. What is more, since the TM effectively has unlimited storage capacity, it turns out to be *more* powerful than any real computer could ever be.

One interesting property of TMs is that we cannot make them any more powerful than they are by adding extra computational facilities. Nevertheless, TMs use only one data structure, a potentially infinite one-dimensional array of symbols, called a tape, and only one type of instruction. A TM can simply read a symbol from its tape, replace that symbol by another, and then move to the next symbol on the right or left in the tape. We find that if a TM is designed so that it carries out many processes simultaneously, or uses many additional tapes, it cannot perform any more computational tasks than the basic serial one-tape version. In fact, it has been established that no other formalisms we might create for representing computation give us any more functional power than does a TM. This assertion is known as *Turing's thesis*.

We see that we can actually take any TM, code it and its data structure (tape) as a sequence of zeros and ones, and then let another TM run the original TM as if it were a program. This TM can carry out the computation described by any TM. We thus call it the universal TM (UTM). However, it is simply a standard TM, which, because of the coding scheme used, only needs to expect either a zero or a one every time it examines a symbol on its tape. The UTM is an abstract counterpart of the real computer.

The UTM has unlimited storage, so no real computer can exceed its power. We use this fact as the basis of some extremely important results of computer science, one of which is to see that the following problem cannot be solved, in the general case:

> Given any program, and appropriate input values to that program, will that program terminate when run on that input?

In terms of TMs, rather than programs, this problem is known as the *halting problem*. We see that the halting problem is unsolvable,[1] which has implications both for the real world of computing, and for the nature of formal languages. We also see that the halting problem can be used to convince us that there are formal languages that cannot be processed by any TM, and thus by any program. This enables us to finally define the relationship between computable languages and abstract machines.

At the end of Part 2, our discussion about abstract machines leads us on to a topic of computer science, algorithm complexity, that is very relevant to programming. In particular, we discuss techniques that enable us to predict the running time of algorithms, and we see that dramatic savings in running time can be made by certain cleverly defined algorithms.

Figure 1.2 shows the chapters of Part 2 of the book.

The final part of the book, Part 3: Computation and Logic considers formal logical systems as tools for representation, reasoning and computation. We first consider the system known as *Boolean logic*. This form of logic is important because it forms the foundation of digital computer circuitry. We show how Boolean logic can be used to solve problems, and we introduce the *truth table* as a tool for checking the validity of logical statements.

In the same chapter, we consider *propositional logic*. This logic includes the basic operators of Boolean logic, along with an additional operator that represents implication ("if x is true then so is y"). We consider logical rules of inference, which enable us to solve problems by applying the rules to a propositional representation of the problem. We discover that propositional logic is not sufficiently powerful to represent certain things, which leads us on to what is known as FOPL.

FOPL occupies the final two chapters of the book. We see how FOPL can be used to represent and reason about properties that apply to all objects in a domain, and properties that apply to one or more objects in that domain. We consider what is called *classical* FOPL reasoning, when we represent a domain as a set of FOPL

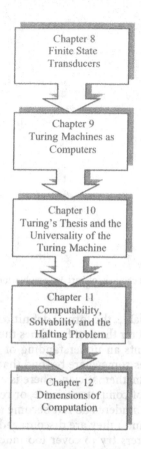

Figure 1.2. The chapters of Part 2: Machines and Computation

statements and then use rules of inference to attempt to derive a statement we believe to be true, and are thus trying to prove. In the final chapter we consider a computational technique for reasoning in FOPL known as *resolution*. We also relate the discussion of resolution to the real world of programming, by briefly considering the logic programming language PROLOG.

Throughout Part 3, we relate our discussions to issues from the first two parts of the book. We discuss the linguistic nature of logic and the time and space requirements of logical reasoning. We end the book by discussing the relationship between two key results of twentieth century mathematics (Gödel's theorem and Turing's thesis) and their implications for computer science.

Figure 1.3 shows the titles of the chapters of Part 3.

1.3 What This Book Tries to Do

This book attempts to present formal computer science in a way that you, the student, can understand. This book does not assume that you are a mathematician. It assumes that you are not particularly familiar with formal notation, set theory, and so on. As far as possible, excessive formal notation is avoided. When formal

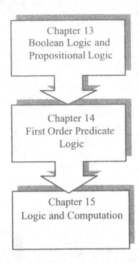

Figure 1.3. The chapters of Part 3: Computation and Logic

notation or jargon is unavoidable, the formal definitions are usually accompanied by short explanations, at least for the first few times they appear.

Overall, this book represents an understanding of formal languages, abstract machines and logic which lies somewhere beyond that of a layperson, but is considerably less than that of a mathematician. There is a certain core of material in the subject that any student of computer science, or related disciplines, should be aware of. Nevertheless, many students do not become aware of this important and often useful material. Sometimes they are discouraged by the way it is presented. Sometimes, books and lecturers try to cover too much material, and the fundamentals get lost along the way.

Above all, this book tries to highlight the connections between what might initially appear to be distinct topics within computer science.

1.4 What This Book Tries Not to Do

This book tries not to be too formal. References to a small number of formal books are included in the section "Further Reading", at the end of the book. In these books you will find many theorems, usually proved conclusively by logical proof techniques. The "proofs" appearing in this book are included usually to establish something absolutely central, or that we need to assume for subsequent discussion. Where possible, such "proofs" are presented in an intuitive way.

This is not to say that proofs are unimportant. I assume that, like me, you are the type of person who has convinced themselves that something is right before you really accept it. However, the proofs in this book are presented in a different way from the way that proofs are usually presented. Many results are established "beyond a reasonable doubt" by presenting particular examples and encouraging the reader to appreciate ways in which the examples can be generalised. Of course, this is not really sound logic, as it does not argue throughout in terms of the general case. Some of the end of chapter exercises present an opportunity to practice or complete proofs. Such exercises often include hints, and/or sample answers.

1.5 The Exercises

At the end of each of Chapters 2–15, you will find a small number of exercises to test and develop your knowledge of the material covered in that chapter. Most of the exercises are of the "pencil and paper" type, though some of them are medium scale programming problems. Any exercise marked with a dagger (†) has a sample solution in the "Solutions to Selected Exercises" section near the end of the book. You do not need to attempt any of the exercises to fully understand the book. However, although the book attempts to make the subject matter as informal as possible, in one very important respect it is very much like maths: *you need to practice applying the knowledge and skills you learn or you do not retain them.*

Finally, some of the exercises give you an opportunity to investigate additional material that is not covered in the chapters themselves.

1.6 Further Reading

A small section called "Further Reading" appears towards the end of the book. This is not meant to be an exhaustive list of reading material. There are many other books on formal computer science than are cited here. The further reading list also refers to books concerned with other fields of computer science (for example, computer networks) where certain of the formal techniques in this book have been applied.

Brief notes accompany each title cited.

1.7 Some Advice

Most of the material in this book is very straightforward, though some requires a little thought the first time it is encountered. Students of limited formal mathematical ability should find most of the subject matter of the book reasonably accessible. You should use the opportunity to practice provided by the exercises, if possible. If you find a section really difficult, ignore it and go on to the next. You will probably find that an appreciation of the overall result of a section will enable you to follow the subsequent material. Sections you omit on first reading may become more comprehensible when studied again later.

You should not allow yourself to be put off if you cannot see immediate applications of the subject matter. There have been many applications of grammar, abstract machines and logic in computing and related disciplines, some of which are referred to by books in the "Further Reading" section.

This book should be interesting and relevant to any intelligent reader who has an interest in Computer science and approaches the subject matter with an open mind. Such a reader may then see the subject of languages, machines and logic as an explanation of the simple yet powerful and profound abstract computational processes beneath the surface of the digital computer.

Notes

1. You have probably realised that the halting problem is solvable in certain cases.

Languages and Machines

Chapter 2
Elements of Formal Languages

2.1 Overview

In this chapter, you learn about:

- the building blocks of formal languages: *alphabets* and *strings*
- *grammars* and *languages*
- a way of classifying grammars and languages: the *Chomsky hierarchy*
- how formal languages relate to the definition of programming languages.

... and you are introduced to:

- writing definitions of sets of strings
- producing *sentences* from grammars
- using the notation of formal languages.

2.2 Alphabets

An alphabet is a finite collection (or set) of symbols. The symbols in the alphabet are entities which cannot be taken apart in any meaningful way, a property which leads to them being sometimes referred to as *atomic*. The symbols of an alphabet are simply the "characters", from which we build our "words". As already said, an alphabet is *finite*. That means we could define a program that would print out its elements (or members) one by one, and (this last part is very important) the program would terminate sometime, having printed out each and every element.

For example, the small letters you use to form words of your own language (for example, English) could be regarded as an alphabet, in the formal sense, if written down as follows:

$$\{a, b, c, d, e, ..., x, y, z\}.$$

The digits of the (base 10) number system we use can also be presented as an alphabet:

$$\{0, 1, 2, 3, 4, 5, 6, 7, 8, 9\}.$$

2.3 Strings

A string is a finite sequence of zero or more symbols taken from a formal alphabet. We write down strings just as we write the words of this sentence, so the word "strings" itself could be regarded as a *string* taken from the alphabet of letters, above. Mathematicians sometimes say that a string taken from a given alphabet is a string *over* that alphabet, but we will say that the string is *taken from* the alphabet. Let us consider some more examples. The string *abc* is one of the many strings which can be taken from the alphabet $\{a, b, c, d\}$. So is *aabacab*. Note that duplicate symbols are allowed in strings (unlike in sets). If there are no symbols in a string it is called the *empty string*, and we write it as ε (the Greek letter *epsilon*), though some write it as λ (the Greek letter *lambda*).

2.3.1 Functions that Apply to Strings

We now know enough about strings to describe some important functions that we can use to manipulate strings or obtain information about them. Table 2.1 shows the basic string operations (note that x and y stand for *any* strings).

You may have noticed that strings have certain features in common with *arrays* in programming languages such as Pascal, in that we can index them. To index a string, we use the notation x_i, as opposed to something like $x[i]$. However, strings actually have more in common with the list data structures of programming languages such as LISP or PROLOG, in that we can concatenate two strings together, creating a new string. This is like the *append* function in LISP, with strings corresponding to lists, and the empty string corresponding to the empty list. It is only possible to perform such operations on arrays if the programming language allows

Table 2.1. The basic operations on strings

Operation	Written as	Meaning	Examples and comments
Length	$\|x\|$	The number of symbols in the string x	$\|abcabca\| = 7$ $\|a\| = 1$ $\|\varepsilon\| = 0$
Concatenation	xy	The string formed by writing down the string x followed immediately by the string y concatenating the empty string to any string makes no difference	let $x = abca$ let $y = ca$ then: $xy = abcaca$ let $x = $ <any string> then: $x\varepsilon = x$ $\varepsilon x = x$
Power	x^n, where n is a whole number $\geqslant 0$	The string formed by writing down n copies of the string x	let $x = abca$ then: $x^3 = abcaabcaabca$ $x^1 = x$ Note: $x^0 = \varepsilon$
Index	x_i, where i is a whole number	The ith symbol in the string x (i.e. treats the string as if it were an array of symbols)	let $x = abca$ then: $x_1 = a$ $x_2 = b$ $x_3 = c$ $x_4 = a$

arrays to be of dynamic size (which Pascal, for example, does not). However, many versions of Pascal now provide a special dynamic "string" data type, on which operations such as concatenation can be carried out.

2.3.2 Useful Notation for Describing Strings

As described above, a string is a sequence of symbols taken from some alphabet. Later, we will need to say such things as:

> suppose x stands for some string taken from the alphabet A.

This is a rather clumsy phrase to have to use. A more accurate, though even clumsier, way of saying it is to say:

> x is an element of the set of all strings which can be formed using zero or more symbols of the alphabet A.

There is a convenient and simple notational device to say this. We represent the latter statement as follows:

$$x \in A^*,$$

which relates to the English version as shown in Figure 2.1.

On other occasions, we may wish to say something like:

> x is an element of the set of all strings which can be formed using *one* or more symbols of the alphabet A,

for which we write:

$$x \in A^+,$$

which relates to the associated verbal description as shown in Figure 2.2.

Suppose we have the alphabet $\{a, b, c\}$. Then $\{a, b, c\}^*$ is the set

$$\{\varepsilon, a, b, c, aa, ab, ac, ba, bb, bc, ca, cb, cc, aaa, aab, aac, aba, abb, abc, \ldots\}.$$

Clearly, for any non-empty alphabet (i.e. an alphabet consisting of one or more symbols), the set so defined will be *infinite*.

Earlier in the chapter, we discussed the notion of a program printing out the elements of a *finite* set, one by one, terminating when all of the elements of the set had been printed. If A is some alphabet, we could write a program to print out all the strings in A^*, one by one, such that each string only gets printed out once. Obviously, such a program would *never* terminate (because A^* is an *infinite* set), but we could design the program so that any string in A^* would appear within a finite period of time. Table 2.2 shows a possible method for doing this (as an

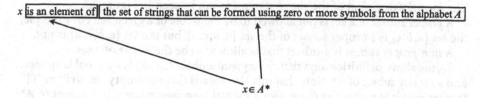

Figure 2.1. How we specify an unknown, possibly empty, string

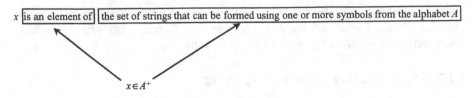

Figure 2.2. How we specify an unknown, non-empty, string

exercise, you might like to develop the method into a program in your favourite programming language). The method is suggested by the way the first few elements of the set A^*, for $A = \{a, b, c\}$ were written down, above.

An infinite set for which we can print out any given element within a finite time of starting the program is known as a *countably infinite* set. I suggest you think carefully about the program in Table 2.2, as it may help you to appreciate just what is meant by the terms "infinite" and "finite". Clearly, the program specified in Table 2.2 would never terminate. However, on each iteration of the loop, i would have a finite value, and so any string printed out would be finite in length (a necessary condition for a string). Moreover, any string in A^* would appear after a finite period of time.

Table 2.2. Systematically printing out all strings in A^*

```
begin
    <print some symbol to represent the empty string>
    i := 1
    while i >= 0 do
        <print each of the strings of length i>
        i := i + 1
    endwhile
end
```

2.4 Formal Languages

Now we know how to express the notion of all of the strings that can be formed by using symbols from an alphabet, we are in a position to describe what is meant by the term *formal language*. Essentially, a formal language is simply any set of strings formed using the symbols from any alphabet. In set parlance, given some alphabet A,

 a *formal language* is "any (proper or non-proper) subset of the set of all strings which can be formed using zero or more symbols of the alphabet A".

The formal expression of the above statement can be seen in Figure 2.3.

A *proper* subset of a set is not allowed to be the *whole* of a given set. For example, the set $\{a, b, c\}$ is a proper subset of the set $\{a, b, c, d\}$, but the set $\{a, b, c, d\}$ is not.

A *non-proper* subset is a subset that is allowed to be the whole of a set.

So, the above definition says that, for a given alphabet, A, A^* is a formal language, and so is any subset of A^*. Note that this also means that the empty set, written "{ }" (sometimes it is written as $\varnothing$) is also a formal language, since it is a subset of A^* (the empty set is a subset of *any* set).

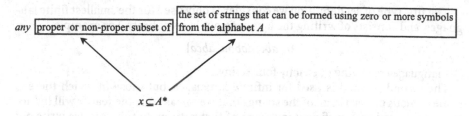

Figure 2.3. The definition of a *formal language*

A formal language, then, is any set of strings. To indicate that the strings are part of a language, we usually call them *sentences*. In some books, sentences are called *words*. However, while the strings we have seen so far are similar to English words, in that they are unbroken sequences of alphabetic symbols (for example, *abca*), later we will see strings that are statements in a programming language, such as

```
if i > 1 then x := x + 1.
```

It seems peculiar to call a statement such as this a "word".

2.5 Methods for Defining Formal Languages

Our definition of a formal language as being a set of strings that are called *sentences* is extremely simple. However, it does not allow us to say anything about the form of sentences in a particular language. For example, in terms of our definition, the Pascal programming language, by which we mean "the set of all syntactically correct Pascal programs", is a subset of the set of all strings which can be formed using symbols found in the character set of a typical computer. This definition, though true, is not particularly helpful if we want to write Pascal programs. It tells us nothing about what makes one string a Pascal program, and another string not a Pascal program, except in the trivial sense that we can immediately rule out any strings containing symbols that are not in the character set of the computer. You would be most displeased if, in attempting to learn to program in Pascal, you opened the Pascal manual to find that it consisted entirely of one statement which said: "Let *C* be the set of all characters available on the computer. Then the set of compilable Pascal programs, *P*, is a subset of C^*".

One way of informing you what constitutes "proper" Pascal programs would be to write all the proper ones out for you. However, this would also be unhelpful, albeit in a different way, since such a manual would be infinite, and thus could never be completed. Moreover, it would be a rather tedious process to find the particular program you required.

In this section we discover three approaches to defining a formal language. Following this, every formal language we meet in this book will be defined according to one or more of these approaches.

2.5.1 Set Definitions of Languages

Since a language is a *set* of strings, the obvious way to describe some language is by providing a set definition. Set definitions of the formal languages in which we are interested are of three different types, as now discussed.

The first type of set definition we consider is only used for the smallest finite languages, and consists of writing the language out in its entirety. For example,

$$\{\varepsilon, abc, abbba, abca\}$$

is a language consisting of exactly four strings.

The second method is used for infinite languages, but those in which there is some obvious pattern in all of the strings that we can assume the reader will induce when presented with sufficient instances of that pattern. In this case, we write out sufficient sentences for the pattern to be made clear, then indicate that the pattern should be allowed to continue indefinitely, by using three dots "...". For example,

$$\{ab, aabb, aaabbb, aaaabbbb, \ldots\}$$

suggests the infinite language consisting of all strings which consist of one or more as followed by one or more bs and in which the number of as equals the number of bs.

The final method, used for many *finite* and *infinite* languages, is to use a set definition to specify how to construct the sentences in the language, i.e. provide a function to deliver the sentences as its output. In addition to the function itself, we must provide a specification of how many strings should be constructed. Such set definitions have the format shown in Figure 2.4.

For the "function to produce strings", of Figure 2.4, we use combinations of the string functions we considered earlier (*index*, *power*, and *concatenation*). A language that was defined immediately above,

all strings which consist of one or more as followed by one or more bs and in which the number of as equals the number of bs

can be defined using our latest method as:

$$\{a^i b^i : i \geq 1\}.$$

The above definition is explained in Table 2.3.

From Table 2.3 we can see that $\{a^i b^i : i \geq 1\}$ means:

the set of all strings consisting of i copies of a followed by i copies of b such that i is is allowed to take on the value of each and every whole number value greater than or equal to 1.

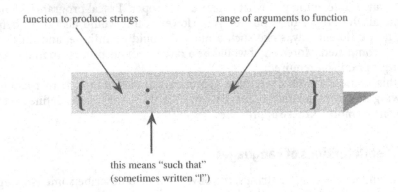

function to produce strings   range of arguments to function

this means "such that"
(sometimes written "|")

Figure 2.4. Understanding a set definition of a formal language

Table 2.3. What the set definition $\{a^i b^i : i \geqslant 1\}$ means

Notation	String function	Meaning
a	N/A	The string a
b	N/A	The string b
a^i	Power	The string formed by writing down i copies of the string a
b^i	Power	The string formed by writing down i copies of the string b
$a^i b^i$	Concatenation	The string formed by writing down i copies of a followed by i copies of b
$: i \geqslant 1$	N/A	Such that i is allowed to take on the value of each and every whole number value greater than or equal to 1 (we could have written $i > 0$)

Changing the right-hand side of the set definition can change the language defined. For example, $\{a^i b^i : i \geqslant 0\}$ defines:

the set of all strings consisting of i copies of a followed by i copies of b such that i is is allowed to take on the value of each and every whole number value greater than or equal to 0.

This latter set is our original set, along with the empty string (since $a^0 = \varepsilon$, $b^0 = \varepsilon$, and therefore $a^0 b^0 = \varepsilon\varepsilon = \varepsilon$). In set parlance, $\{a^i b^i : i \geqslant 0\}$ is the *union* of the set $\{a^i b^i : i \geqslant 1\}$ with the set $\{\varepsilon\}$, which can be written as:

$$\{a^i b^i : i \geqslant 0\} = \{a^i b^i : i \geqslant 1\} \cup \{\varepsilon\}.$$

The immediately preceding example illustrates a further useful feature of sets. We can often simplify the definition of a language by creating several sets and using the union, intersection and set difference operators to combine them into one. This sometimes removes the need for a complicated expression in the right-hand side of our set definition. For example, the definition

$$\{a^i b^j c^k : i \geqslant 1,\ j \geqslant 0,\ k \geqslant 0,\quad if\ i \geqslant 3\ then\ j = 0\ else\ k = 0\},$$

is probably better represented as

$$\{a^i c^j : i \geqslant 3,\ j \geqslant 0\} \cup \{a^i b^j : 1 \leqslant i < 3,\ j \geqslant 0\},$$

which means

the set of strings consisting of three or more as followed by zero or more cs, or consisting of one or two as followed by zero or more bs.

2.5.2 Decision Programs for Languages

We have seen how to define a language by using a formal set definition. Another way of describing a language is to provide a program that tells us whether or not any given string of symbols is one of its sentences. Such a program is called a *decision program*. If the program always tells us, for any string, whether or not the string is a sentence, then the program in an implicit sense defines the language, in that the language is the set containing each and every string that the program would tell us is a sentence. That is why we use a special term, "sentence", to describe *a string that belongs to a language*. A *string* input to the program may or may not be a *sentence* of the language; the program should tell us. For an alphabet A, a

Table 2.4. A decision program for a formal language

```
1:  read(sym)                {assume read just gives us the next symbol in the string being examined}
       case sym of
          eos:  goto N       {assume read returns special symbol "eos" if at end of string}
          "a":  goto 2
          "b":  goto N
          "c":  goto N
       endcase               {case statement selects between alternatives as in Pascal}
2:  read(sym)
       case sym of
          eos:  goto Y       {if we get here we have a string of one a which is ok}
          "a":  goto 3
          "b":  goto 6       {we can have a b after one a}
          "c":  goto N       {any cs must follow three or more as – here we've only had one}
       endcase
3:  read(sym)
       case sym of
          eos:  goto Y       {if we get here we've read a string of two as which is OK}
          "a":  goto 4
          "b":  goto 6       {we can have a b after two as}
          "c":  goto N       {any cs must follow three or more as – here we've only had two}
       endcase
4:  read(sym)
       case sym of
          eos:  goto Y       {if we get here we've read a string of three or more as which is OK}
          "a":  goto 4       {we loop here because we allow any number of as ≥ 3}
          "b":  goto N       {b can only follow one or two as}
          "c":  goto 5       {cs are OK after three or more as}
       endcase
5:  read(sym)
       case sym of
          eos:  goto Y       {if we get here we've read ≥3 as followed by ≥1 cs which is OK}
          "a":  goto N       {as after cs are not allowed}
          "b":  goto N       {bs are only allowed after one or two as}
          "c":  goto 5       {we loop here because we allow any number of cs after ≥3 as}
       endcase
6:  read(sym)
       case sym of
          eos:  goto Y       {we get here if we've read 1 or 2 as followed by ≥1 bs – OK}
          "a":  goto N       {no as allowed after bs}
          "b":  goto 6       {we loop here because we allow any number of bs after 1 or 2 as}
          "c":  goto N       {no cs are allowed after bs}
       endcase
Y:  write("yes")
       goto E
N:  write("no")
       goto E
E:  {end of program}
```

language is any *subset* of A^*. For any interesting language, then, there will be many strings in A^* that are not sentences.

Later in this book we will be more precise about the form these decision programs take, and what can actually be achieved with them. For now, however, we will consider an example to show the basic idea.

If you have done any programming at all, you will have used a decision program on numerous occasions. The decision program you have used is a component of the *compiler*. If you write programs in a language such as Pascal, you submit your program text to a compiler, and the compiler tells you if the text is a syntactically correct Pascal program. Of course, the compiler does a lot more than this, but a very important part of its job is to tell us if the source text (string) is a syntactically correct Pascal program, i.e. a *sentence* of the *language* called "Pascal".

Consider again the language

$$\{a^i c^j : i \geqslant 3, \ j \geqslant 0\} \cup \{a^i b^j : 1 \leqslant i < 3, \ j \geqslant 0\},$$

i.e.

the set of strings consisting of three or more *a*s followed by zero or more *c*s, or consisting of one or two *a*s followed by zero or more *b*s.

Table 2.4 shows a decision program for the language.

The program of Table 2.4 is purely for illustration. In the next chapter we consider formal languages for which the above type of decision program can be created automatically. For now, examine the program to convince yourself that it correctly meets its specification, which can be stated as follows:

given any string in $\{a, b, c\}^*$, tell us whether or not that string is a sentence of the language

$$\{a^i c^j : i \geqslant 3, \ j \geqslant 0\} \cup \{a^i b^j : 1 \leqslant i < 3, \ j \geqslant 0\}.$$

2.5.3 Rules for Generating Languages

We have seen how to describe formal languages by providing set definitions, and we have encountered the notion of a decision program for a language. The third method, which is the basis for the remainder of this chapter, defines a language by providing a set of rules to generate sentences of a language. We require that such rules are able to generate every one of the sentences of a language, and no others. Analogously, a set definition describes every one of the sentences, and no others, and a decision program says "yes" to every one of the sentences, and to no others.

There are several ways of specifying rules to generate sentences of a language. One popular form is the *syntax diagram*. Such diagrams are often used to show the structure of programming languages, and thus inform you how to write syntactically correct programs (*syntax* is considered in more detail in Chapter 3).

Figure 2.5 shows a syntax diagram for the top level syntax of the Pascal "program" construct.

The diagram in Figure 2.5 tells us that the syntactic element called a "program" consists of

the string "PROGRAM" (entities in rounded boxes and circles represent actual strings that are required at a given point),

program

Figure 2.5. Syntax diagram for the Pascal construct "program"

followed by something called

> an "identifier" (entities in rectangles are those which need elaborating in some way that is specified in a further definition),

followed by

> an open bracket "(",

followed by

> a list of one or more "identifiers", in which *every one except the last* is followed by a comma, ",", followed by a semicolon, ";",

followed by

> a close bracket, ")",

followed by

> something called a "block",

followed by

> a full stop, ".".

In Figure 2.6 we see the syntax diagram for the entity "identifier".

Figure 2.6 shows us that an "identifier" consists of a letter followed by zero or more letters and/or digits.

The following fragment of Pascal:

```
program calc(input, output, infile26, outfile23);
```

associates with the syntax diagram for "program" as shown in Figure 2.7.

Of course, the diagrams in Figures 2.5 and 2.6, together with all of the other diagrams defining the syntax of Pascal, cannot tell us how to write a program to solve a given problem. That is a *semantic* consideration, relating to the *meaning* of the program text, not only its *form*. The diagrams merely describe the *syntactic structure* of constructs belonging to the Pascal language.

An alternative method of specifying the syntax of a programming language is to use a notation called Backus–Naur form (BNF).[1] Table 2.5 presents a BNF version of our syntax diagrams from above.

The meaning of the notation in Table 2.5 should be reasonably clear when you see its correspondence with syntax diagrams, as shown in Figure 2.8.

Formalisms such as syntax diagrams and BNF are excellent ways of defining the syntax of a language. If you were taught to use a programming language, you may never have looked at a formal definition of its syntax. Analogously, you probably did not learn your own "natural" language by studying a book describing its grammar. However, many programming languages are similar to each other in

Table 2.5. BNF version of Figures 2.5 and 2.6

<program> ::= <program heading> <block>
<program heading> := <u>program</u> <identifier> (<identifier> { , <identifier>})
<identifier> ::= <letter> {<letter or digit>}
<letter or digit> ::= <letter>	<digit>

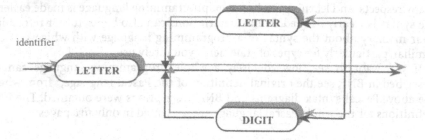

Figure 2.6. Syntax diagram for a Pascal "identifier"

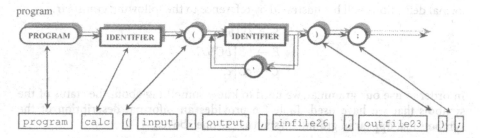

Figure 2.7. How a syntax diagram describes a Pascal statement

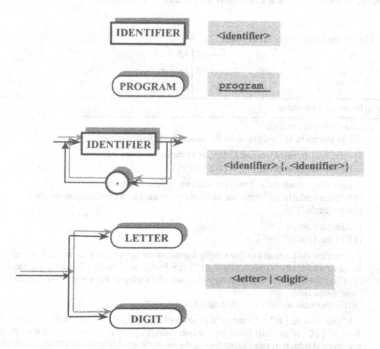

Figure 2.8. How syntax diagrams and BNF correspond

many respects, and learning a subsequent programming language is made easier if the syntax is clearly defined. Syntax descriptions can also be useful for refreshing your memory about the syntax of a programming language with which you are familiar, particularly for types of statements you rarely use.

If you want to see how concisely a whole programming language can be described in BNF, see the original definition of the Pascal language,[2] from where the above Pascal syntax diagrams and BNF descriptions were obtained. The BNF definitions for the whole Pascal language are presented in only *five* pages.

2.6 Formal Grammars

A *grammar* is a set of rules for generating strings. The grammars we will use in the remainder of this book are known as *phrase structure grammars* (PSGs). Here, our formal definitions will be illustrated by reference to the following grammar:

$$S \rightarrow aS \,|\, bB$$
$$B \rightarrow bB \,|\, bC \,|\, cC$$
$$C \rightarrow cC \,|\, c.$$

In order to use our grammar, we need to know something about the status of the symbols that we have used. Table 2.6 provides an informal description of the symbols that appear in grammars such as the one above.

2.6.1 Grammars, Derivations, and Languages

Table 2.7 presents an informal description, supported by examples using our grammar above, of how we use a grammar to generate a sentence.

Table 2.6. The symbols that make up the PSG:

$$S \rightarrow aS \,|\, bB$$
$$B \rightarrow bB \,|\, bC \,|\, cC$$
$$C \rightarrow cC \,|\, c$$

Symbols	Name and meaning
S, B, C	*Non-terminal* symbols [BNF: things in angled brackets, for example <identifier>]
S	Special non-terminal, called a *start*, or *sentence*, symbol [BNF: in our example above, <program>]
a, b, c	*Terminal* symbols: only these symbols can appear in sentences [BNF: the underlined terms (for example program) and punctuation symbols (for example ";")]
$\rightarrow$	*Production arrow* [BNF: the symbol "::="]
$S \rightarrow aS$	*Production rule*, usually called simply a *production* (or sometimes we will just use the word *rule*). Means "*S produces aS*", or "*S can be replaced by aS*". The string to the left of $\rightarrow$ is called the *left-hand side* of the production, the string to the right of $\rightarrow$ is called the *right-hand side* [BNF: this rule would be written as <S> ::= a <S>]
|	"Or", so $B \rightarrow bB \,|\, bC \,|\, cC$ means "*B produces bB* <u>or</u> *bC* <u>or</u> *cC*". Note that this means that $B \rightarrow bB \,|\, bC \,|\, cC$ is really *three* production rules, i.e. $B \rightarrow bB$, $B \rightarrow bC$, and $B \rightarrow cC$. So there are *seven* production rules altogether in the example grammar above [BNF: exactly the same]

Table 2.7. Using a PSG

Action taken	Resulting string	Production applied
Start with S, the *start* symbol	S	
If a substring of the resulting string matches the left-hand side of one or more productions, replace that substring by the right-hand side of any one of those productions	aS	$S \rightarrow aS$
Same as above	aaS	$S \rightarrow aS$
Same as above	$aabB$	$S \rightarrow bB$
Same as above	$aabcC$	$B \rightarrow cC$
Same as above	$aabccC$	$C \rightarrow cC$
Same as above	$aabccc$	$C \rightarrow c$
If the result string consists entirely of terminals, then stop		

As you can see from Table 2.7, there is often a choice as to which rule to apply at a given stage. For example, when the resulting string was aaS, we could have applied the rule $S \rightarrow aS$ as many times as we wished (adding another a each time). A similar observation can be made for the applicability of the $C \rightarrow cC$ rule when the resulting string was $aabcC$, for example.

Here are some other strings we could create, by applying the rules in various ways:

$$abcc, \quad bbbbc, \quad \text{and} \quad a^3b^2c^5.$$

You may like to see if you can apply the rules yourself to create the above strings. You must always begin with a rule that has S on its left-hand side (that is why S is called the *start* symbol).

We write down the S symbol to start the process, and we merely repeat the process described in Table 2.7 as

if a substring of the resulting string matches the left-hand side of one or more productions, replace that substring by the right-hand side of any one of those productions,

until the following becomes true:

if the result string consists entirely of terminals,

at which point we:

stop.

You may wonder why the process of matching the substring was not presented as:

if a <u>non-terminal symbol</u> *in the resulting string matches the left-hand side of one or more productions, replace that* <u>non-terminal symbol</u> *by the right-hand side of any one of those productions.*

This would clearly work for the example grammar given. However, as discussed in the next section, grammars are not necessarily restricted to having single non-terminals on the left-hand sides of their productions.

The process of creating strings using a grammar is called *deriving* them, so when we show how we have used the grammar to *derive* a string (as was done in Table 2.7), we are showing a *derivation* for (or of) that string.

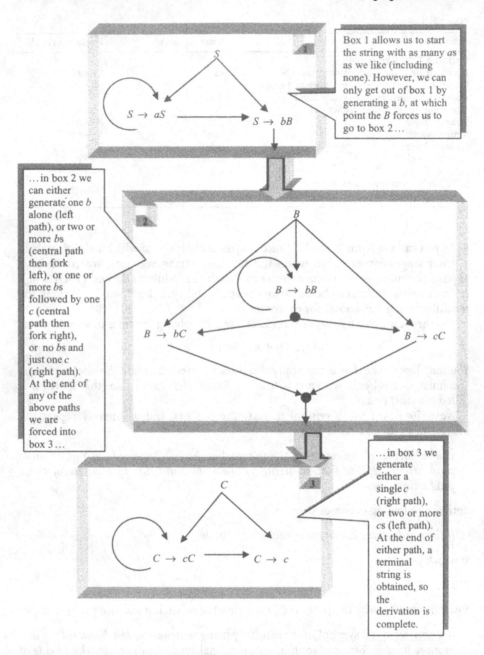

Box 1 allows us to start the string with as many *a*s as we like (including none). However, we can only get out of box 1 by generating a *b*, at which point the *B* forces us to go to box 2 ...

... in box 2 we can either generate one *b* alone (left path), or two or more *b*s (central path then fork left), or one or more *b*s followed by one *c* (central path then fork right), or no *b*s and just one *c* (right path). At the end of any of the above paths we are forced into box 3 ...

... in box 3 we generate either a single *c* (right path), or two or more *c*s (left path). At the end of either path, a terminal string is obtained, so the derivation is complete.

Figure 2.9. Working out all of the terminal strings that a grammar can generate

Let us now consider all of the "terminal strings" – strings consisting entirely of terminal symbols, also known as *sentences* – we can use the example grammar to derive. As this is a simple grammar, it is not too difficult to work out what they are. Figure 2.9 shows the choice of rules possible for deriving terminal strings from the example grammar.

Table 2.8. The language generated by a grammar

Box in diagram of Figure 2.9	Informal description of derived strings	Formal description of derived strings
① i.e. productions $S \to aS \mid S \to bB$ ... is expanded in box 2 ...	"Any non-zero number of as followed by bB" or "just bB" which is the same as saying "zero or more as followed by bB"	$a^i bB, \; i \geq 0$ ... the B at the end ...
② i.e. *productions* $B \to bB \mid bC \mid cC$ ... is expanded in box 3 ...	"Any non-zero number of bs followed by either bC or cC" or "just bC" or "just cC"	$b^j C, \; j \geq 1$ or $b^i cC, \; j \geq 0$ ... the C at the end ...
③ i.e. productions $C \to cC \mid c$	"Any non-zero number of cs followed by one c" or "just one c" which is the same as saying "one or more cs"	$c^k, \; k \geq 1$

Any "legal" application of our production rules, starting with S, the start symbol, alone, and resulting in a *terminal string*, would involve us in following a path through the diagram in Figure 2.9, starting in box 1, passing through box 2, and ending up in box 3. The boxes in Figure 2.9 are annotated with the strings produced by taking given options in applying the rules. Table 2.8 summarises the strings described in Figure 2.9.

We now define a set that contains all of the terminal strings (and only those strings) that can be derived from the example grammar. The set will contain all strings defined as follows:

A string taken from the set $\{a^i b: i \geq 0\}$ concatenated with a string taken from the set $\{b^j: j \geq 1\} \cup \{b^j c: j \geq 0\}$ concatenated with a string taken from the set $\{c^k: k \geq 1\}$.

The above can be written as:

$$\{a^i bb^j c^k: i \geq 0, \; j \geq 1, \; k \geq 1\} \cup \{a^i bb^j cc^k: i \geq 0, \; j \geq 0, \; k \geq 1\}.$$

Observe that $bb^j, j \geq 1$ is the same as $b^j, j \geq 2$, and $bb^j, j \geq 0$ is the same as b^j, $j \geq 1$, and $cc^k, k \geq 1$ is the same as $c^k, k \geq 2$ so we could write:

$$\{a^i b^j c^k: i \geq 0, \; j \geq 2, \; k \geq 1\} \cup \{a^i b^j c^k: i \geq 0, \; j \geq 1, \; k \geq 2\}.$$

This looks rather complicated, but essentially there is only one awkward case, which is that if there is only one b then there must be two or more cs (any more than one b and we can have 1 or more cs). So we could have written:

$$\{a^i b^j c^k: i \geq 0, \; j \geq 1, \; k \geq 1, \; if \, j = 1 \; then \, k \geq 2 \; else \, k \geq 1\}.$$

Whichever way we write the set, one point should be made clear: the set is a set of strings formed from symbols in the alphabet $\{a, b, c\}$, that is to say, the set is a *formal language*.

2.6.2 The Relationship Between Grammars and Languages

We are now ready to give an intuitive definition of the relationship between grammars and languages:

> The *language generated by a grammar* is the set of all *terminal strings* that can be *derived* using the *productions* of that grammar, each derivation beginning with the *start symbol* of that grammar.

Our example grammar, when written like this:

$$S \rightarrow aS \mid bB$$
$$B \rightarrow bB \mid bC \mid cC$$
$$C \rightarrow cC \mid c$$

is not fully defined. A grammar is fully defined when we know which symbols are terminals, which are non-terminals, and which of the non-terminals is the start symbol. In this book, we will usually see only the productions of a grammar, and we will assume the following:

- capitalised letters are *non-terminal symbols*
- non-capitalised letters are *terminal symbols*
- the capital letter S is the *start symbol*.

The above will always be the case unless explicitly stated otherwise.

2.7 Phrase Structure Grammars and the Chomsky Hierarchy

The production rules of the example grammar from the preceding section are simple in format. For example, the left-hand sides of all the productions consist of lone non-terminals. As we see later in the book, restricting the form of productions allowed in a grammar in certain ways simplifies certain language processing tasks, but it also reduces the sophistication of the languages that such grammars can generate. For now, we will define a scheme for classifying grammars according to the "shape" of their productions which will form the basis of our subsequent discussion of grammars and languages. The classification scheme is called the Chomsky hierarchy, named after Noam Chomsky, an influential American linguist.

2.7.1 Formal Definition of PSGs

To prepare for specifying the Chomsky hierarchy, we first need to precisely define the term PSG. Table 2.9 does this.

Formally, then, a PSG, G, is specified as (N, T, P, S). This is what mathematicians call a "tuple" (of four elements).

The definition in Table 2.9 makes it clear that the empty string, ε, cannot appear alone on the left-hand side of any of the productions of a PSG. Moreover, the definition tells us that ε *is* allowed on the right-hand side. Otherwise, any strings of terminals and/or non-terminals can appear on either side of productions. However, in most grammars we usually find that there are one or more non-terminals on the *left-hand side* of each production.

Table 2.9. The formal definition of a PSG

Any PSG, G, consists of the following:	
N, a set of *non-terminal* symbols	An alphabet, containing no symbols that can appear in sentences
T, a set of *terminal* symbols	Also an alphabet, containing only symbols that *can* appear in sentences
P, a set of *production rules* of the form	This specification uses the notation for specifying strings from an alphabet we looked at earlier
$x \rightarrow y$, where $x \in (N \cup T)^+$, and $y \in (N \cup T)^*$	x is the left-hand side of a production, y the right-hand side
	The definition of y means: *the right-hand side of each production is a possibly empty string of terminals and/or non-terminals*
	The only difference between the specification above and the one for x (the left-hand side) is that the one for x uses "+" rather than "*"
	So the specification for x means: *the left-hand side of each production is a non-empty string of terminals and/or non-terminals*
S, a member of N, designated as the *start*, or *sentence* symbol	The non-terminal symbol with which we always begin a derivation

As we always start a derivation with a lone S (the *start* symbol), for a grammar to derive anything it must have at least one production with S alone on its left-hand side. This last piece of information is not specified in the definition above, as there is nothing in the formal definition of PSGs that says they *must* generate anything. To refer back to our earlier example grammar, its full formal description would be as shown in Table 2.10.

Table 2.10. The (N, T, P, S) form of a grammar

Productions	(N, T, P, S)	
	({S, B, C},	---N
$S \rightarrow aS \mid bB$	{a, b, c},	---T
$B \rightarrow bB \mid bC \mid cC$	{S → aS, S → bB, B → bB, B → bC,	---P
$C \rightarrow cC \mid c$	B → cC, C → cC, C → c},	
	S	---S
	)	

2.7.2 Derivations, Sentential Forms, Sentences, and "$L(G)$"

We have formalised the definition of a PSG. We now formalise our notion of derivation, and introduce some useful terminology to support subsequent discussion. To do this, we consider a new grammar:

$$S \rightarrow aB \mid bA \mid \varepsilon$$
$$A \rightarrow aS \mid bAA$$
$$B \rightarrow bS \mid aBB.$$

Using the conventions outlined earlier, we know that S is the start symbol, $\{S, A, B\}$ is the set of non-terminals (N), and $\{a, b\}$ is the set of terminals (T). So we need not provide the full (N, T, P, S) definition of the grammar.

As in our earlier example, the left-hand sides of the above productions all consist of single non-terminals. We see an example grammar that differs from this later in the chapter.

Here is a string in $(N \cup T)^+$ that the above productions can be used to derive, as you might like to verify for yourself:

$$abbbaSA.$$

This is not a terminal string, since it contains non-terminals (S and A). Therefore it is not a sentence. The next step could be, say, to apply the production $A \rightarrow bAA$, which would give us

$$abbbaSbAA,$$

which is also not a sentence.

We now have two strings, *abbbaSA* and *abbbaSbAA* that are such that the *former can be used as a basis for the derivation of the latter by the application of one production rule of the grammar*. This is rather a mouthful, even if we replace *"by the application of one production rule of the grammar"* by the phrase *"in one step"*, so we introduce a symbol to represent this relationship. We write:

$$abbbaSA \Rightarrow abbbaSbAA.$$

To be absolutely correct, we should give our grammar a name, say G, and write

$$abbbaSA \Rightarrow^G abbbaSbAA$$

to denote which particular grammar is being used. Since it is usually clear in our examples which grammar is being used, we will simply use $\Rightarrow$. We now use this symbol to show how our example grammar derives the string *abbbaSbAA*:

$$S \Rightarrow aB$$
$$aB \Rightarrow abS$$
$$abS \Rightarrow abbA$$
$$abbA \Rightarrow abbbAA$$
$$abbbAA \Rightarrow abbbaSA$$
$$abbbaSA \Rightarrow abbbaSbAA.$$

As it is tedious to write out each intermediate stage twice, apart from the first (S) and the last (*abbbaSbAA*), we allow an abbreviated form of such a derivation as follows:

$$S \Rightarrow aB \Rightarrow abS \Rightarrow abbA \Rightarrow abbbAA \Rightarrow abbbaSA \Rightarrow abbbaSbAA.$$

We now use our new symbol as the basis of some additional useful notation, as shown in Table 2.11.

A new term is now introduced, to simplify references to the intermediate stages in a derivation. We call these intermediate stages *sentential forms*. Formally, given any grammar, G, a *sentential form* is any string that can be derived in zero or more steps from the start symbol, S. By "any string", we mean exactly that; not only terminal strings, but any string of terminals and/or non-terminals. Thus, *a sentence is a sentential form*, but *a sentential form is not necessarily a sentence*. Given the simple grammar

$$S \rightarrow aS \mid a,$$

some sentential forms are: S, $aaaaaaS$, and a^{10}. Only one of these sentential forms (a^{10}) is a sentence, as it is the only one that consists entirely of *terminal symbols*.

Table 2.11. Useful notation for discussing derivations, and some example true statements for grammar:

$$S \rightarrow aB \mid bA \mid \varepsilon$$
$$A \rightarrow aS \mid bAA$$
$$B \rightarrow bS \mid aBB$$

Notation	Meaning	Example true statements
$x \Rightarrow y$	The application of one production rule results in the string x becoming the string y Also expressed as "x generates y <u>in one step</u>", or "x produces y <u>in one step</u>", or "y is derived from x <u>in one step</u>"	$aB \Rightarrow abS$ $S \Rightarrow \varepsilon$ $abbbaSA \Rightarrow abbbaSbAA$
$x \Rightarrow^* y$	x generates y <u>in zero or more steps</u>, or just "x generates y", or "x produces y", or "y is derived from x"	$S \Rightarrow^* S$ $S \Rightarrow^* abbbaSA$ $aB \Rightarrow^* abbbaa$
$x \Rightarrow^+ y$	x generates y <u>in one or more steps</u>, or just "x generates y", or "x produces y", or "y is derived from x"	$S \Rightarrow^+ abbbaSA$ $abbbaSbAA \Rightarrow^+ abbbabaa$

Formally, using our new notation,

if $S \Rightarrow^* x$, then x is a sentential form;
if $S \Rightarrow^* x$, and x is a terminal string, then x is a *sentence*.

We now formalise a definition given earlier, this being the statement that

the language generated by a grammar is the set of all terminal strings that can be derived using the productions of that grammar, each derivation beginning with the start symbol of that grammar.

Using various aspects of the notation introduced in this chapter, this becomes:

given a PSG, G, $L(G) = \{x : x \in T^* \text{ and } S \Rightarrow^* x\}$.

(Note that the definition assumes that we have specified the set of terminals and the start symbol of the grammar, which as we said earlier is done implicitly in our examples.)

So, if G is some PSG, $L(G)$ means *the language generated by G*. As the set definition of $L(G)$ clearly states, the set $L(G)$ contains all of the terminal strings generated by G, but only the strings that G generates. It is very important to realise that this is what it means when we say *the language generated by the grammar*.

We now consider three examples, to reinforce these notions. The first is an example grammar encountered above, now labelled G_1:

$$S \rightarrow aS \mid bB$$
$$B \rightarrow bB \mid bC \mid cC$$
$$C \rightarrow cC \mid c.$$

We have already provided a set definition of $L(G_1)$; it was:

$$L(G_1) = \{a^i b^j c^k : i \geq 0, \ j \geq 1, \ k \geq 1, \ \text{if } j = 1 \ \text{then } k \geq 2 \ \text{else } k \geq 1\}.$$

Another grammar we have already encountered, which we now call G_2, is:

$$S \rightarrow aB \mid bA \mid \varepsilon$$
$$A \rightarrow aS \mid bAA$$
$$B \rightarrow bS \mid aBB.$$

This is more complex than G_1, in the sense that some of G_2's productions have more than one non-terminal on their right-hand sides:

$$L(G_2) = \{x: x \in \{a, b\}^* \text{ and the number of } as \text{ in } x \text{ equals the number of } bs\}.$$

I leave it to you to establish that the above statement is true.

Note that $L(G_2)$ is not the same as a set that we came across earlier, i.e.

$$\{a^i b^i : i \geqslant 1\},$$

which we will call set A. In fact, set A is a proper subset of $L(G_2)$. G_2 can generate all of the strings in A, but it generates many more besides (such as ε, *bbabbbaaaaab*, and so on). A grammar, G_3, such that $L(G_3) = A$ is:

$$S \rightarrow ab \mid aSb.$$

2.7.3 The Chomsky Hierarchy

This section describes a classification scheme for PSGs, and the corresponding phrase structure languages (PSLs) that they generate, which is of the utmost importance in determining certain of their computational features. PSGs can be classified in a hierarchy, the location of a PSG in that hierarchy being an indicator of certain characteristics required by a decision program for the corresponding language. We saw above how one example language could be processed by an extremely simple decision program. Much of this book is devoted to investigating the computational nature of formal languages. We use as the basis of our investigation the classification scheme for PSGs and PSLs called the *Chomsky hierarchy*.

Classifying a grammar according to the Chomsky hierarchy is based solely on the presence of certain patterns in the productions. Table 2.12 shows how to make the classification. The types of grammar in the Chomsky hierarchy are named types 0 to 3, with 0 as the most general type. Each type from 1 to 3 is defined according to one or more restrictions on the definition of the type numerically preceding it, which is why the scheme qualifies as a *hierarchy*.

If you are observant, you may have noticed an anomaly in Table 2.12. Context sensitive grammars are not allowed to have the empty string on the right-hand side of productions, whereas all of the other types are. This means that, for example, our grammar G_2, which can be classified as unrestricted and as context free (but not as regular), cannot be classified as context sensitive. However, every grammar that can be classified as regular can be classified as context free, and every grammar that can be classified as context free can be classified as unrestricted.

When classifying a grammar according to the Chomsky hierarchy, you should remember the following:

> For a grammar to be classified as being of a certain type, *each and every production of that grammar must match the pattern specified for productions of that type.*

Table 2.12. The Chomsky hierarchy

Type no.	Type name	Patterns to which ALL productions must conform	Informal description and examples
0	Unrestricted	$x \rightarrow y$, $x \in (N \cup T)^+$, $y \in (N \cup T)^*$	The definition of PSGs we have already seen. Anything allowed on the left-hand side (except for ε), anything allowed on the right. All of our example grammars considered so far conform to this. Example type 0 production: $aXYpq \rightarrow aZpq$ (all productions of G_1, G_2 and G_3 conform – but see below).
1	Context sensitive	$x \rightarrow y$, $x \in (N \cup T)^+$, $y \in (N \cup T)^+$, $|x| \leqslant |y|$	As for *type 0*, but we are not allowed to have ε on the left *or the right-hand-sides*. Note that the example production given for *type 0* is <u>not</u> a context sensitive production, as the length of the right-hand side is less than the length of the left. Example type 1 production: $aXYpq \rightarrow aZwpq$ (all productions of G_1 and G_3 conform, but not all of those of G_2 do).
2	Context free	$x \rightarrow y$, $x \in N$, $y \in (N \cup T)^*$	Single non-terminal on left, any mixture of terminals and/or non-terminals on the right. Also, ε is allowed on the right. Example type 2 production: $X \rightarrow XapZQ$ (all productions of G_1, G_2, and G_3 conform).
3	Regular	$w \rightarrow x$, or $w \rightarrow yz$ $w \in N$, $x \in T \cup \{\varepsilon\}$, $y \in T$, $z \in N$	Single non-terminal on left, and either ε or a single terminal or a single terminal followed by a single non-terminal, on the right. Example type 3 productions: $P \rightarrow pQ$, $F \rightarrow a$ (all of the productions of G_1 conform to this, but G_2 and G_3 do not).

Which means that the following grammar:

$$S \rightarrow aS \mid aA \mid AA$$
$$A \rightarrow aA \mid a$$

is classified as *context free*, since the production $S \rightarrow AA$ does not conform to the pattern for regular productions, even though all of the other productions do.

Table 2.13. The order in which to attempt the classification of a grammar, G, in the Chomsky hierarchy

---

if G is regular then
  return("regular")
else
  if G is context free then
    return("context free")
  else
    if G is context sensitive then
      return("context sensitive")
    else
      return("unrestricted")
    endif
  endif
endif

So, given the above rule that all productions must conform to the pattern, you classify a grammar, G, according to the procedure in Table 2.13.

Table 2.13 tells us to begin by attempting to classify G according to the most restricted type in the hierarchy. This means that, as indicated by Table 2.12, G_1 is a *regular* grammar, and G_2 and G_3 are *context free* grammars (CFGs). Of course, we know that as *all* regular grammars are CFGs, G_1 is also context free. Similarly, we know that they can *all* be classified as unrestricted. But we make the classification as specific as possible.

From the above, it can be seen that classifying a PSG is done simply by seeing if its productions match a given pattern. As we already know, grammars generate languages. In terms of the Chomsky hierarchy, *a language is of a given type if it is generated by a grammar of that type*. So, for example,

$$\{a^i b^i : i \geqslant 1\} \quad \text{(set } A \text{ mentioned above)}$$

is a context free language (CFL), since it is generated by G_3, which is classified as a CFG. However, how can we be sure that there is not a *regular grammar* that could generate A? We see later on that the more restricted the language (in the Chomsky hierarchy), the simpler the decision program for the language. It is therefore useful to be able to define the simplest possible type of grammar for a given language. In the meantime, you might like to see if you can create a *regular* grammar to generate set A (clue: do not devote too much time to this!).

From a theoretical perspective, the immediately preceding discussion is very important. If we can establish that there are languages that can be generated by grammars at some level of the hierarchy and cannot be generated by more restricted grammars, then we are sure that we do indeed have a genuine *hierarchy*. However, there are also *practical* issues at stake, for as mentioned above, and discussed in more detail in Chapters 4, 5 and 7, each type of grammar has associated with it a type of decision program, in the form of an abstract machine. The more restricted a language is, the simpler the type of decision program we need to write for that language.

In terms of the Chomsky hierarchy, our main interest is in CFLs, as it turns out that the syntactic structure of most programming languages is represented by CFGs. The grammars and languages we have looked at so far in this book have all been context free (remember that any *regular* grammar or language is, by definition, also context free).

2.8 A Type 0 Grammar: Computation as Symbol Manipulation

We close this chapter by considering a grammar that is more complex than our previous examples. The grammar, which we label G_4, has productions as follows (each row of productions has been numbered, to help us to refer to them later):

$$S \rightarrow AS \mid AB \tag{2.1}$$
$$B \rightarrow BB \mid C \tag{2.2}$$
$$AB \rightarrow HXNB \tag{2.3}$$
$$NB \rightarrow BN \tag{2.4}$$
$$BM \rightarrow MB \tag{2.5}$$
$$NC \rightarrow Mc \tag{2.6}$$
$$Nc \rightarrow Mcc \tag{2.7}$$
$$XMBB \rightarrow BXNB \tag{2.8}$$
$$XBMc \rightarrow Bc \tag{2.9}$$
$$AH \rightarrow HA \tag{2.10}$$
$$H \rightarrow a \tag{2.11}$$
$$B \rightarrow b \tag{2.12}$$

G_4 is a type 0, or unrestricted grammar. It would be context sensitive, but for the production $XBMc \rightarrow Bc$, which is the only production with a right-hand side shorter than its left-hand side.

Table 2.14 represents the derivation of a particular sentence using this grammar. It is presented step by step. Each sentential form, apart from the *sentence* itself, is followed by the number of the row in G_4 from which the production used to achieve the next step was taken. Table 2.14 should be read row by row, left to right.

The sentence derived is $a^2b^3c^6$. Notice how, in Table 2.14, the grammar replaces each A in the sentential form $AABBBC$ by H, and each time it does this it places one

Table 2.14. A type 0 grammar is used to derive a sentence

Stage	Row	Stage	Row	Stage	Row
S	(1)	AS	(1)	AAB	(2)
AABB	(2)	AABBB	(2)	AABBBC	(3)
AHXNBBBC	(4)	AHXBNBBC	(4)	AHXBBNBC	(4)
AHXBBBNC	(6)	AHXBBBMc	(5)	AHXBBMBc	(5)
AHXBMBBc	(5)	AHXMBBBc	(8)	AHBXNBBc	(4)
AHBXBNBc	(4)	AHBXBBNc	(7)	AHBXBBMcc	(5)
AHBXBMBcc	(5)	AHBXMBBcc	(8)	AHBBXNBcc	(4)
AHBBXBNcc	(7)	AHBBXBMccc	(9)	AHBBBccc	(10)
HABBBccc	(3)	HHXNBBBccc	(4)	HHXBNBBccc	(4)
HHXBBNBccc	(4)	HHXBBBNccc	(7)	HHXBBBMcccc	(5)
HHXBBMBcccc	(5)	HHXBMBBcccc	(5)	HHXMBBBcccc	(8)
HHBXNBBcccc	(4)	HHBXBNBcccc	(4)	HHBXBBNcccc	(7)
HHBXBBMccccc	(5)	HHBXBMBccccc	(5)	HHBXMBBccccc	(8)
HHBBXNBccccc	(4)	HHBBXBNccccc	(7)	HHBBXBMcccccc	(9)
HHBBBcccccc	(11)	aHBBBcccccc	(11)	aaBBBcccccc	(12)
aabBBcccccc	(12)	aabbBcccccc	(12)	aabbbcccccc	

c at the rightmost end for each B. Note also how the grammar uses non-terminals as "markers" of various types:

- H is used to replace the As that have been accounted for
- X is used to indicate how far along the Bs we have reached
- N is used to move right along the Bs, each time ending in a c being added to the end of the sentential form
- M is used to move left back along the Bs.

You may also notice that at many points in the derivation several productions are applicable. However, many of these productions lead eventually to "dead ends", i.e. sentential forms that cannot lead eventually to sentences.

The language generated by G_4, i.e. $L(G_4)$, is $\{a^i b^j c^{i \times j}: i, j \geqslant 1\}$. This is the set:

all strings of the form one or more as followed by one or more bs followed by cs in which the number of cs is the number of as *multiplied* by the number of bs.

You may wish to convince yourself that this is the case.

G_4 is rather a complicated grammar compared to our earlier examples. You may be wondering if there is a simpler *type* of grammar, perhaps a CFG, that can do the same job. In fact there is not. However, while the grammar is comparatively complex, the method it embodies in the generation of the sentences is quite simple. Essentially, like all grammars, it simply replaces one string by another at each stage in the derivation.

An interesting way of thinking about G_4 is in terms of it performing a kind of *computation*. Once a sentential form like $A^i B^j C$ is reached, the productions then ensure that $i \times j$ cs are appended to the end by essentially modelling the simple algorithm in Table 2.15.

The question that arises is: what range of computational tasks can we carry out using such purely syntactic transformations? We see from our example that the type 0 grammar simply specifies string substitutions. If we take our strings of as and bs as representing numbers, so that, say, a^6 represents the *number* 6, we see that G_4 is essentially a model of a process for multiplying together two arbitrary length numbers.

Later in this book, we encounter an abstract machine, called a *Turing machine* (TM), that specifies string operations, each operation involving the replacing of only one symbol by another, and we see that the machine is actually as powerful as the type 0 grammars. Indeed, the machine is capable of performing a wider range of computational tasks than even the most powerful real computer.

However, we will not concern ourselves with these issues until later. In the next chapter, we encounter more of the fundamental concepts of formal languages: *syntax*, *semantics*, and *ambiguity*.

Table 2.15. The "multiplication" algorithm embodied in grammar G_4

```
for each A do
  for each B do
    put a c at the end of the sentential form
  endfor
endfor
```

2.9 Exercises

For exercises marked †, solutions, partial solutions, or hints to get you started appear in "Solutions to Selected Exercises" at the rear of the book.

2.1. Classify the following grammars according to the Chomsky hierarchy. In all cases, briefly justify your answer:

(a)† $S \rightarrow aA$
$A \rightarrow aS \mid aB$
$B \rightarrow bC$
$C \rightarrow bD$
$D \rightarrow b \mid bB.$

(b)† $S \rightarrow aS \mid aAbb$
$A \rightarrow \varepsilon \mid aAbb.$

(c) $S \rightarrow XYZ \mid aB$
$B \rightarrow PQ \mid S$
$Z \rightarrow aS.$

(d) $S \rightarrow \varepsilon.$

2.2†. Construct set definitions of each of the languages generated by the four grammars in Exercise 2.1.

Hint: the language generated by 2.1(c) is not the same as that generated by 2.1(d), as one of them contains no strings at all, whereas the other contains exactly one string.

2.3†. It was pointed out above that we usually insist that one or more non-terminals must be included in the left-hand side of type 0 productions. Write down a formal expression representing this constraint. Assume that N is the set of non-terminals, and T the set of terminals.

2.4. Construct regular grammars, G_v, G_w, and G_x, such that

(a) $L(G_v) = \{c^j : j > 0,$ and j does not divide exactly by 3$\}$.

(b) $L(G_w) = \{a^i b^j [cd]^k : i, k \geqslant 0, 0 \leqslant j \leqslant 1\}$

Note: as we are dealing only with whole numbers, the expression $0 \leqslant j \leqslant 1$, which is short for $0 \leqslant j$ and $j \leqslant 1$, is the same as writing: $j = 0$ or $j = 1$.

(c) $L(G_x) = \{a, b, c\}^*.$

2.5†. Use your answer to 2.4(c) as the basis for sketching out an intuitive justification that A^* is a regular language, for any alphabet, A.

2.6. Use the symbol $\Rightarrow$ in showing the step by step derivation of the string c^5 using (a) G_v and (b) G_x from Exercise 2.4.

2.7.   Construct CFGs, G_y and G_z, such that

(a) $L(G_y) = \{a^{2i+1}c^jb^{2i+1} : i \geq 0, \ 0 \leq j \leq 1\}$

   Note: if $i \geq 0$, a^{2i+1} means "all odd numbers of as".

(b)[†] $L(G_z) = $ all Boolean expressions in your favourite programming language (Boolean expressions are discussed in Chapter 13).

2.8.   Use the symbol $\Rightarrow$ in showing the step-by-step derivation of a^3b^3 using

(a) G_y from Exercise 2.7,

and the grammar

(b) G_3 from Chapter 2, i.e. $S \rightarrow ab \mid aSb$

2.9.   Provide a regular grammar to generate the language $\{ab, abc, cd\}$.

   Hint: make sure your grammar generates only the three given strings, and no others.

2.10[†].   Use your answer to Exercise 2.9 as the basis for sketching out an intuitive justification that any finite language is regular.

   Note: the converse, i.e. that every regular language is finite, is certainly not true. Consider, for example, the languages specified in Exercise 2.4.

Notes

1. The formalism we describe here is actually *Extended* BNF (EBNF). The original BNF did not include the repetition construct found in Table 2.5.
2. Jensen and Wirth (1975), see Further Reading section.

Chapter 3
Syntax, Semantics, and Ambiguity

3.1 Overview

In this chapter, we consider the following aspects of formal languages, and their particular relevance to programming languages:

- the *syntax*, or *grammatical structure*, of a sentence
- the *semantics*, or *meaning*, of a sentence
- the graphical representation of *derivations* by structures called *derivation* trees
- *parsing*, or trying to discover the grammatical structure of a given sentence
- *ambiguity*, when a sentence in a formal language has more than one possible meaning.

3.2 Syntax Vs. Semantics

The *phrase structure grammars* (PSGs), as introduced in the preceding chapter, describe the *syntax* of languages. The syntax of a language is the set of rules by which "well-formed" phrases, which we call *sentences*, come about. The Backus–Naur form (BNF) definitions or syntax charts for Pascal to which we referred in the previous chapter tell us about the correct *form* of Pascal statements, but do not tell us what the statements cause the machine to do, after they have been compiled. They do not tell us how to write a program to compute a particular function, or solve a given problem. To write programs to do these things requires us to know about the *meaning* of the program constructs, i.e. their *semantics*.

Some of the languages we looked at in Chapter 2 had no semantics whatsoever (or at least none that we referred to). For semantics is not form alone, but also "interpretation", and, like syntax, requires that we have access to a set of rules which tell us how to make this interpretation. For natural languages such as English these rules of interpretation are extremely complex, and not completely understood. Moreover, the rules of natural languages are not necessarily universal, and are subject to constant revision, especially in artistic usage.

The semantics of a sentence is its meaning. For a program, the term "semantics" refers to the computation carried out after the program source code has been compiled, when the program actually runs. In *formal languages*, meaning is inextricably linked to form. The grammatical structure of a sentence, by which we mean an account of how the *productions* were applied to obtain the sentence, is assumed to determine the meaning that we attribute to that sentence. We shall see certain implications of this below, but first we provide a foundation for our discussion by investigating a very useful way of representing the structure of context free *derivations*.

3.3 Derivation Trees

As we know, *context free grammars* (CFGs) have productions where every left-hand side consists of a single *non-terminal*, a property also shared by the *regular grammars*. In each step (application of one production) of the derivation, therefore, a single non-terminal in the *sentential form* will have been replaced by a string of zero or more *terminals* and/or non-terminals to yield the next sentential form. Any derivation of a sentential form begins with S, the *start symbol*, alone. These features mean that we can represent the derivation of any sentential form as a *tree*, and a very useful feature of trees is that they can be represented graphically, and therefore be used to express the derivation in a visually useful way. Trees used to represent derivations are called *derivation trees* (or sometimes *parse trees*, as we see later).

Let us describe how *derivation trees* are constructed by using examples. First, a simple tree, featuring grammar G_3 from Chapter 2:

$$S \rightarrow ab \mid aSb.$$

Consider the derivation:

$$S \Rightarrow aSb \Rightarrow aaSbb \Rightarrow aaabbb.$$

This can be represented as the tree in Figure 3.1. Figure 3.2 shows how the derivation tree relates to the derivation itself.

In a tree, the circles are called *nodes*. The nodes that have no nodes attached beneath them are called *leaf* nodes, or sometimes *terminal nodes*. In such a tree, the resulting string is taken by reading the *leaf* nodes in left to right order.

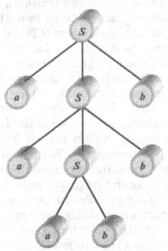

Figure 3.1. A derivation tree

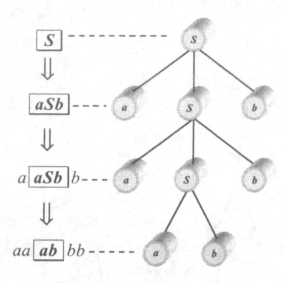

Figure 3.2. A derivation and its correspondence with the derivation tree

A slightly more complex example is based on grammar G_2, also from Chapter 2:

$$S \to aB \mid bA \mid \varepsilon$$
$$A \to aS \mid bAA$$
$$B \to bS \mid aBB.$$

The derivation

$$S \Rightarrow aB \Rightarrow aaBB \Rightarrow aabSB \Rightarrow aabB \Rightarrow aabaBB \Rightarrow aababSB$$
$$\Rightarrow aababB \Rightarrow aababbS \Rightarrow aababb$$

is represented by the tree in Figure 3.3.

Dashed lines have been included in Figure 3.3 to clarify the order in which we read the terminal symbols from the tree.

Note how, for the purposes of drawing the derivation tree, the *empty string, ε,* is treated exactly as any other symbol. However, remember that the final string, *aababb*, does not show the εs, as they disappear when *concatenated* into the resulting string.

3.4 Parsing

The *derivation* of a string, as described in Chapter 2, is the process of applying various productions to produce that string. In the immediately preceding example, we *derived* the string *aababb* using the productions of grammar G_2. *Parsing*, on the other hand, is the creation of a structural account of the string according to a grammar. The term *parsing* relates to the Latin for the phrase *parts of speech*. As for *derivation trees*, we also have *parse trees*, and the *derivation* trees shown above are also *parse* trees for the corresponding strings (i.e. they provide an account of the *grammatical structure* of the strings).

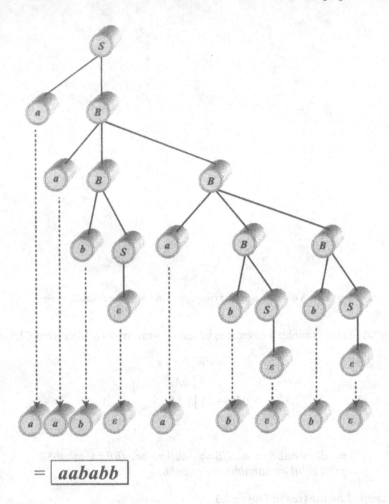

Figure 3.3. A more complex derivation tree

Parsing is an important part of (formal and natural) language understanding. For the compilation of source programs it is absolutely crucial, since unless the compiler can arrive at an appropriate parse of a statement, it cannot be expected to produce the appropriate object code.

For now we look at the two overall approaches to parsing, these being *top-down* and *bottom-up*. The treatment given here to this subject is purely abstract, and takes a somewhat extreme position. There are actually many different approaches to parsing, and it is common to find methods that embody elements of both top-down and bottom-up approaches.

The essence of parsing is that we start with a grammar, G, and a *terminal string*, x, and we say:

- construct a derivation (parse) tree for x that can be produced by the grammar G.

Now, while x is a *terminal string*, it may not be a *sentence*, so we may need to precede the above instruction with "*find out if x is a sentence and if so ...*". As previously stated, there are two overall approaches to parsing, these being *top-down* and *bottom-up*. We now briefly describe the two approaches.

In *top-down parsing*, we build the parse tree for the terminal string *x*, starting from the top, i.e. from *S*, the start symbol. The most purely top-down approach would be to use the grammar to repeatedly generate different sentences until *x* was produced. *Bottom-up parsing* begins from the terminal string *x* itself, trying to build the parse tree from the sentence upwards, finishing when the tree is fully formed, with *S*, the start symbol at the top.

To clarify bottom-up parsing, we will consider one bottom-up approach, known as the *reduction* method of parsing. We begin with our grammar, *G*, and make a grammar G_{red}, by swapping over the left- and right-hand sides of all of the productions in *G*. We call the resulting "productions" *reductions*. The goal then becomes to use the *reductions* to eventually arrive at *S* after starting with the string *x*.

We again consider grammar G_3, i.e.

$$S \rightarrow aSb \mid ab$$

and the problem of producing the parse tree for *aaabbb*. The "reductions" version of G_3, called G_{3red}, is as follows:

$$aSb \rightarrow S$$
$$ab \rightarrow S.$$

We start with our sentence *x*, and seek a left-hand side of one of the reductions that matches some *substring* of *x*. We replace that substring by the right-hand side of the chosen reduction, and so on. We terminate when we reach a string consisting only of *S*. Parsing is simple in our example, which is shown in Figure 3.4, since only one reduction will be applicable at each stage.

In the example in Figure 3.4, there was never a point at which we had to make a choice between several applicable reductions. In general, however, a grammar may be such that at many points in the process there may be several applicable reductions, and several substrings within the current string that match the left-hand sides of various reductions.

As a more complex example, consider using the reduction method to produce a parse tree for the string *aababb*. This is a sentence we derived using grammar G_2 above. For G_2, the reductions G_{2red} are:

$$\varepsilon \rightarrow S$$
$$aB \rightarrow S$$
$$bA \rightarrow S$$
$$aS \rightarrow A$$
$$bAA \rightarrow A$$
$$bS \rightarrow B$$
$$aBB \rightarrow B.$$

We have to begin with the $\varepsilon \rightarrow S$ reduction, to place an *S* somewhere in our string so that one of the other reductions will apply. However, if we place the *S* after the first *a*, for example, we will be unable to complete the tree (you should convince yourself of this). The same applies if we place the *S* after the second *a* in the string. In fact, you will appreciate that a reduction rule with an ε on its left-hand side can

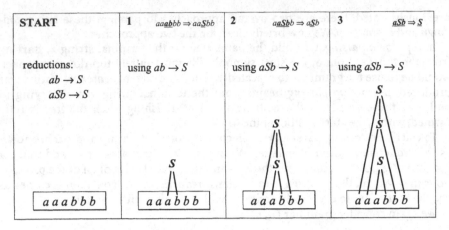

Figure 3.4. Reduction parsing

be applied *at any stage* in the parse. Moreover, suppose that later on in the parse we reach, say, *aabSaBB*? How do we choose between *aBB* → *B* and *aB* → *S*?

We shall see in Chapters 4 and 5 how some of the problems of parsing can be simplified for certain classes of grammars, i.e. the *regular* grammars and (some of) the CFGs. We do this by establishing that, associated with these grammars, are *abstract machines*. We then see that such machines can be used as a basis for designing parsing programs for the languages in question. For now, however, we look at an extremely important concept that relates to languages in general.

3.5 Ambiguity

Now that we have familiarised ourselves with the notion, and some of the problems, of parsing, we are going to consider one of the most important related concepts of formal languages, i.e. that of *ambiguity*. Natural language permits, and indeed in some respects actually thrives on, ambiguity. In talking of natural language, we usually say a statement is ambiguous if it has more than one possible meaning. The somewhat contrived example often used to demonstrate this is the phrase:

>   Fruit flies like a banana.

Does this mean "the insects called fruit flies are positively disposed towards bananas"? Or does it mean "something called fruit is capable of the same type of trajectory as a banana"? These two potential meanings are partly based on the (at least) two ways in which the phrase can be *parsed*. The former meaning assumes that "Fruit flies" is a single noun, while the latter assumes that the same phrase is a noun followed by a verb.

The point of the above example is to show that while our intuitive description of what constitutes ambiguity makes reference to *meaning*, the different meanings can in fact be regarded as the *implication* of the ambiguity. Formally, ambiguity in a language is reflected in the existence of more than one parse tree for one or more sentences of that language. To illustrate this, let us consider an example that relates to a programming language called Algol60, a language that influenced the designers of Pascal. The example also applies to Pascal itself, and is based on the

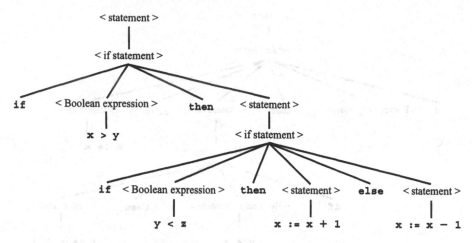

Figure 3.5. PARSE TREE 1: A derivation tree for a Pascal "if statement"

following fragment of a programming language definition, for which we revert, for convenience, to the *BNF* notation introduced in Chapter 2:

<statement> ::= <if statement> | <arithmetic expression> | ...
<if statement> ::= *if* <Boolean expression> *then* <statement> |
                 if <Boolean expression> *then* <statement> *else*
                 <statement>

Remember that the terms in angled brackets ("<statement>", etc.) are *non-terminals*, and that other items (*if, then* and *else*) are regarded as *terminals*.

Now consider the following statement:

```
if x > y then if y < z then x := x + 1 else x := x − 1.
```

Suppose that x > y and y < z are <Boolean expression>s, and that x := x + 1 and x := x − 1 are <arithmetic expression>s. Then two different parse trees can be constructed for our statement. The first is shown in Figure 3.5, and is labelled "PARSE TREE 1".

"PARSE TREE 2", an alternative parse tree for the same statement is shown in Figure 3.6.

We have two distinct structural accounts of a single sentence. This tells us that the grammar is *ambiguous*.

Now, suppose that the compiler for our language used the structure of the parse tree to indicate the order in which the parts of the statement were executed. Let us write out the statement again, but this time indicate (by inserting "[" and "]" into the statement) the interpretation suggested by the structure of each of the two trees.

PARSE TREE 1 (Figure 3.5) suggests

```
if x > y then [if y < z then x := x + 1 else x := x − 1].
```

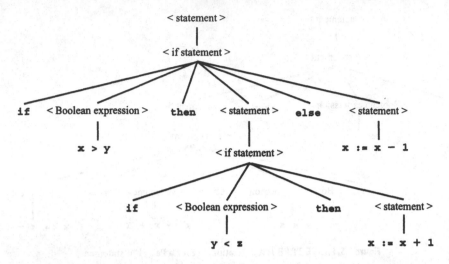

Figure 3.6. PARSE TREE 2: A different derivation tree for the same "if statement" as that in Figure 3.5

However, PARSE TREE 2 (Figure 3.6) suggests

```
if x > y then [if y < z then x := x + 1] else x := x − 1.
```

The "else part" belongs to different "if parts" in the two interpretations. This would clearly yield different results according to which interpretation our compiler made. For example, if $x = 1, y = 1$, and $z = 2$, then in case 1 the execution of our statement would result in x still having the value 1, while after execution in case 2, x would have the value 0.

More seriously, a compiler for our language on one manufacturer's machine may create PARSE TREE 1 (Figure 3.5), while a compiler for another type of machine creates PARSE TREE 2 (Figure 3.6). What is more, a note would probably have to be added to the syntax descriptions of our programming language, for example, in the manual, to explain which interpretation was assumed by the particular compiler. This is not a good policy to adopt in programming language definition, as the syntax for the programming language ought to be specified in a way that is subject to a single interpretation, and that interpretation should be obvious from looking at the formal definition of the syntax. That is to say, the language should be *unambiguous*.

Now, our example statement above is actually syntactically correct Pascal, as an "if" statement is not a compound statement in Pascal, and therefore the subordinate if statement ("if $y < z$ then ...") does not need to be bracketed with begin and end. Thus, there is ambiguity in the standard definition of Pascal. The solution to this in the original Pascal manual was to inform the reader (in an additional note) which interpretation would be taken. The solution adopted for Algol was to change a later version of the language (Algol68) by introducing "bracketing tokens"; sometimes called "guards" (similar to the *endif*s used in the algorithms in this book).

If we have bracketing tokens, and we would like the interpretation specified by PARSE TREE 1 (Figure 3.5), we write:

```
if x > y then if y < z then x := x + 1 else
    x := x − 1 endif endif.
```

If, on the other hand, we desire the PARSE TREE 2 (Figure 3.6) interpretation we write:

```
if x > y then if y > z then x := x + 1 endif else
    x := x - 1 endif.
```

The difference in meaning between the two statements is now very clear.

While ambiguity is, as mentioned above, an accepted part of natural languages, it *cannot* be accepted in programming languages, where the consequences of ambiguity can be serious. However, ambiguity in natural language also leads to problems on occasions. Krushchev, a president of the former Soviet Union, spoke a phrase to western politicians that was translated into English literally as "Communism will bury you". This was interpreted to be a threat that the Soviet Union would destroy the west, but in Russia the original phrase also has the meaning "Communism will outlast you!".

To complete this introduction to ambiguity, we will precisely state the definition of ambiguity in formal grammars.

A PSG, G, is *ambiguous* if L(G) contains *any* sentence which has two or more distinct derivation trees.

Correspondingly:

A PSG, G, is *unambiguous* if L(G) contains *no* sentences which have two or more distinct derivation trees.

If we find any sentence in L(G) which has two or more derivation trees, we have established that our grammar is ambiguous. Sometimes, an ambiguous grammar can be replaced by an *un*ambiguous grammar which does the same job (i.e. generates exactly the same language). This was not done to solve the Algol60 problem described above, since in that case new terminal symbols were introduced (the bracketing terms), and so the new grammar generated many different sentences from the original grammar, and therefore the *language* was changed.

Ambiguity is problematic, as there is no general solution to the problem of determining whether or not an arbitrary PSG is ambiguous. There are some individual cases where we can show that a grammar is unambiguous, and there are some cases where we can show that a given grammar is ambiguous (exactly what happened in the case of Algol60). However, there is no general solution to the ambiguity problem, since establishing that a grammar is unambiguous could require that we make sure that there is absolutely no sentence that the grammar generates for which we can construct more than one derivation tree.

In programming languages, ambiguity is a great cause for concern, as stated above. The main reason for this is that the compiler is usually designed to use the parse tree of a program to derive the structure of the object code that is generated. It is obvious therefore, that the programmer must be able to write his or her program in the knowledge that there is only one possible interpretation of that program. Being unable to predict the behaviour of a statement in a program directly from the text of that statement could result in dire consequences both for the programmer and the user.

3.6 Exercises

For exercises marked †, solutions, partial solutions, or hints to get you started appear in "Solutions to Selected Exercises" at the rear of the book.

3.1. Given the following grammar, *G*:

$$S \rightarrow XC \mid AY$$
$$X \rightarrow aXb \mid ab$$
$$Y \rightarrow bYc \mid bc$$
$$A \rightarrow a \mid aA$$
$$C \rightarrow c \mid cC.$$

   (a)   Classify *G* according to the *Chomsky* hierarchy.

   (b)† Write a *set definition* of *L(G)*.

   (c)† Using the sentence $a^3 b^3 c^3$ in *L(G)*, show that *G* is an *ambiguous* grammar.

   Note: L(G) is known as an "inherently ambiguous language", as any CFG that generates it is necessarily ambiguous. You might like to justify this for yourself (hint: think about the derivation of sentences of the form $a^i b^i c^i$).

3.2. Given the following productions of a grammar, *G*:

$$S \rightarrow E$$
$$E \rightarrow T \mid E + T \mid T - E$$
$$T \rightarrow 1 \mid 2 \mid 3$$

   (note that 1, 2, 3, +, and − are terminal symbols);
   and the following sentence in *L(G)*:

$$3 - 2 + 1.$$

   (a)   Use the sentence to show that *G* is an *ambiguous grammar*.

   (b)† Assuming the standard arithmetic interpretation of the terminal symbols, and with particular reference to the example sentence, discuss the *semantic* implications of the ambiguity.

3.3. Discuss the semantic implications of the ambiguity in the grammar:

$$P \rightarrow P \, or \, P \mid P \quad and \quad P \mid x \mid y \mid z$$

   (where "or", and "and" are terminals),

assuming the usual Boolean logic interpretation of "or" and "and" (Boolean logic is discussed in Chapter 13). Introduce new terminals "(" and ")" in such a way to produce a grammar that generates similar, but unambiguous logical expressions.

3.4. Given the following ambiguous grammar:

$$S \rightarrow T \mid A \mid T \text{ or } S \mid A \text{ or } S$$
$$A \rightarrow T \text{ and } T \mid T \text{ and } A \mid \text{not } A$$
$$T \rightarrow a \mid b \mid c \mid \text{not } T$$

(where *or, and, not, a, b,* and *c* are *terminal symbols*).

Discuss possible problems that the ambiguity might cause when interpreting the sentence *a or not b and c* (assuming the usual Boolean logic interpretation of the symbols *and, or* and *not*).

Chapter 4
Regular Languages and Finite State Recognisers

4.1 Overview

This chapter is concerned with the most restricted class of languages in the *Chomsky hierarchy* (as introduced in Chapter 2): the *regular* languages. In particular, we encounter our first, and most simple, type of abstract machine, the *finite state recogniser* (FSR).

We discover:

- how to convert a *regular grammar* into an FSR
- how any *FSR* can be made *deterministic* so it never has to make a choice when *parsing* a regular language
- how a *deterministic* FSR (DFSR) converts directly into a simple *decision program*
- how to make an FSR as small as possible
- the *limitations* of the regular languages.

4.2 Regular Grammars

As we saw in Chapter 2, a *regular grammar* is one in which every production conforms to one of the following patterns:

$$X \to xY \qquad X \to y$$

where X and Y are each *single non-terminals*, x is a terminal, and y is either the *empty string* (ε), or a *single terminal*.

Here are some productions that conform to this specification (A and B are non-terminals, a and b are terminals):

$$A \to \varepsilon$$
$$A \to aB$$
$$B \to bB$$
$$A \to bA$$
$$A \to b$$
$$B \to a.$$

49

When we discussed *reduction parsing* in Chapter 3, we saw that productions such as $A \rightarrow \varepsilon$, with ε, the empty string on the right-hand side can cause problems in parsing. However, for reasons to be discussed later, we can ignore them for the present, and assume for now that regular grammars contain no productions with ε on the right-hand side.

Here is an example of (the productions of) a regular grammar, G_5:

$$S \rightarrow aS \mid aA \mid bB \mid bC$$
$$A \rightarrow aC$$
$$B \rightarrow aC$$
$$C \rightarrow a \mid aC$$

and here is an example *derivation* of a *sentence*:

$$S \Rightarrow aS \Rightarrow aaS \Rightarrow aaaA \Rightarrow aaaaC \Rightarrow aaaaa.$$

An equivalent *derivation tree* as introduced in Chapter 3, can be seen in Figure 4.1.

You may have observed that the above grammar is also *ambiguous* (ambiguity was discussed in Chapter 3), since there is at least one alternative derivation tree for the given sentence. Try to construct one, to practice creating derivation trees.

In this chapter, we use G_5 to illustrate some important features of regular languages. For now, you might like to try to write a *set definition* of the language generated by the grammar. You can assess your attempt later in the Chapter, when $L(G_5)$ is defined.

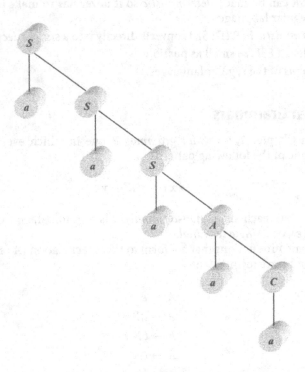

Figure 4.1. A derivation tree produced by a regular grammar

4.3 Some Problems with Grammars

It is not always straightforward to define a language by examining the corresponding grammar. Take, for example, grammar G_5 above. One problem is that various symbols representing the same entity are scattered throughout the productions (for example, the non-terminal C). Another problem is that, as computer scientists, we are not usually satisfied by merely designing and writing down a grammar, we usually want to write a *parser* for the language it generates. This leads to questions that grammars do not conveniently allow us to address. Will our grammar yield an efficient parser (in terms of space and time)? If we design another grammar with fewer productions, say, to generate the same language, is there a way of being sure that it does indeed do this? What about the design of grammars themselves: is there a more convenient design notation which we can use in place of the textual form? Then there are productions, such as $S \rightarrow aS$ and $S \rightarrow aA$, that may cause problems in the parsing of the language, as was discussed in Chapter 3. Can we eliminate situations where there is a choice, such as these?

For the regular grammars, the answer to all the above questions is "yes".

4.4 Finite State Recognisers and Finite State Generators

For the regular grammars there is a corresponding *abstract machine* that can achieve the same tasks (parsing and generating) as those supported by the grammar. Moreover, as the structure of the regular grammar is so simple, the corresponding machine is also simple. We call the machine the *FSR*. It is called a *recogniser* because it "recognises" if a string presented to it as input belongs to a given language. If we were using the machine to *produce*, rather than *analyse*, strings, we would call it a *finite state generator*.

4.4.1 Creating an FSR

We can devise an *equivalent* FSR from any regular grammar, by following the simple rules in Table 4.1.

Figure 4.2 demonstrates the application of part (a) of the instructions in Table 4.1 to our example grammar G_5.

Figure 4.3 shows the machine that is produced after parts (b) and (c) of the procedure in Table 4.1. have been carried out.

To show its correspondence with the grammar from which it was constructed (G_5), we will call our new machine M_5.

To support a discussion of how our machines operate, we need some new terminology, which is provided by Table 4.2.

4.4.2 The Behaviour of the FSR

A machine is of little use if it does not do something. Our machines are abstract, so we need to specify their rules of operation. Table 4.3 provides us with these rules.

It can be seen from Table 4.3 that there are two possible conditions that cause the halting of the machine:

1. the input is *exhausted*, so no transitions can be made
2. the input is *not exhausted*, but no transitions are possible.

Table 4.1. Constructing an FSR from the productions of a regular grammar

The construction rules (X and Y denote any non-terminals, y denotes any terminal)

(a) Convert all productions using whichever of the following three rules applies, in each case:

1. a production of the form
   $$X \to yY \ (Y \neq X)$$
   becomes a drawing of the form

2. a production of the form
   $$X \to yX$$
   becomes a drawing of the form

3. a production of the form
   $$X \to y$$
   becomes a drawing of the form

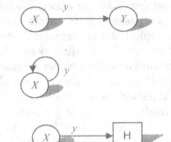

H is not a non-terminal in the grammar. We use the label H each time the rule applies.

(b) Take all the drawings produced in (a), and join all of the circles and squares with the same labels together.
   (Once the basic concept is appreciated, one usually finds the whole machine can be drawn directly)

(c) The circle ⓢ (where S is the *start symbol* of the grammar) becomes ⤳ⓢ
   (i.e. an arrow is added that comes from nowhere and points to S)

The result is a pictorial representation of a *directed graph* ("*digraph*") that we call a *finite state recogniser* (FSR)

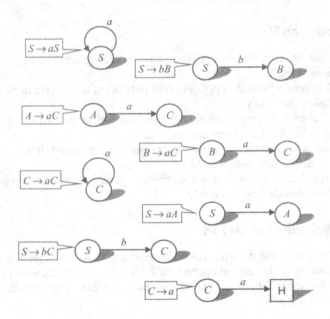

Figure 4.2. The first part of expressing a regular grammar as an FSR. This represents the application to grammar G_5 of part (a) of Table 4.1

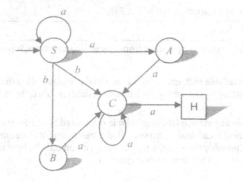

Figure 4.3. The FSR M_5 that represents regular grammar G_5

It can also be seen that one of two possible additional conditions must prevail when the machine halts:

A. it is in an *acceptance* state
B. it is in a *non-acceptance* state.

Table 4.2. Terminology for referring to FSRs

Term	Pictorial representation (from M_5 – Figure 4.3)
states (also called *nodes*)	S A B C H
arcs Note: this is a *labelled* arc. The label is the terminal *a*. The arc that enters state S from nowhere is an *unlabelled* arc	X_1 ──*a*──▶ X_2
ingoing arcs	C
outgoing arcs	C
Special states *start state* (has an *ingoing arc* coming from nowhere)	──▶ S
halt or *acceptance* state	H

Table 4.3. The rules governing the behaviour of the FSR

Rule no.	Description
1	The FSR begins its operation from any of its *start states*. It is said to be *in* that start state.
	It has available to it any string of terminal symbols, called the *input string*. If the input string is non-empty, its leftmost symbol is called the *current input symbol*
2	When in any state, q (the symbol q is often used to denote states in abstract machines) it can make a move, called a *transition* (or *state transition*) to any other (possibly the same) single state, q_1, if and only if there is an arc as follows (x represents the *current input symbol*):

Note that either or both of q or q_1 can be halt states

After making a transition, the following conditions hold:

- the machine is *in* state q_1
- the symbol in the input string immediately to the right of the current input symbol now becomes the current input symbol. If there is no such symbol we say that the input is *exhausted*

If, at any point, any transition is possible, the machine *must* make one of them. However, at any point where more than one transition is possible, the choice between the applicable transitions is completely arbitrary

Note: when the machine makes a transition and moves to the next symbol in the input string, we say that the machine is *reading* the input

| 3 | When the machine is in a given state, and can make no transition whatsoever, the machine is said to have *halted* in that state |

Our machine is very simple, and does not produce output as a program might, and so, if we imagine that we are blessed with the power to observe the internal workings of the machine, the situation that we observe at the time of halting is of critical importance. What we will observe when the machine stops is the situation "inside" the machine with respect to the combination of (1) or (2) with (A) or (B) from above, as represented in Table 4.4.

Table 4.5 defines the *acceptance* and *rejection* conditions of the FSR.

We now consider examples of the machine M_5 in operation. These examples are shown in Figures 4.4–4.6. In each case, we show the initial configuration of the machine and input, followed by the sequence of transitions. In each of Figures 4.4–4.6, the emboldened state in a transition represents the state that the

Table 4.4. Possible state $\times$ input string configurations when the FSR halts

	Condition of input	
State in which FSR halts	Input exhausted	Input not exhausted
Acceptance (halt) state	Accepted	Rejected
Non-acceptance state	Rejected	Rejected

Table 4.5. String acceptance and rejection conditions for the FSR

When the FSR halts …
… if the FSR is in an *acceptance* (*halt*) state and the input is *exhausted*, we say that the input string is *accepted* by the FSR
… if the FSR is not in an *acceptance* (*halt*) state, or is in a halt state and the input is not *exhausted*, we say that the input string is *rejected* by the FSR

machine reaches on making that transition. The arrow beneath the input string shows the new current input symbol, also after a transition is made.

Example 1. Input string: *aaba* (Figure 4.4)

Example 1 (Figure 4.4) shows how the string *aaba* can be *accepted* by M_5. You can observe that several choices were made, as if by magic, as to which transition to take in a given situation. In fact, you could show a sequence of transitions that lead to *aaba* being *rejected*. This appears to be a problem with rule 2 in Table 4.3, to which we return later. First, another example.

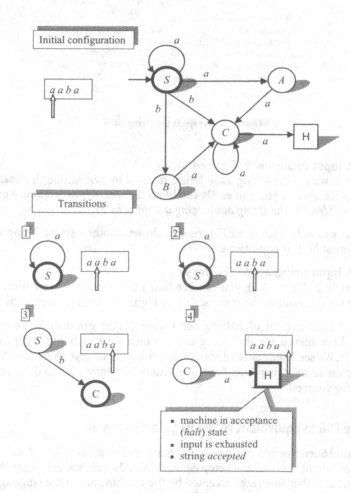

Figure 4.4. M_5 accepts the string *aaba*

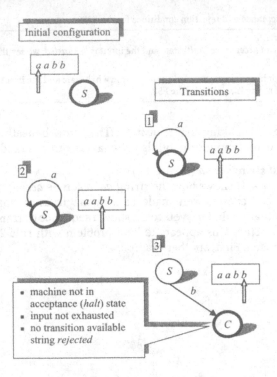

Figure 4.5. M_5 rejects the string *aabb*

Example 2. Input string: *aabb* (Figure 4.5)
In Figure 4.5, we see the string *aabb* being *rejected* by M_5. Although choices were made at several points, you can easily establish that *there is no sequence of transitions that will lead to the string aabb being accepted by M_5*.

The next example, shown in Figure 4.6, shows another string being rejected under different *halting* conditions.

Example 3. Input string: *aaabb* (Figure 4.6)
As in example 2 (Figure 4.5), you can see that no sequence of transitions can be found that would result in the input string in Figure 4.6 being accepted by M_5.

The final combination of halting conditions (*input exhausted, not in accept state*) can be achieved for the string *aaba*, which we saw being accepted above (example 1). We see a sequence of transitions that will achieve these rejection conditions for the same string later. For the moment, you may wish to discover such a sequence for yourself.

4.4.3 The FSR as Equivalent to the Regular Grammar

Let us remind ourselves why we created the machine in the first place. Recall the earlier statement about the machine being *equivalent* to the original grammar. What this ought to mean is that all strings accepted by the machine are in the language generated by the grammar (and vice versa). The problem is that, according to the particular

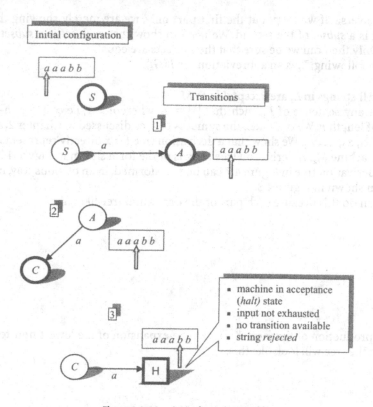

Figure 4.6. M_5 rejects the string *aaabb*

transition chosen in the application of rule 2 (Table 4.3), the machine can sometimes reject a string that (like *aaba* in example 1, above) *is* generated by the grammar. We want to say something such as: "all strings that *can* be accepted by the machine are in the language generated by the grammar, and all strings that are generated by the grammar *can* be accepted by the machine". That is exactly what we do next.

First we say this:

- a string is *acceptable* if there is some sequence of *transitions* which lead to that string being *accepted*.

Thus, *aaba*, from example 1, is acceptable, even though we could provide an alternative sequence of transitions that would lead to the string being rejected. On the other hand, the strings from examples 2 and 3 are not acceptable, and also cannot be generated by the grammar. You may wish to convince yourself that the preceding statement is true.

Suppose that G_z is any regular grammar, and we call the FSR derived from G_z's productions M_z, we can say that *the set of acceptable strings of M_z is exactly the same as* $L(G_z)$, *the language generated by the grammar G_z.* The rules we followed to produce M_z from G_z ensure that this is the case, but to convince ourselves, let us argue the case thoroughly.

To prove that two sets (in this case, *the set of acceptable strings of M_z*, and $L(G_z)$) are equal, we have to show that all members of one set are members of the other, and vice versa. If we carry out the first part only, we are merely showing that the first set is a *subset* of the second. We need to show that *each set is a subset of the other*. Only then can we be sure that the two sets are equal.

In the following, L_z is an abbreviation for $L(G_z)$.

Part 1. All strings in L_z are acceptable to M_z

Let x be any sentence of L_z, such that $|x| = n$, where $n \geqslant 1$, i.e. x is a non-empty string of length n. We can *index* the symbols of x, as discussed in Chapter 2, and we write it $x_1, x_2, \ldots, x_n$. We show that a derivation tree for x directly represents a part of the machine M_z. A derivation tree for x is of the form shown in Figure 4.7.

The derivation tree in Figure 4.7 can be transformed, in an obvious way, into the diagram shown in Figure 4.8.

We can do this because each part of the derivation tree like this:

used a production of the form $X \to xY$. The expansion of the lowest non-terminal node in the tree will look like this:

which used a production of the form $X \to x$ (single non-terminal to single terminal). The sequence of productions used in Figure 4.7 become, if the conversion algorithm is applied, the FSR of Figure 4.8, since $X_1 = S$ (the start symbol of the grammar), and a production of the form $X \to x$ becomes an arc:

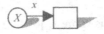

Some of the non-terminals may, of course, appear more than once in the derivation tree, and may have to be joined together (as specified in part (b) of Table 4.1). For any sentence, x, of the language there will thus be a path in the FSR that will accept that sentence – x is an *acceptable* string to M_z.

Part 2. All acceptable strings of M_z are in L_z

Let x be any acceptable string of M_z. Choose any path through M_z that accepts x. We can write out this path so that it has no *loops* in it (as in Figure 4.8). We then convert this version into a tree using the converse of the rules applied to the derivation tree in Figure 4.7. You can easily justify that the resulting tree will be a derivation tree for x that uses only productions of G_z. Thus, any *acceptable* string of M_z can be generated by G_z and is thus in L_z.

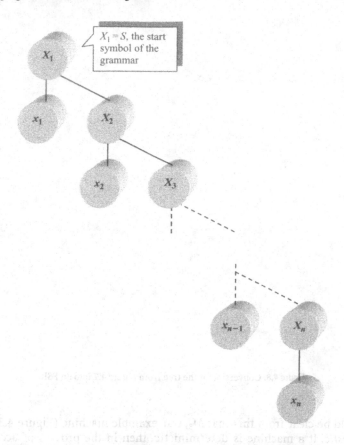

Figure 4.7. A derivation tree for an arbitrary regular grammar derivation

We have shown (*part* 1) that every string generated by G_z is acceptable to M_z, and (*part* 2) that every string acceptable to M_z is generated by G_z. We have thus established the equivalence of M_z and G_z.

4.5 Non-determinism in FSRs

As was pointed out above, rule 2 of the rules governing the behaviour of an FSR (see Table 4.3) implies that there may be a choice of transitions to make when the machine is in a given state, with a given current input symbol. As we saw in the case of example 1, a wrong choice of transition in a given state may result in an acceptable string being rejected. It is not rule 2 itself that is at fault, it is the type of machine that we may produce from a given set of productions. We will now state this more formally.

- If a FSR/generator contains any state with two or more identically labelled *outgoing arcs*, we call it a *non-DFSR*.
- If a FSR/generator contains no states with two or more identically labelled *outgoing arcs*, we call it a *DFSR*.

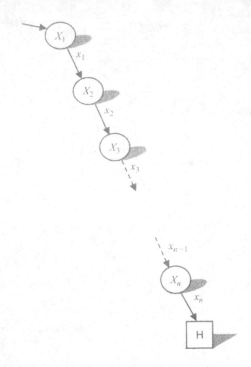

Figure 4.8. Conversion of the tree from Figure 4.7 into an FSR

It should be clear from this that M_5, our example machine (Figure 4.3), is *non-deterministic*. If a machine is deterministic, then in the process of accepting or rejecting a string, it *never has to make a choice as to which transition to take, for a given current input symbol, when in a given state; there is only one possible transition which can be made at that point.* This considerably simplifies the processes of parsing, as we shall see in the next section. Incidentally, we can still use the same rules for FSR behaviour as those above, for DFSRs; the "choice" of transitions will at each point consist of only one option.

In following our rules to produce an FSR from the productions of a regular grammar, we may create a *non*-deterministic machine. Remember our machine M_5 (Figure 4.3), and in particular our discussion about input string *aaba*. If the wrong sequence of transitions is chosen, for example:

S to S reading an a,

S to S reading an a,

S to B reading a b,

B to C reading an a,

the machine's halting conditions would be *input exhausted, not in halt state*; indicating a rejection of what we know to be an *acceptable* string. This implies that a parser that behaves in a similar way to a non-DFSR might have to try many sequences of applicable transitions to establish whether or not a string was acceptable. Worse than this, for a non-acceptable string, the parser might have to try *all possible sequences of applicable transitions of length equal to the length of the string* before being sure that the string could indeed be rejected.

The main problem associated with non-determinism, therefore, is that it may result in parsers that have to *backtrack*. Backtracking means returning to a state of affairs (for FSRs this means to a given *state* and *input symbol*) at which a choice between applicable transitions was made, in order to try an alternative. Backtracking is undesirable (and as we shall see, for regular languages it is also *unnecessary*), since

- a backtracking parser would have to save arbitrarily long input strings in case it made the wrong choice at any of the choice points (as it may have made several transitions *since* the choice point)
- backtracking uses up a lot of time.

If an FSR is deterministic there is, in any given state, only one choice of transition for any particular input symbol, which means that a corresponding parser *never* has to backtrack.

4.5.1 Constructing Deterministic FSRs

It may surprise you to learn that *for every set of regular productions we can produce an equivalent DFSR*. As we will see, we can then use that FSR to produce a simple, but efficient, deterministic decision program for the language. First, we consider a procedure to convert any non-DFSR into an *equivalent* DFSR. We will call the old machine M and the new deterministic machine M^d.

Table 4.6 presents the procedure, which is called the *subset construction algorithm*. Our machine M_5 will be used to illustrate the steps of the procedure. Table 4.6, followed by Figures 4.9–4.14 show how the new machine M_5^d is built up.

Continuing the execution of step 2, we will choose the state of M_5^d with A and S in it (see the note that follows step 2 of Table 4.6). This results in the partial machine shown in Figure 4.9.

We still have states of M_5^d with no *outgoing arcs* in Figure 4.9, so we choose one (in this case, the one with B and C inside it). The result is shown in Figure 4.10.

Now, from Figure 4.10 we choose the state containing A, S, and C, resulting in Figure 4.11.

From Figure 4.11, we choose the state that contains A, S, C and H. The result can be seen in Figure 4.12.

Then we choose the state in Figure 4.12 that contains only C and H. Figure 4.13 shows the outcome.

Finally, from Figure 4.13 we choose the empty state of M_5^d. This state represents unrecognised input in certain states of M_5. Thus, if M_5^d finds itself in this state it is reading a string that is not in the language. To make M_5^d totally deterministic, we need to cater for the remainder (if any) of the string that caused M_5^d to reach this "*null*" state. Thus we draw an arc, for each terminal in M_5's alphabet, leaving and immediately re-entering the null state. The final machine is as shown in Figure 4.14.

If the machine ever finds itself in the null state, it will never leave it, but will continue to process the input string until the input is *exhausted*. It will not accept such a string, however, since the null state is not a halt state.

Table 4.6. The *subset algorithm*

Step	Description
1	(a) Copy out all of the start states of M Note: If M was constructed from a regular grammar it will have only one start state However, in general, an FSR can have several start states (see later in this chapter)
	(b) If all of the states you just drew are non-acceptance states, draw a circle (or ellipse) around them all, otherwise draw a square (or rectangle) around them all. We call the resulting object a *state* of M^d
	(c) Draw an *unlabelled ingoing* arc (see Table 4.2) to the shape you drew in (b)

Note: The result of step 1 for M_5 is:

2	Choose any state of M^d that has no outgoing arcs

	for each terminal, t, in the alphabet of M do
	    for each state, q, of M inside the chosen state of M^d do
	       copy out each state of M which has an ingoing arc labelled t leaving q
	       (but only copy out any given state once)
	    endfor

	if you just copied out NO states, draw an empty circle else do step 1(b) then return here
	if the state of M^d you have just drawn contains exactly the same states of M as an existing state of M^d, rub out the state of M^d you have just drawn

	Draw an arc labelled t from the chosen state of M^d to either
	    (i) the shape you just drew at (b)
	    or
	    (ii) the state of M^d which was the same as the one you rubbed out
	 endfor

Note: the result of the first complete application of step 2 to M is:

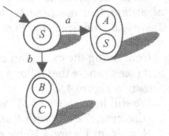

| 3 | Repeat step 2 until every single state of M^d has exactly one outgoing arc for each terminal in M's alphabet |

The subset algorithm now terminates (see step 3 of Table 4.6), leaving us with a totally deterministic machine, M_5^d (shown in Figure 4.14) which is equivalent to M_5 (Figure 4.3).

We can now tidy up M_5^d by renaming its states so that the names consist of single symbols, resulting in the machine we see in Figure 4.15.

This renaming of states illustrates that the names of states in an FSR are insignificant (as long as each label is distinct). What is important in this case is that there is a one-to-one association between the states and arcs of the old M_5^d and the states and arcs of the renamed machine.

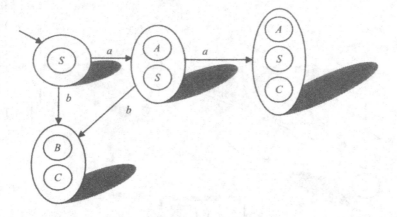

Figure 4.9. The subset algorithm in action (1)

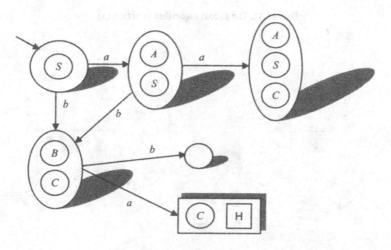

Figure 4.10. The subset algorithm in action (2)

The null state and its ingoing and outgoing arcs are sometimes removed from the deterministic machine, their presence being assumed. Later, we will see that the null state in the DFSR has its uses.

4.5.2 The DFSR as Equivalent to the Non-DFSR

We will now convince ourselves that M_5 and M_5^d are equivalent. From M_5^d, we observe the following properties of acceptable strings:

- if they consist of *a*s alone, then they consist of three or more *a*s
- they can consist of any number (≥ 0) of *a*s, followed by exactly one *b*, followed by one or more *a*s.

This can be described in set form:

$$\{a^i : i \geq 3\} \cup \{a^i b a^j : i \geq 0, j \geq 1\}$$

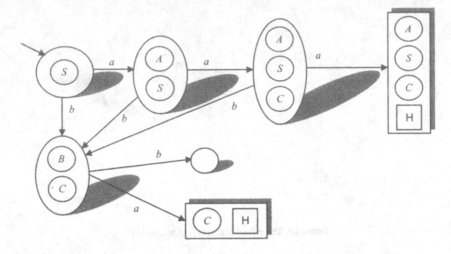

Figure 4.11. The subset algorithm in action (3)

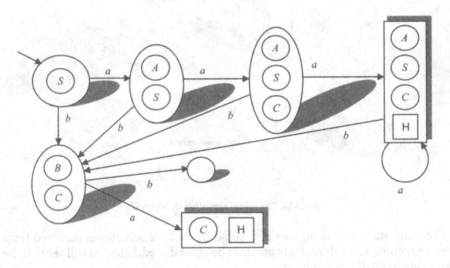

Figure 4.12. The subset algorithm in action (4)

You can verify for yourself that all strings accepted by the original machine (M_5), of Figure 4.3, are also described by one of the above statements.

If we establish that for any non-DFSR, M, we can produce an equivalent deterministic machine, M^d, we will, for the remainder of our discussion about regular languages, be able to assume that any FSR we discuss is deterministic. This is extremely useful, as we will see.

Suppose that M is any FSR, and that M^d is the machine produced when we applied our subset method to M. We will establish two things: (1) every string which is *acceptable* to M is *accepted* by M^d; (2) every string which is *accepted* by M^d is *acceptable* to M.[1]

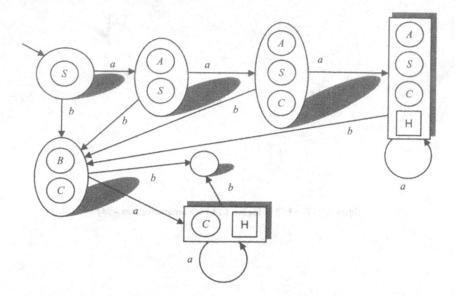

Figure 4.13. The subset algorithm in action (5)

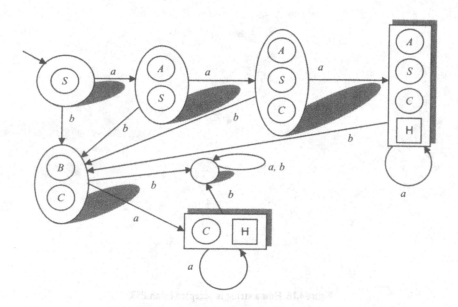

Figure 4.14. The result of the subset algorithm

Part 1. Every acceptable string of M is accepted by M^d

Let x be an *acceptable* string of M. Suppose x has length n, where $n \geqslant 1$. We can *index* the symbols in x, as discussed in Chapter 2: $x_1, x_2, \ldots, x_n$. x is acceptable to M, so there is a path through M, from a start state, S, to a halt state, H, as represented in Figure 4.16.

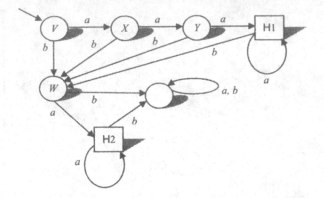

Figure 4.15. The FSR of Figure 4.14 with renamed states – M_5^d

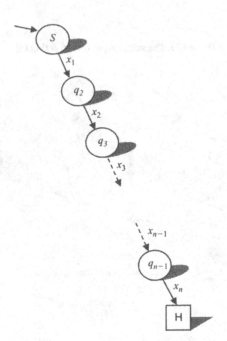

Figure 4.16. How a string is accepted by an FSR

The states on the path in Figure 4.16 are not necessarily all distinct. Also, if $n = 1$, the path looks like this:

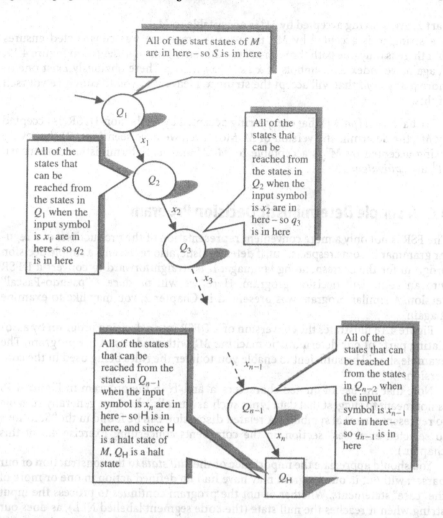

All of the start states of M are in here – so S is in here

Q_1

x_1

All of the states that can be reached from the states in Q_1 when the input symbol is x_1 are in here – so q_2 is in here

Q_2

x_2

All of the states that can be reached from the states in Q_2 when the input symbol is x_2 are in here – so q_3 is in here

Q_3

x_3

All of the states that can be reached from the states in Q_{n-1} when the input symbol is x_n are in here – so H is in here, and since H is a halt state of M, Q_H is a halt state

x_{n-1}

Q_{n-1}

x_n

All of the states that can be reached from the states in Q_{n-2} when the input symbol is x_{n-1} are in here – so q_{n-1} is in here

Q_H

Figure 4.17. How a deterministic version of an FSR (M) accepts the string of Figure 4.16. The states Q_i represent the composite states produced by the subset algorithm. The states q_i represent the states of the original non-deterministic machine from Figure 4.16

With respect to the above path through M, M^d will be as represented in Figure 4.17.

The string x, which was acceptable to M, is thus accepted by M^d. The above representation of the path taken by the string x through M^d may also incorporate any paths for the same string which result in the situations:

- *input exhausted, not in halt state* (the non-halt state of M reached will be incorporated into the halt state of M^d),
- *input not exhausted, in halt state* (other states of M^d on the path in Figure 4.17 may be halt states),
- *input not exhausted, not in halt state* (M would have stopped in one of the states contained in the non-halt states of Figure 4.17, but M^d will be able to proceed to its halt state with the remainder of the string).

Part 2. Every string accepted by M^d is acceptable to M

If a string, x, is accepted by M^d, the way the machine was constructed ensures that there is only one path through M^d for that string, as represented in Figure 4.17, if, again, we index the symbols of x as $x_1, x_2, \ldots, x_n$. There obviously exist one or more paths in M that will accept the string x. I leave it to you to convince yourself of this.

We have seen (*part 1*) that every string accepted by M, the non-DFSR, is accepted by M^d, the deterministic version of M. Moreover, we have seen (*part 2*) that every string accepted by M^d is acceptable to M. M and its deterministic counterpart, M^d, are *equivalent* FSRs.

4.6 A Simple Deterministic Decision Program

The FSR is not only a more convenient representation of the productions of a regular grammar in some respects, but, if deterministic, also represents a simple decision program for the corresponding language. It is straightforward to convert a DFSR into an equivalent decision program. Here, we will produce a "pseudo-Pascal" version. A similar program was presented in Chapter 2; you may like to examine it again.

Figure 4.18 illustrates the conversion of a DFSR into a decision program by associating parts of our deterministic machine M_5^d with modules of the program. The example should be sufficient to enable you to infer the rules being used in the conversion.

Note that the use of quoted characters 'a' and 'b' in the program in Figure 4.18 is not meant to suggest that data types such as Pascal's "char" are generally suitable to represent terminal symbols. (A related discussion can be found in the "Solutions to selected exercises" section, in the comments concerning Exercise 4.3 of this chapter.)

You should appreciate the importance of the *null state* to the construction of our parser: without it, our program may have had no defined action in one or more of the "case" statements. Whether or not the program continues to process the input string when it reaches the null state (the code segment labelled *NLL*), as does our program in Figure 4.18, is a matter of preference. One may require, in certain circumstances, for the program to abort immediately (with a suitable message), as soon as this segment is entered, not bothering with any remaining symbols in the input stream (as do some compilers). On the other hand, it may be desirable to clear the input stream of the remainder of the symbols of the rejected string.

4.7 Minimal FSRs

Each boxed segment of the program in Figure 4.18 represents one *state* in the corresponding FSR, M_5^d. In general, if we were to build such a parser for a regular language, we would like to know that our parser did not feature an unnecessarily large number of such segments.

Now, our program (Figure 4.18) *is* structured, notwithstanding misgivings often expressed about the use of unconditional jumps. If an equally structured program, consisting of fewer such segments, but doing the same job, could be derived, then we would prefer, for obvious and sound reasons, to use that one. In terms of a given

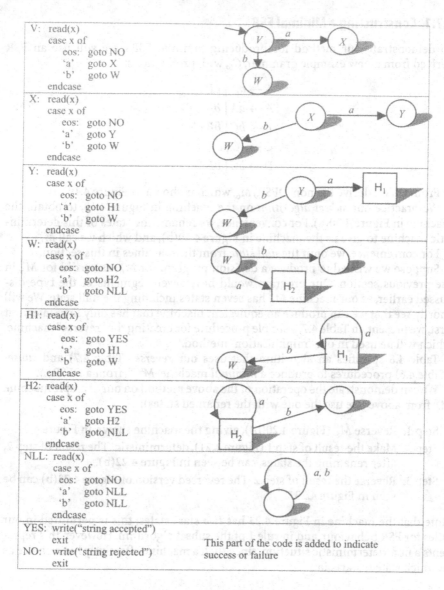

```
V:  read(x)
    case x of
        eos:  goto NO
        'a'   goto X
        'b'   goto W
    endcase
X:  read(x)
    case x of
        eos:  goto NO
        'a'   goto Y
        'b'   goto W
    endcase
Y:  read(x)
    case x of
        eos:  goto NO
        'a'   goto H1
        'b'   goto W
    endcase
W:  read(x)
    case x of
        eos:  goto NO
        'a'   goto H2
        'b'   goto NLL
    endcase
H1: read(x)
    case x of
        eos:  goto YES
        'a'   goto H1
        'b'   goto W
    endcase
H2: read(x)
    case x of
        eos:  goto YES
        'a'   goto H2
        'b'   goto NLL
    endcase
NLL: read(x)
    case x of
        eos:  goto NO
        'a'   goto NLL
        'b'   goto NLL
    endcase
YES: write("string accepted")
     exit
NO:  write("string rejected")
     exit
```

This part of the code is added to indicate success or failure

Figure 4.18. The correspondence between a DFSR (M_5^d of Figure 4.15) and a decision program for the equivalent regular language

FSR, the question is this: can we find an equivalent machine that features fewer states? We will go beyond this, however, we will discover a method for producing a machine, from a given FSR, which is deterministic and uses *the smallest number of states necessary for a DFSR to recognise the same language as that recognised by the original machine*. We call such a machine the *minimal FSR*.

If we produce such a minimal FSR, we can be confident that the corresponding decision program uses the smallest number of segments possible to perform the parsing task. Of course, there may be other methods for reducing the amount of code, but these are not of interest here.

4.7.1 Constructing a Minimal FSR

To demonstrate our method for producing minimal FSRs, we will use an FSR derived from a new example grammar, G_6, with productions:

$$S \rightarrow aA \mid bB$$
$$A \rightarrow aA \mid bB \mid bD$$
$$B \rightarrow bC \mid bB \mid bD$$
$$C \rightarrow cC \mid c$$
$$D \rightarrow bC \mid bD.$$

From G_6, we derive the non-DFSR, M_6, which is shown in Figure 4.19.

We practice our *subset algorithm* on the machine in Figure 4.19, to obtain the machine in Figure 4.20(a). For convenience, we rename the states of this deterministic machine to give us the machine in Figure 4.20(b), and which we call M_6^d:

For convenience, we omit the *null state* from the machines in this section.

Suppose we wished to produce a decision program for M_6^d, as we did for M_5^d in the previous section. Our program would have seven segments of the type discussed earlier, as our machine M_6^d has seven states including the null state. We will shortly see that we can produce an equivalent machine that has only five states, but first, we present, in Table 4.7, a simple procedure for creating the "reverse" machine, which will be used in our "minimisation" method.

Table 4.8 specifies an algorithm that uses our *reverse* (Table 4.7) and *subset* (Table 4.6) procedures to produce a *minimal* machine, M^{min}, from a DFSR, M.

We can demonstrate the operation of the above method on our example machine M_6^d from above (we use the one with the renamed states):

Step 1. Reverse M_6^d (Figure 4.20(b)), giving the machine shown in Figure 4.21.

Step 2. Make the result of step 1 (Figure 4.21), deterministic. The result of step 2, after renaming the states, can be seen in Figure 4.22(b).

Step 3. Reverse the result of step 2. The reversed version of Figure 4.22(b) can be seen in Figure 4.23.

Note that the machine in Figure 4.23 has *two* start states; this is permitted in our rules for FSR behaviour, and in rule 1 of the subset algorithm. However, this represents a non-deterministic situation, since such a machine effectively has a choice as to which state it starts in.

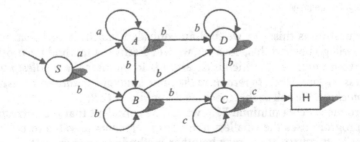

Figure 4.19. The non-DFSR, M_6

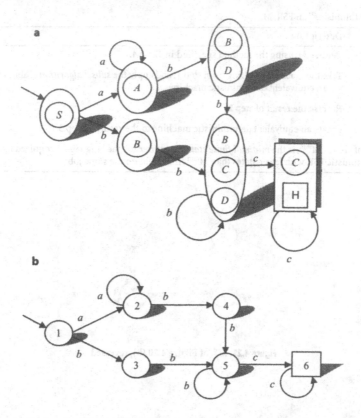

Figure 4.20. a M_6^d, the output of the subset algorithm for input M_6; and **b** the machine in (a) with renamed states

Table 4.7. Deriving M^{rev}, the "reversed" form of an FSR, M

Step	Description	
1	Let M^{rev} be an exact copy of M	
2	Objects of this form in M ...	... become objects of this form in M^{rev}

Note that all other parts of M^{rev} remain unchanged

If L is the language recognised by M, then the language L^{rev}, recognised by M^{rev}, is L *with each and every sentence reversed*

Table 4.8. "Minimising" an FSR, M

Step	Description
1	*Reverse M*, using the method specified in Table 4.7
2	Take the machine resulting from step 1, and apply the *subset algorithm* (Table 4.6) to create an equivalent deterministic machine
3	*Reverse* the result of step 2
4	Create an equivalent deterministic machine to the result of step 3

The result of step 4, $M^{\min}$, is the *minimal finite state recogniser* for the language recognised by M. There is no deterministic FSR with fewer states than $M^{\min}$ that could do the same job

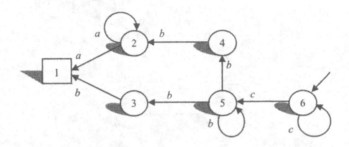

Figure 4.21. M_6^d, of Figure 4.20 (b), reversed

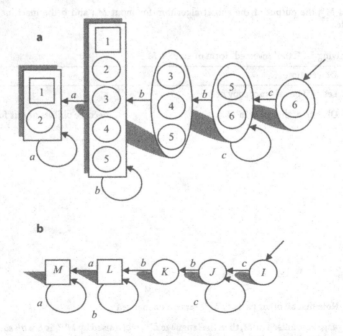

Figure 4.22. a The deterministic version of Figure 4.21; and **b** the machine in (a) with renamed states

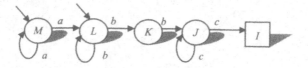

Figure 4.23. The reverse construction of Figure 4.22 (b)

Now for the final step of the minimisation process:

Step 4. Make the result of step 3 (Figure 4.23) deterministic. We obtain the machine depicted in Figure 4.24(a).

The machine in Figure 4.24(b), resulting from renaming the states of the machine in Figure 4.24(a), has been called M_6^{min}. It is the *minimal DFSR* for the language recognised by M_6^d.

We now see clearly that the language recognised by M_6^{min} (and, of course, M_6 and M_6^d), is:

$$\{a^i b^j c^k : i \geqslant 0, j \geqslant 2, k \geqslant 1\},$$

i.e. all sentences are of the form:

zero or more *a*s, followed by two or more *b*s, followed by one or more *c*s.

With a little thought, we can easily justify that the machine is minimal. We must have one state for the "zero or more *a*s" part, a further *two* states for the "two or more *b*s" part, and an extra state to ensure the "one or more *c*s" part.

4.7.2 Why Minimisation Works

The minimisation algorithm of Table 4.8 simply uses the *subset algorithm* twice, first applying it to the reversed form of the original machine, and then to the

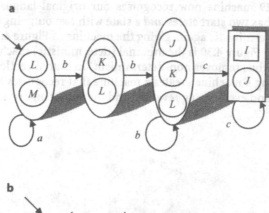

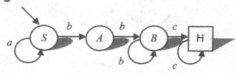

Figure 4.24. a The deterministic version of Figure 4.23; and b the machine in (a) with renamed states, which is M_6^{min}, the minimal version of M_6

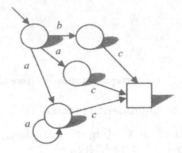

Figure 4.25. An FSR which will have a non-deterministic reverse construction

reversed form of the deterministic reversed form. Let us consider why this approach produces the correct result. To do this, we focus on the example FSR in Figure 4.25.

Making the machine in Figure 4.25 deterministic, by using our subset algorithm, we merge the two paths labelled *a* into one. We then obtain the machine in Figure 4.26.

Now, while the machine in Figure 4.26 is deterministic, it is not *minimal*. It seems that we ought to be able to deal with the fact that, regardless of whether the first symbol is an *a* or a *b*, the machine could get to its halt state given one *c*. However, as this is not a non-deterministic situation, the subset algorithm has ignored it.

Now, reversing the machine in Figure 4.26 yields the machine in Figure 4.27.

The "three *cs* situation" has become non-deterministic in Figure 4.27. The subset algorithm *will* now deal with this, resulting in the machine of Figure 4.28.

While the machine in Figure 4.28 now consists of only four states, and is deterministic, it unfortunately recognises our original language in reverse, so we must reverse it again, giving Figure 4.29.

The Figure 4.29 machine now recognises our original language, but is non-deterministic (it has two start states *and* a state with two outgoing arcs labelled *a*). So, we make it deterministic again, giving the machine of Figure 4.30.

The machine in Figure 4.30 is the minimal (deterministic) machine.

The algorithm thus removes non-determinism "in both directions" through the machine. Reversing a machine creates a machine that recognises the original language with each and every string *reversed*, and the subset algorithm always results

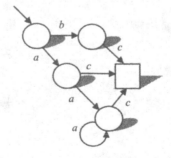

Figure 4.26. A deterministic version of Figure 4.25

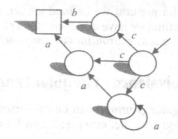

Figure 4.27. The reverse construction of Figure 4.26

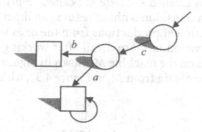

Figure 4.28. A deterministic version of Figure 4.27

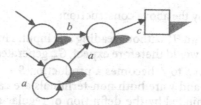

Figure 4.29. The reverse construction of Figure 4.28. (Note that this is also an example of an FSR with two start states)

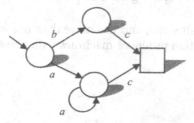

Figure 4.30. Deterministic version of Figure 4.29, the minimal version of Figure 4.25

in a machine that accepts exactly the same strings as the machine to which it is applied. Thus, making the reversed machine deterministic and then reversing the result will not change the language recognised by the machine.

There are other ways of creating a minimal FSR. One of the most popular is based on *equivalence sets*. The beauty of the method we use here is that it is very simple to specify, and makes use of a method we defined earlier, i.e. the *subset algorithm*. Thus, the minimisation method we have used here has a pleasing modularity. It uses the *subset* and *reverse* methods as *subroutines* or *components*, so to speak.

4.8 The General Equivalence of Regular Languages and FSRs

We saw above that any *regular grammar* can be converted into an equivalent FSR. In fact, the converse is also true, i.e. every FSR can be converted into a regular grammar.

We will consider an example showing the conversion of an FSR into regular *productions*. The interesting thing about our example is that the particular machine in question has transitions labelled ε. These are called *empty moves* or ε *moves*, and enable an FSR to make a transition without reading an input symbol. As we will see, they produce slightly different productions from the ones we have seen up to now. We consider this observation in more detail after working through the following example, which is based on the machine M shown in Figure 4.31.

The grammar that we obtain from M, in Figure 4.31, which we will call G_m, is as follows:

$$S \to aA \mid A$$
$$A \to bA \mid bX \mid Y$$
$$X \to aX \mid \varepsilon$$
$$Y \to aX \mid \varepsilon.$$

Some issues are raised by the above construction:

- M can get to its halt state $H1$ without reading any input. Thus, ε is in the language accepted by M. As we would therefore expect, G_m generates ε ($S \Rightarrow A \Rightarrow Y \Rightarrow \varepsilon$).
- The *empty move* from S to A becomes a production $S \to A$. Productions of the form $x \to y$ where x and y are both non-terminals are called *unit productions*, and are actually prohibited by the definition of regular grammar productions with which we began this chapter. However, we see in Chapter 5 that there is a method for removing these productions, leaving a grammar that *does* conform to our specification of regular grammar productions, and generates the same language as the original.
- M has arcs *leaving* halt states, and has more than one halt state. We saw above that the subset algorithm produces machines with these characteristics. In any

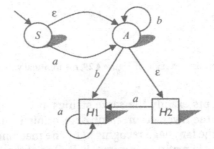

Figure 4.31. An FSR, M, with "empty moves" ("ε *moves*")

case, there is nothing in the definition of the FSR that precludes multiple halt *or* start states. Where there is an arc leaving a halt state, such as the arc labelled *a* leaving *H2*, the machine could either halt or read an *a*. Halting in such a situation is exactly the same as an empty move, so in the grammar the *H2* situation becomes $Y \to aX \mid \varepsilon$, where *Y* represents *H2*, and *X* represents *H1*.

I leave it to you to convince yourself that $L(G_m)$ is identical to the language accepted by *M*, and that the way the productions of G_m were obtained from *M* could be applied to any FSR, no matter how complex, and no matter how many empty moves it has. In the above machine, the start state was conveniently labelled *S*. We can always label the states of an FSR how we please; a point made earlier in this chapter. Note that, for FSRs with more than one start state we can simply introduce a new *single* start state, with outgoing arcs labelled ε leaving it and directly entering each state that was a start state of the original machine (which we then no longer class as start states). Then the machine can be converted into productions, as was *M*, above.

The above should convince you that the following statement is true:

for any FSR, *M*, there exists a regular grammar, *G*, which is such that $L(G)$, the language generated by *G*, is equal to the set of strings accepted by *M*.

Our earlier result that for *every* regular grammar there is an *equivalent* FSR means that if we come across a language that *cannot* be accepted by an FSR, we know that we are wasting our time trying to create a regular grammar to generate that language. As we will see in Chapter 6, there are properties of the FSR that we can exploit to establish that certain languages could not be accepted by the FSR. We now know that if we do this, we have also established that we could not define a regular grammar to generate that language. Conversely, if we can establish that there is no regular grammar to generate a given language, we know it is pointless attempting to create an FSR, or a parser of "FSR-level-power", like the decision program earlier in the chapter, to accept that language.

4.9 Observations on Regular Grammars and Languages

The properties of regular languages described in this chapter make them very useful in a computational sense. We have seen that simple *parsers* can be constructed for languages generated by a regular grammar. We have also seen how we can make our parsers extremely efficient, by ensuring that they are deterministic, reduced to the smallest possible form, and reflect a modular structure. We have also seen that there is a general *equivalence* between the FSRs and the regular languages.

Unfortunately, the simple structure of regular grammars means that they can only be used to represent simple constructs in programming languages, for example, basic objects such as *identifiers* (variable names) in Pascal programs. In fact, the definition of Pascal "identifiers" given in context free form in Chapter 2 could be represented as a regular grammar. The regular languages can also be represented as symbolic statements known as *regular expressions*. These are not only a purely formal device, but are also implemented in text editors such as such as "vi" (Unix), and "emacs".

We will defer more detailed discussion of the limitations of regular grammars until Chapter 6. In that chapter we show that even a language as simple as $\{a^i b^i : i \geq 1\}$, generated by the two-production *context free* grammar (CFG) $S \to aSb \mid ab$

(grammar G_3 of Chapters 2 and 3) is not regular. If such a language cannot be generated by a regular grammar, then a regular grammar is not likely to be capable of specifying syntactically valid Pascal programs, where, for example, a parser has to ensure that the number of *begins* is the same as the number of *ends*. In fact, when it comes to programming languages, the *CFGs* as a whole are much more important.[2] CFGs and context free languages (CFLs) are the subject of the next chapter.

4.10 Exercises

For exercises marked †, solutions, partial solutions, or hints to get you started appear in "Solutions to Selected Exercises" at the rear of the book.

4.1. Construct FSRs *equivalent* to the following grammars. For each FSR briefly justify whether or not the machine is *deterministic*. If the machine is *non-deterministic*, convert it into an equivalent deterministic machine.

(a)[†] $S \rightarrow aA$

$A \rightarrow aS \mid aB$

$B \rightarrow bC$

$C \rightarrow bD$

$D \rightarrow b \mid bB.$

Note: This is the grammar from Chapter 2, Exercise 1(a).

(b) $S \rightarrow aA \mid aB \mid aC$

$A \rightarrow aA \mid aC \mid bC$

$B \rightarrow aB \mid cC \mid c$

$C \rightarrow cC \mid c.$

(c) Grammar G_1 from Chapter 2, i.e.

$S \rightarrow aS \mid bB$

$B \rightarrow bB \mid bC \mid cC$

$C \rightarrow cC \mid c.$

4.2. Given the FSR in Figure 4.32,

(a) Produce a *minimal* version.

(b) Construct an equivalent regular grammar.

Hint: it is often more straightforward to use the original, non-deterministic machine (in this case, the one in Figure 4.32) to obtain the productions of the grammar, i.e. we write down the productions that would have given us the machine according to our conversion rules.

4.3[†]. Design suitable data structures to represent FSRs. Design a program to apply operations such as the *subset method* and the *minimisation algorithm* to FSRs.

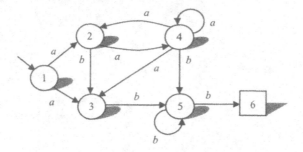

Figure 4.32. A finite state recogniser

4.4. The *subset algorithm* is so called because each new state of M^d, the deterministic machine, produced contains a *subset* of the states of *M*. An *algorithm* is a procedure that always terminates for valid inputs. Consider why we can be sure that the subset procedure will terminate, given any non-deterministic machine as its input, and thus deserves to be called an algorithm.

Note: this has something to do with the necessary limit on the number of new, composite states that can be created by the procedure. This point is explored in more detail in Chapter 13.

Notes

1. Note that this is not the only way of doing this: we could show (1) and then for (2) show that every string that is *non-acceptable* to *M* is *rejected* by M^d. You might like to try this approach.
2. Bear in mind that the class of the CFGs includes the class of the regular grammars.

Chapter 5
Context Free Languages and Pushdown Recognisers

5.1 Overview

This chapter considers the *context free grammars* (CFGs) and *context free languages* (CFLs).

First, we consider two methods for changing the productions of CFGs without changing the language they generate. These two methods are:

- removing the empty string (ε) from a CFG
- converting a CFG into *Chomsky normal form* (CNF).

We find that, as for the *regular languages*, there is an *abstract machine* for the CFLs called the *pushdown recogniser* (PDR), which is like a *finite state recogniser* (FSR) equipped with a storage "device" called a *stack*.

We consider:

- how to create a *non-deterministic* PDR (NPDR) from a CFG
- the difference between *deterministic* PDRs (DPDRs) and NPDRs
- the *deterministic* and *non-deterministic* CFLs
- the implications of the fact that not all CFLs are *deterministic*.

5.2 Context Free Grammars and Context Free Languages

Recall from Chapter 2 that the CFGs (type 2 in the *Chomsky hierarchy*) are *phrase structure grammars* (PSGs) in which each and every production conforms to the following pattern:

$$x \to y, \quad x \in N, \ y \in (N \cup T)^*,$$

i.e. x is a single *non-terminal*, and y is an arbitrary (possibly empty) mixture of *terminals* and or *non-terminals*.

So, the grammar G_2 of Chapter 2, i.e.

$$S \to aB \mid bA \mid \varepsilon$$
$$A \to aS \mid bAA$$
$$B \to bS \mid aBB$$

is context free. If G is a CFG, then $L(G)$, the *language generated by the grammar G*, is a *CFL*.

We will consider a further example CFG, G_7:

$$S \rightarrow ABC$$
$$A \rightarrow aAb \mid B$$
$$B \rightarrow bB \mid \varepsilon$$
$$C \rightarrow ccC \mid B$$

$$L(G_7) = \{a^h b^i c^{2j} b^k : h, j, k \geqslant 0, i \geqslant h\},$$

i.e. 0 or more as followed by at least as many bs, followed by an even number of (or no) cs, followed by 0 or more bs, as you may like to justify for yourself. You may arrive at a different, though equivalent, set definition; if so, check that it describes the same set as the one above.

5.3 Changing *G* Without Changing *L(G)*

The purpose of this section is to show how one can modify the productions of a CFG in various ways without changing the language the CFG generates. There are more ways of modifying CFGs than are featured here, but the two methods below are relatively simple to apply. The results of this section serve also to simplify various proofs in Chapter 6.

5.3.1 The Empty String (ε)

Our example grammar G_7, above, features the *empty string* (ε) on the right-hand side of one of its productions. The empty string often causes problems in parsing and in terms of *empty moves* in FSRs, as was noted in Chapter 3. For parsing purposes, we would prefer the empty string not to feature in our grammars at all. Indeed, you may wonder whether the empty string actually features in any "real world" situations, or if it is included for mathematical reasons. Consider, for example string *concatenation*, where ε can be considered to play a similar part as 0 in addition, or 1 in multiplication.

The empty string has its uses, one of these being in the definition of programming languages. Pascal, for example, allows a <statement> to be empty (i.e. a statement can be an *empty string*). Such "empty" statements have their practical uses, such as when the programmer leaves empty "actions" in a *case* statement, or an empty *else* action in an *if* statement, to provide a placeholder for additional code to be entered later as needed. Note, however, that the empty string is not a *sentence* of the Pascal language. The Pascal syntax does not allow for an empty <program>, and thus, according to the definition of the language, an empty source code file should result in an error message from the Pascal compiler.

It turns out that for CFGs we can deal with the empty string in one of two ways. Firstly, for *any* PSG (see Chapter 2), G, we can, by applying the rules we are about to see, remove ε altogether from the productions, except for one special case which we discuss later. We will need the terminology defined in Table 5.1 to discuss the empty string in CFGs. The table also features examples based on grammar G_7 from above.

Table 5.1. Terminology for discussing the empty string in grammars

Term	Definition	Examples from G_7
ε production	Any production with ε on the right-hand side	$B \rightarrow \varepsilon$
ε generating non-terminal	Any non-terminal, X, which is such that $X \Rightarrow {}^*\varepsilon$ (i.e. any non-terminal from which we can generate the empty string)	S, A, B, and C since we can have: $S \Rightarrow ABC$ and $A \Rightarrow B \Rightarrow \varepsilon$ and $C \Rightarrow B \Rightarrow \varepsilon$
ε free production	Any production which does not have ε on the right-hand side	All productions except $B \rightarrow \varepsilon$

The method for removing ε *productions* from a grammar is described in Table 5.2. On the right of the table, we see the method applied to grammar G_7.

You should verify for yourself that the new grammar, which in Table 5.2 is called G_8, generates exactly the same set of strings as the grammar with which we started (i.e. G_7).

Referring to Table 5.2, at step 3, the productions added to the set Y were necessary to deal with cases where one or more non-terminals in the original productions could have been used to generate ε. For example, we have the production $S \rightarrow ABC$. Now in G_7, the B could have been used to generate ε, so we need to cater for this by introducing the production $S \rightarrow AC$. Similarly, both the A and the C could have been used to generate ε, so we need a production $S \rightarrow B$, and so on. Of

Table 5.2. How to remove from a CFG

Operation number and description	As applied to G_7
1 Put all the ε *generating* non-terminals into a set, call the set X	S, A, B, C — set X
2 Put all the ε *free* productions into another set, call the set Y	$S \rightarrow ABC$ — set Y $A \rightarrow aAb \mid B$ $B \rightarrow bB$ $C \rightarrow ccC \mid B$
3 For each production, p, in Y, add to Y any ε *free* productions we can make by omitting one or more of the non-terminals in X from p's right-hand side	$S \rightarrow ABC \mid AB \mid AC \mid BC \mid A \mid B \mid C$ $A \rightarrow aAb \mid B \mid ab$ $B \rightarrow bB \mid b$ $C \rightarrow ccC \mid cc \mid B$ — new set Y
4 We only do this if S is member of set X:	
4.1 Replace the non-terminal S *throughout the set Y*, by a new non-terminal, A, *which is not in any of the productions of Y*	$S \rightarrow Z \mid \varepsilon$ $Z \rightarrow ABC \mid AB \mid AC \mid BC \mid A \mid B \mid C$ $A \rightarrow aAb \mid B$ $B \rightarrow bB \mid b$ $C \rightarrow ccC \mid cc \mid B$ — new set Y (G_8)
4.2 Put the productions $S \rightarrow A \mid \varepsilon$ into Y	

course, we still need to include $S \rightarrow ABC$ itself, because, in general, the ε generating non-terminals may generate other things apart from ε (as they do in our example).

Note that step 4 of Table 5.2 is only applied if ε is in the language generated by the grammar (in formal parlance, if $\varepsilon \in L(G)$). This occurs only when S itself is an ε generating non-terminal, as it is in the case of G_7. If S is an ε generating non-terminal for a grammar G, then obviously ε is in $L(G)$. Conversely, if S is not such a non-terminal, then ε cannot be in the language. If we applied step 4 to a grammar for which ε is not in the corresponding language, the result would be a new grammar that did not generate the same language (it would generate the original set of strings along with ε, formally $L(G) \cup \{\varepsilon\}$).

Alternatively, if we did not apply step 4 to a grammar in which S *is* an ε generating non-terminal, then we would produce a grammar which generated all the strings in $L(G)$ except for the string ε (formally $L(G) - \{\varepsilon\}$). We must always take care to preserve exactly the set of strings generated by the original grammar.

If we were to apply the above rules to a *regular grammar*, the resulting grammar would also be regular. This concludes a discussion in Chapter 4 suggesting that the empty string in regular grammars need not be problematical.

For *parsing*, the grammars produced by the above rules are useful, since at most they contain one ε production, and if they do it is $S \rightarrow \varepsilon$. This means that when we parse a string x:

- If x is non-empty, our parser need no longer consider the empty string; it will not feature in any of the productions that could have been used in the derivation of x.
- If, on the other hand, $x = \varepsilon$, either our grammar has the one ε production $S \rightarrow \varepsilon$, in which case we accept x, or the grammar has no ε productions, and we simply reject it.

For parsing purposes our amended grammars may be useful, but there may be reasons for leaving the empty string in a grammar showing us the structure of, say, a programming language. One of these reasons is readability. As we can see, G_8 consists of many more productions (16) than does the equivalent G_7 (7), and thus it could be argued that G_7 is more readily comprehensible to the reader. A format for grammars that is useful for a parser may not necessarily support one of the other purposes of grammars, i.e. to show *us* the structure of languages.

To complete this section, we will consider a further example of the removal of ε productions from a CFG. In this case ε is not in the language generated by the grammar. Only a sketch of the process will be given; you should be able to follow the procedure by referring to the rules in Table 5.2.

The example grammar, G_9, is as follows:

$$S \rightarrow aCb$$
$$C \rightarrow aSb \mid cC \mid \varepsilon.$$

In this case, the only ε generating non-terminal is C. S cannot generate ε, as any string it generates must contain at least one a and one b. The new productions we would add are $S \rightarrow ab$ and $C \rightarrow c$, both resulting from the omission of C. Step 4 of Table 5.2 should not be applied in this case, as S is not ε generating. The final amended productions would be:

$$S \rightarrow aCb \mid ab$$
$$C \rightarrow aSb \mid cC \mid c.$$

A set definition of the language generated by this grammar, which, of course, is exactly the same as $L(G_9)$, is:

$$\{a^{2i+1}c^j b^{2i+1}: \ i, j \geq 0\}$$

i.e. a non-zero odd number of as, followed by 0 or more cs, followed by the same number of bs as there were as.

5.3.2 Chomsky Normal Form

Once we have removed the empty string (ε) from a CFG, isolating it if necessary in one production, $S \to \varepsilon$, there are various methods for transforming the ε *free* part of G while leaving $L(G)$ unchanged. One of the most important of these methods leaves the productions in a form we call *CNF*. This section briefly describes the method for converting a CFG into CNF by presenting a worked example. That we can assume that any CFG can be converted into CNF is useful in Chapter 6.

As an example grammar, we use part of G_8 (the ε free version of G_7), and assume in this case that Z is the *start symbol*:

$$Z \to ABC \mid AB \mid AC \mid BC \mid A \mid B \mid C$$
$$A \to aAb \mid B$$
$$B \to bB \mid b$$
$$C \to ccC \mid cc \mid B.$$

The method to "Chomskyfy" a grammar is as follows. There are four steps, as now explained.

Step 1: Unit productions
Unit productions were briefly introduced in Chapter 4. A unit production is a production of the form: $x \to y$, where x and y are single non-terminals. Thus, G_8 has the following unit productions:

$$Z \to A \mid B \mid C$$
$$A \to B$$
$$C \to B.$$

To help us deal with the unit productions of G_8, we draw the structure shown in Figure 5.1. This represents all derivations that begin with the application of a unit production and end with a string that does not consist of a single non-terminal (each path stops the first time such a string is encountered).

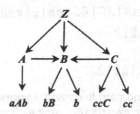

Figure 5.1. The *unit production graph* for grammar G_8

Table 5.3. Example productions and a corresponding unit production graph

Example productions	Resulting graph

$S \rightarrow aS \mid B$

$B \rightarrow bS \mid A \mid C$

$A \rightarrow a \mid C$

$C \rightarrow bA \mid S \mid B$

The structure in Figure 5.1 is not a *tree*, but a *graph* (in fact, like an FSR, it is directed graph, or *digraph*). We can call the structure a *unit production graph*. It could be drawn as a tree if we repeated some *nodes* at lower levels (in the graph in Figure 5.1 we would repeat B beneath A and beneath C). However, such a representation would not be so useful in the procedure for creating new non-unit productions described below.

Some unit production graphs we draw may lead to even more circular ("*cyclical*"), situations, as in the example in Table 5.3.

We now continue with our initial example. From the unit production graph for G_8, (Figure 5.1), we obtain:

- every possible non-unit production we can obtain by following a directed path from every non-terminal to every node that does not consist of a single non-terminal, where each production we make has a left-hand side that is the non-terminal we started at, and a right-hand side that is the string where we stopped.

So, from our unit production graph for G_8 in Figure 5.1, we obtain the following productions (those crossed out are already in G_8, so we need not add them):

$Z \rightarrow bB \mid b \mid aAb \mid ccC \mid cc$ {by following all paths from Z}

$A \rightarrow aAb \mid bB \mid b$ {by following all paths from A}

$B \rightarrow bB \mid b$ {... from B}

$C \rightarrow ccC \mid cc \mid bB \mid b$ {... and from C}.

We remove all of the unit productions from the grammar, and add the new productions to it, giving $G_{8(a)}$ (new parts are underlined):

$Z \rightarrow ABC \mid AB \mid AC \mid BC \mid \underline{bB} \mid \underline{b} \mid \underline{aAb} \mid \underline{ccC} \mid \underline{cc}$

$A \rightarrow aAb \mid \underline{bB} \mid \underline{b}$

$B \rightarrow bB \mid b$

$C \rightarrow ccC \mid cc \mid \underline{bB} \mid \underline{b}$.

You should perhaps take a moment to convince yourself that the terminal strings we can generate from $G_{8(a)}$ are exactly the same as those we can generate starting from Z in G_8 (remember that Z is the start symbol, in this case). Observe, for example, that from the initial grammar, we could perform the derivation $Z \Rightarrow A \Rightarrow B \Rightarrow C \Rightarrow ccC \Rightarrow cccc$. Using $G_{8(a)}$, this derivation is simply $Z \Rightarrow ccC \Rightarrow cccc$.

In Chapter 4, with reference to the conversion of an arbitrary *FSR* into *regular grammar* productions, we noted that *empty moves* in an FSR (transitions between states that require no input to be read, and are thus labelled with ε) correspond to unit productions. You can now see that if we applied step 1 of "Chomskyfication" to the ε free part of a regular grammar containing unit productions, we would obtain a grammar that was also regular.

Step 1 is now complete, as unit productions have been removed. We turn now to step 2.

Step 2: Productions with terminals and non-terminals on the right-hand side, or with more than one terminal on the right-hand side

Our grammar $G_{8(a)}$ now contains no unit productions. We now focus on the productions in $G_{8(a)}$ described in the title to this step. These productions we call *secondary productions*. In $G_{8(a)}$ they are:

$$Z \to bB \mid aAb \mid ccC \mid cc$$
$$A \to aAb \mid bB$$
$$B \to bB$$
$$C \to ccC \mid cc \mid bB.$$

To these productions we do the following:

- replace each terminal, t, by a new non-terminal, N, and introduce a new production $N \to t$, in each case.

So, for example, from $A \to aAb \mid bB$, we get:

$$A \to JAK \mid KB$$
$$J \to a$$
$$K \to b.$$

Now we can use J and K to represent a and b, respectively, throughout our step 2 productions, adding all our new productions to $G_{8(a)}$ in place of the original step 2 productions. Then we replace c by L in all the appropriate productions, and introduce a new production $L \to c$. The result of this we will label $G_{8(b)}$:

$$Z \to ABC \mid AB \mid AC \mid BC \mid b \mid \underline{KB} \mid \underline{JAK} \mid \underline{LLC} \mid \underline{LL}$$
$$A \to \underline{JAK} \mid \underline{KB} \mid b$$
$$B \to \underline{KB} \mid b$$
$$C \to \underline{LLC} \mid \underline{LL} \mid \underline{KB} \mid b$$
$$\underline{J \to a}$$
$$\underline{K \to b}$$
$$\underline{L \to c}.$$

The new productions are again underlined. You can verify that what we have done has not changed the terminal strings that can be generated from Z.

We are now ready for the final step.

Step 3: Productions with more than two non-terminals on the right-hand side
After steps 1 and 2, all our productions will be of the form:

$$x \to y \quad \text{or} \quad x \to z$$

where y consists of two or more *non-terminals*, and z is a single *terminal*.

As the step 3 heading suggests, we now focus on the following productions of $G_{8(b)}$:

$$Z \to ABC \mid JAK \mid LLC$$
$$A \to JAK$$
$$C \to LLC.$$

As an example consider $Z \to ABC$. It will not affect things if we introduce a new non-terminal, say P, and replace $Z \to ABC$ by:

$$Z \to AP$$
$$P \to BC,$$

the $G_{8(b)}$ derivation $Z \Rightarrow ABC$ would then involve doing: $Z \Rightarrow AP \Rightarrow ABC$. As P is not found anywhere else in the grammar, this will not change any of the terminal strings that can be derived. We treat the rest of our step 3 productions analogously, again replacing the old productions by our new ones, and using a new non-terminal in each case. This gives us the grammar $G_{8(c)}$:

$$Z \to \underline{AP} \mid AB \mid AC \mid BC \mid b \mid KB \mid \underline{JQ} \mid \underline{LR} \mid LL$$
$$P \to BC$$
$$Q \to AK$$
$$R \to LC$$
$$A \to \underline{IT} \mid KB \mid b$$
$$T \to AK$$
$$B \to KB \mid b$$
$$C \to \underline{LU} \mid LL \mid KB \mid b$$
$$U \to LC$$
$$J \to a$$
$$K \to b$$
$$L \to c.$$

(Changed and new productions have once again been indicated by underlining.)
The resulting grammar, $G_{8(c)}$, is in *CNF*.

It is important to appreciate that, unlike in G_8, step 3 productions may include right-hand sides of *more than three* non-terminals. For example, we may have a production such as

$$E \to DEFGH.$$

In such cases, we split the production up into productions as follows:

$$E \to DW$$
$$W \to EX$$
$$X \to FY$$
$$Y \to GH.$$

where W, X, and Y are new non-terminals, not found anywhere else in the grammar. A derivation originally expressed $E \Rightarrow DEFGH$ would now involve the following:

$$E \Rightarrow DW \Rightarrow DEX \Rightarrow DEFY \Rightarrow DEFGH.$$

From this example, you should be able to induce the general rule for creating several "two-non-terminals-on-the-right-hand-side" productions from one "more-than-two-non-terminals-on-the-right-hand-side" production.

If we apply steps 1, 2 and 3 to any ε free CFG, we obtain a CFG where all productions are of the form

$$x \to y \quad \text{or} \quad x \to uv,$$

where y is a single *terminal*, and x, u, and v are single non-terminals.

A CFG like this is said to be in *CNF*.

It should therefore be clear that any ε *free* CFG can be converted into CNF.

Once the ε free part of the grammar has been converted into CNF, we need to include the two productions that resulted from step 4 of the method for removing the empty string (Table 5.2), *but only if* step 4 *was carried out for the grammar*. In the case of G_8, step 4 *was* carried out. To arrive at a grammar equivalent to G_8 we need to add the productions $S \to \varepsilon$ and $S \to Z$ to $G_{8(c)}$, and once again designate S as the start symbol.

It is very useful to know that any CFG can be converted into CNF. For example, we will use this "CNF assumption" to facilitate an important proof in the next chapter. In a CFG in CNF, all *reductions* will be considerably simplified, as the *right-hand sides* of productions consist of single terminals or two non-terminals. A *parser* using such a grammar need only examine two adjacent non-terminals or a single terminal symbol in the *sentential form* at a time. This simplifies things considerably, since a *non-CNF* CFG could have a large number of terminals and or non-terminals on the right-hand side of some of its productions.

A final point is that if the CFG is in CNF, any *parse tree* will be such that every non-terminal node expands into either two non-terminal nodes or a single terminal node. Derivations based on CNF grammars are therefore represented by *binary* trees. A binary tree is a tree in which no node has more than two outgoing arcs. A parser written in Pascal could use a three field *record* to hold details of each non-terminal node that expands into two non-terminals (one field for the node's label, and two pointers to the two non-terminals into which the non-terminal expands). A single type of record structure would therefore be sufficient to represent all of the non-terminal nodes of the parse tree.

5.4 Pushdown Recognisers

In Chapter 4, we saw that from any *regular grammar* we can build an *abstract machine* called an FSR. The machine is such that it recognises every string (and no others), that is generated by the corresponding grammar. Moreover, for any FSR we invent we can produce an equivalent regular grammar. It was also said (though it will not be *proved* until Chapter 6) that there is at least one language, $\{a^i b^i : i \geqslant 1\}$, that cannot be recognised by any FSR (and therefore cannot be generated by a regular grammar), which nevertheless we have seen to be context free.

In terms of the processing of the CFLs, the problem with the FSR is that it has no memory device: it has no way of "remembering" details about the part of the string it has processed at a given moment in time. This observation is crucial in several results considered in this book. The FSR's only memory is in terms of the state it is in, which provides limited information. For example, if a FSR wishes to "count" n symbols, it has to have n states with which to do it. This immediately makes it unsuitable even for a simple CFL such as $\{a^i b^i : i \geq 1\}$, that requires a recogniser to be able to "count" a number of as that cannot be predicted in advance, so that it can make sure that an equal number of bs follows.

It turns out that with the addition of a rudimentary memory device called a *stack*, the FSR can do the job for CFLs that the FSR can do for regular languages. We call such an augmented FSR a *pushdown recogniser* (PDR). Before we examine these machines in detail, we need to be clear about what is meant by a *stack*.

5.4.1 The Stack

A stack is simply a linear data structure that stores a sequence of objects. In terms of this chapter, these "objects" are symbols from a formal alphabet. The stack can only be accessed by putting an object on, or removing an object from, one end of the stack that is usually referred to as the *top* of the stack. The stack has a special marker ($\perp$), called the *stack bottom marker* which is always at the *bottom* of the stack. We use a diagrammatic representation of a stack as shown in Figure 5.2. Formally, our stacks are really strings that have the symbol $\perp$ at the rightmost end, and to which we can add or remove symbols from the leftmost end.

Figure 5.3 specifies the operations *PUSH* and *POP* that are needed by our machines.

However, note that we cannot *POP* anything from an empty stack. We have defined a *destructive POP*, in that its result is the top element of the stack, but its *side effect* is to remove that element from the stack. Finally, observe from Figure 5.4 that pushing the *empty string* onto the stack leaves the stack unchanged.

The stacks in this chapter are used to store strings of terminals and non-terminals, such that each symbol in the string constitutes one element of the stack. For convenience, we allow *PUSH(abc)*, for example, as an abbreviation of *PUSH(c)* followed by *PUSH(b)* followed by *PUSH(a)*. Given that, as stated above, our stacks are really strings, we could have defined *POP* and *PUSH* as formal string functions.

Stacks are important data structures, and have many applications throughout computer science. A related data structure you may encounter (though we have no cause to use it in this book) is called a *queue*. A queue is like a stack, except that, as its name suggests, items are added to one end, but taken from the other.

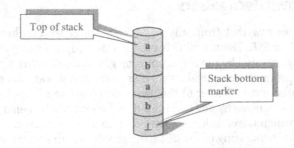

Figure 5.2. A *stack*. Symbols can only be added to, and removed from, the *top*

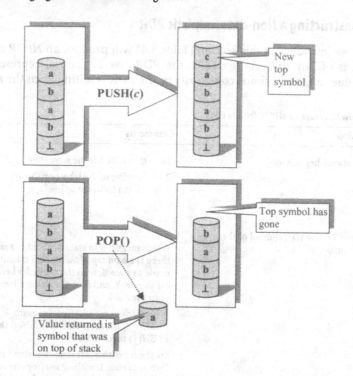

Figure 5.3. "push" and "pop" applied to the stack of Figure 5.2

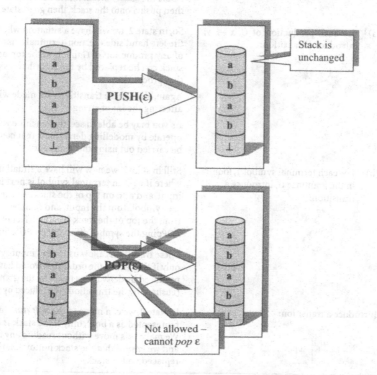

Figure 5.4. "push" and "pop" and the empty string (ε)

5.4.2 Constructing a Non-deterministic PDR

The rules we introduce in this section (Table 5.4) will produce an *NPDR* from the productions of any CFG. As for FSRs, the PDR has a pictorial representation. However, due to the additional complexity of the PDR resulting from the addition

Table 5.4. How to create an NPDR from a CFG

Step	Actions	Comments
1	We always begin with $\varepsilon / \bot / S \bot$ (state 1 → state 2) where S is the start symbol of the grammar	Note that state 1 is the *start state* The meaning of the labels on the *arcs* can be appreciated from the following: $a/b/x$ (A → B) which means: if in state A, about to read a, and there is a b *on top of the stack*, a transition can be made to state B, with the symbol b being popped off the stack, and then the string x being pushed onto the stack When *strings* are PUSHed, for example, $PUSH(efg)$ is interpreted as $PUSH(g)$ then $PUSH(f)$ then $PUSH(e)$ So, the *transition* introduced in step 1 means: "when in state 1, *without reading any input*, and POPping the stack bottom marker off the stack, PUSH the stack bottom marker back on the stack, then push S onto the stack, then go to state 2"
2	(a) For each production of G, $x \rightarrow y$, introduce a transition: $\varepsilon/x/y$ (self-loop on state 2)	So, in state 2, we will have a situation where the left-hand side (i.e. non-terminal symbol) of any production of G that is at the top of the stack can be replaced by the right-hand side of that production. Again, all of these transitions are made without affecting the input string As you may be able to see, the machine will operate by modelling derivations that could be carried out using the grammar
	(b) For each terminal symbol, t, found in the grammar G, introduce a transition: $t/t/\varepsilon$ (self-loop on state 2)	Still in state 2, we now will have a situation where if a given terminal symbol is next in the input, and also on top of the stack, we can read the symbol from the input, and remove the one from the top of the stack (i.e. the effect of POPping the symbol off and then PUSHing ε) These transitions allow us to read input symbols only if they are in the order that would have been produced by derivations made by the grammar (ensured by the transitions introduced by rule 2a)
3	Introduce a transition: $\varepsilon/\bot/\bot$ (state 2 → H)	In state 2, we can move to the *halt state* (again represented as a box), only if the stack is empty. We make this move without reading any input, and we ensure that the stack bottom marker is replaced on the stack

Table 5.5. The acceptance conditions for the NPDR

The machine begins in the *start state*, with the stack containing only the *stack bottom marker*, $\perp$

If there is a sequence of transitions by which the machine can reach its *halt state*, H, and on doing so, the input is *exhausted* and the stack contains only the stack bottom marker, $\perp$, then the string x is said to be *acceptable*

of the stack, the pictorial form does not provide such a succinct description of the language *accepted* by a PDR as it does in the case of the FSR. The rules in Table 5.4 convert any CFG, G, into a three-state NPDR.

The *acceptance conditions* for the NPDR, given some input string, x, are specified in Table 5.5.

As for FSRs (Chapter 4), we assert that the set of all *acceptable* strings of an NPDR derived from a grammar, G, is exactly the same as $L(G)$. We are not going to prove this, however, because as you will be able to see from the examples we consider next, it is reasonably obvious that any PDR constructed by using rules in Table 5.4 will be capable of reproducing any derivation that the grammar could have produced. Moreover, the PDR will reach a halt state only if its input string is a sentence that the grammar could derive. However, if you want to try to *prove* this, it makes a useful exercise.

5.4.3 Example NPDRs, M_3 and M_{10}

Let us now produce an NPDR from the grammar G_3 that we encountered earlier, i.e.

$$S \rightarrow aSb \mid ab.$$

Figure 5.5 shows the transitions produced by each rule (the rule number applied is in the upper right-hand corner of each box).

Figure 5.6 shows the connected machine (we have simply joined the appropriate circles together).

As in the previous chapter, we show the correspondence between the grammar and the machine by using the same subscript number in the name of both. We now investigate the operation of such a machine, by presenting M_3 with the input string a^3b^3. G_3 can generate this string, so M_3 should accept it. The operation of M_3 on the

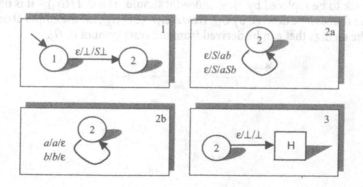

Figure 5.5. Creating a non-deterministic PDR for grammar G_3

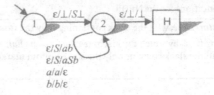

Figure 5.6. The NPDR, M_3

example string is shown in Table 5.6. Table 5.6(a) shows M_3's initial configuration, while Tables 5.6(b) and (c) show the machine's behaviour, eventually accepting the string in Table 5.6(c).

Notice in Tables 5.6(b) and (c) how either of the transitions derived from the productions $S \rightarrow aSb$ and $S \rightarrow ab$ can be applied any time we have S on the top of the stack. There are in fact (infinitely) many sequences of transitions that can be followed, but most of them will lead to a non-acceptance state.

We complete this section by considering a further example, this time a grammar that is equivalent to grammar G_2, of Chapters 2 and 3. G_2 is:

$$S \rightarrow aB \mid bA \mid \varepsilon$$
$$A \rightarrow aS \mid bAA$$
$$B \rightarrow bS \mid aBB.$$

However, our example, G_{10}, below, has been constructed by applying our rules for removing the empty string (Table 5.2) to G_2. You should attempt this construction as an exercise.

G_{10}'s productions are as follows:

$$S \rightarrow \varepsilon \mid X$$
$$X \rightarrow aB \mid bA$$
$$A \rightarrow aX \mid a \mid bAA$$
$$B \rightarrow bX \mid b \mid aBB.$$

We now apply the rules for creating the *equivalent* NPDR, resulting in the machine M_{10}, shown in Figure 5.7.

Note that, unlike M_3, M_{10} will accept the empty string (since it allows S at the top of the stack to be replaced by ε), as indeed it should, as $\varepsilon \in L(G_{10})$ – it is one of the sentences that can be derived by G_{10}. However, $\varepsilon \notin L(G_3)$, i.e. the empty string is not one of the strings that can be derived from the start symbol of G_3.

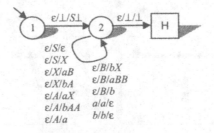

Figure 5.7. The NPDR, M_{10}

Table 5.6a. An initial configuration of the NPDR, M_3

Initial state	Initial stack	Input
		a a a b b b

Table 5.6b. Trace of M_3's behaviour (part 1), starting from the initial configuration in Table 5.6(a)

Transition	Resulting stack	Input
$\varepsilon/\perp/S\perp$	S $\perp$	a a a b b b
$\varepsilon/S/aSb$	a S b $\perp$	a a a b b b
$a/a/\varepsilon$	S b $\perp$	a a a b b b
$\varepsilon/S/aSb$	a S b b $\perp$	a a a b b b
$a/a/\varepsilon$	S b b $\perp$	a a a b b b

Table 5.6c. Trace of M_3's behaviour (part 2), continuing from Table 5.6(b)

Transition	Resulting stack	Input

Input *exhausted*, in halt state, therefore input string *accepted*.

I now leave it to you to follow M_{10}'s behaviour on some example strings. You are reminded that $L(G_{10})$ is the set of all strings of *a*s and *b*s in any order, but in which the number of *a*s is equal to the number of *b*s.

5.5 Deterministic PDRs

For the FSRs, *non-determinism* is not a critical problem. We discovered in Chapter 4 that for any non-deterministic FSR (non-DFSR) we could construct an equivalent, DFSR. The situation is different for PDRs, as we will see. First, let us clarify how non-determinism manifests itself in PDRs. In PDRs, two things determine the next

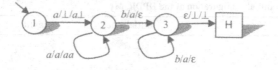

Figure 5.8. The DPDR, M_3^d

move that can be made from a given state:

1. the next symbol to be read from the input, and
2. the symbol currently on top of the stack.

If, given a particular instance of 1 and 2, the machine has *transitions* that would allow it to:

A. make a choice as to which state to move to, or
B. make a choice as to which string to push onto the stack (or both), then the PDR is non-deterministic.

On the other hand, a *deterministic PDR* (DPDR), is such that

- in any state, given a particular input symbol and particular stack top symbol, the machine has only *one* applicable transition.

The question we consider here is: *for any NPDR, can we always construct an equivalent DPDR?* We saw in Chapter 4 that the answer to the corresponding question for FSRs was "yes". For the PDRs, however, the answer is more like "in general no, but in many cases, yes". We shall see that the "many cases" includes all of the regular languages and some of the *proper* CFLs (i.e. languages that are context free *but not regular*). However, we shall also see that there are some languages that are certainly context free, but *cannot* be recognised by a DPDR.

5.5.1 M_3^d, a Deterministic Version of M_3

First, we consider CFLs that *can* be recognised by a DPDR. The language,

$$A = \{a^i b^i : i \geqslant 1\},$$

is not regular, as was stated earlier and will be proved in Chapter 6. However, A is certainly context free, since it is generated by the CFG, G_3. Above, we constructed an NPDR (M_3) for A, from the productions of G_3 (see Figure 5.6). However, we can see that any sentence in the language A consists of a number of as followed by the same number of bs. A deterministic method for accepting strings of this form might be:

- PUSH each a encountered in the input string onto the *stack* until a b is encountered, then POP an a off the stack for each b in the input string.

The DPDR M_3^d, shown in Figure 5.8, uses the above method.

In Table 5.6 we traced the operation of the NPDR M^3 for the input $a^3 b^3$. We now trace the operation of M_3^d for the same input. Table 5.7(a) shows the initial configuration of M_3^d, Table 5.7(b) shows the behaviour of the machine until it accepts the string.

Table 5.7a. An initial configuration of the DPDR, M_3^d

Initial state	Initial stack	Input
1	$\bot$	a a a b b b

Table 5.7b. Trace of M_3^d's behaviour, starting from the initial configuration in Table 5.7(a)

Transition	Resulting stack	Input
1 $\xrightarrow{a/\bot/a\bot}$ 2	a $\bot$	a a a b b b
2 $a/a/aa$	a a $\bot$	a a a b b b
2 $a/a/aa$	a a a $\bot$	a a a b b b
2 $\xrightarrow{b/a/\varepsilon}$ 3	a a $\bot$	a a a b b b
3 $b/a/\varepsilon$	a $\bot$	a a a b b b
3 $b/a/\varepsilon$	$\bot$	a a a b b b
3 $\xrightarrow{\varepsilon/\bot/\bot}$ H	$\bot$	a a a b b b

Input *exhausted*, in halt state, stack empty, therefore string *accepted*.

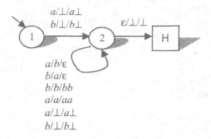

Figure 5.9. A DPDR for the language $L(G_{10}) - \{\varepsilon\}$, i.e. the language recognised by M_{10} but without the empty string

5.5.2 More DPDRs

Two more example DPDRs follow (Figures 5.9 and 5.10). It is suggested that you trace their behaviour for various valid and invalid input strings.

First, Figure 5.9 shows a DPDR for the language $L(G_{10})$, the set of all strings of as and bs in which the number of as is equal to the number of bs. This is the same language as $L(G_2)$, as we obtained the productions for G_{10} by applying our procedure for removing ε productions to G_2. Let us consider $L(G_{10})$ without the empty string (using the *set difference* operator, $L(G_{10}) - \{\varepsilon\}$), which is obviously the set of all *non-empty* strings of as and bs in which the number of as is equal to the number of bs.

The DPDR shown in Figure 5.10 *recognises* the language $L(G_{10})$, which is:

$$\{a^{2i+1}c^jb^{2i+1}: i, j \geqslant 0\}.$$

i.e. a non-zero odd number of as, followed by 0 or more cs, followed by the same number of bs as there were as.

Clearly, then, there are CFLs that can be recognised by DPDRs. Here, we have encountered but a few. However, there are, as you can probably imagine, an infinite number of them. If a CFL can be recognised by a DPDR, we call it a *deterministic CFL*.

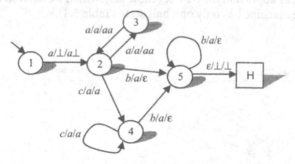

Figure 5.10. A DPDR for the language $L(G_9)$

5.6 Deterministic and Non-deterministic CFLs

In this section, we discover that there are CFLs that cannot be recognised by a DPDR. There is no general equivalence between the NPDRs and the DPDRs, as there is for FSRs. We begin our investigation with the regular languages, which we find to be wholly contained within the deterministic CFLs.

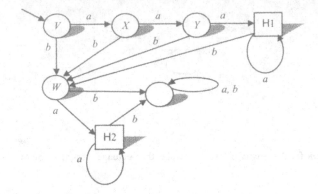

Figure 5.11. The DFSR, M_5^d, from Chapter 4

5.6.1 Every Regular Language Is a Deterministic CFL

Any *regular language* is generated by a regular grammar without ε *productions* (apart from the one ε production $S \rightarrow \varepsilon$, if it is necessary). Such a grammar can be represented as a non-DFSR, which can, if necessary, be represented as a *DFSR*.

Now, any DFSR can be represented as a DPDR that effectively does not use its stack. We will use a single example that suggests a method by which any DFSR could be converted into an equivalent DPDR. The example DFSR is M_3^d, from Chapter 4, shown in Figure 5.11.

This produces the DPDR of Figure 5.12.

The DPDR in Figure 5.12 has two *halt states* (there is nothing wrong with this). We have also ignored the *null state* of the DFSR. We could clearly amend any DFSR to produce an equivalent DPDR as we have done above.

However, for the DPDR, we do need to amend our definition of the *acceptance conditions* as they applied to the NPDR (Table 5.5), which we derived by using rules that ensure the machine has only one halt state (Table 5.4).

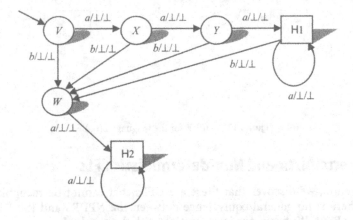

Figure 5.12. A DPDR equivalent to the DFSR, M_5^d, of Figure 5.11

Table 5.8. The acceptance conditions for the DPDR

The machine begins in the *start state*, with the stack containing only the *stack bottom marker*, ⊥

If there is a sequence of transitions by which the machine can reach one of its *halt states*, and on doing so, the input is *exhausted*, then the string x is said to be *acceptable*

The *acceptance conditions* for the DPDR, for an input string x, are specified in Table 5.8.

You may notice that the condition "the stack contains *only the stack bottom marker* ⊥", that applied to NPDRs (Table 5.5) does not appear in Table 5.8. An exercise asks you to justify that the language $\{a^i b^j : i \geq 1, j = i \text{ or } j = 0\}$ can only be recognised by a DPDR if that machine is not required to have an empty stack every time it accepts a string.

So far, we have established the existence of the *deterministic CFLs*. Moreover, we have also established that this class of languages includes the regular languages. We have yet to establish that there are some CFLs that are necessarily *non-deterministic*, in that they can be accepted by an NPDR but not a DPDR.

5.6.2 The Non-deterministic CFLs

Consider the following CFG, G_{11}:

$$S \rightarrow a \mid b \mid c \mid aSa \mid bSb \mid cSc.$$

$L(G_{11})$ is the set of all possible non-empty odd length *palindromic* strings of *a*s and/or *b*s and/or *c*s. A *palindromic* string is the same as the reverse of itself. We could use our procedure from Table 5.4 to produce an NPDR to accept $L(G_{11})$. However, could we design a DPDR to do the same? We cannot, as we now establish by intuitive argument. General palindromic strings are of the form specified in Figure 5.13.

The central box of Figure 5.13 could be empty if even length palindromic strings were allowed. Note that G_{11} does not generate even length palindromes, though it could be easily amended to do so, as you might like to establish for yourself. Figure 5.14 shows how an example sentence from $L(G_{11})$ is partitioned according to Figure 5.13.

Let us hypothesise that a DPDR could accept $L(G_{11})$. It would have to read the input string from left to right, as that is how it works. An obvious deterministic way

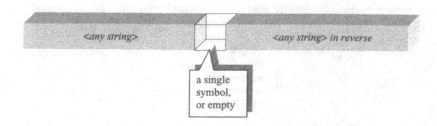

Figure 5.13. A specification of arbitrary palindromic strings

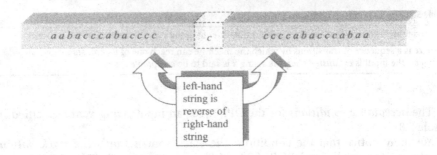

Figure 5.14. A palindrome, partitioned according to Figure 5.13.

to do the task on that basis is as follows:

1. Read, and push onto the stack, each input symbol encountered until the "centre" of the string is reached.
2. Read the rest of the string, comparing each symbol to the next symbol on the stack. If the string is valid, the string on the stack following (1) will be the part of the string after the centre *in reverse*.

The problem with the above is in the statement *"until the 'centre' of the string is reached"*, in step (1). How can the DPDR detect the centre of the string? Of course, it cannot. If there is a symbol in the centre it will be one of the symbols – *a*, *b*, or *c* – that can feature *anywhere* in the input string. The machine must push every input symbol onto the stack, on the assumption that it has not yet reached the centre of the input string. This means that the machine must be allowed to *backtrack*, once it reaches the end of the input, to a situation where it can see if it is possible to empty its stack by comparing the stack contents with the remainder of the input. By similar argument, any such language of arbitrary palindromic strings defined with respect to an alphabet of two or more symbols is necessarily non-deterministic, while nevertheless being context free.

We have argued that the problem *vis a vis* palindromic strings lies in the PDR's inability to detect the "centre" of the input. If a language of palindromic strings is such that every string is of the form shown in Figure 5.15, then it can be accepted by a DPDR, and so is deterministic.

As an example of a deterministic palindromic language, consider $L(G_{12})$, where G_{12} (a slightly amended G_{11}) is as follows:

$$S \to d \mid aSa \mid bSb \mid cSc.$$

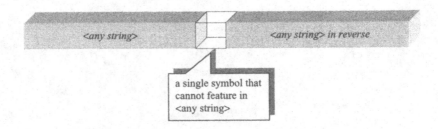

Figure 5.15. Palindromes with a distinct central "marker"

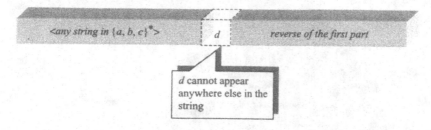

Figure 5.16. Sentences of $L(G_{12})$

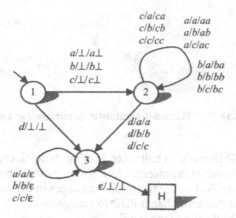

Figure 5.17. The DPDR, M_{12}, which deterministically accepts a language of palindromic strings that feature a central "marker"

$L(G_{12})$ is the set of all palindromic strings of a form described in Figure 5.16. $L(G_{12})$ is accepted by the DPDR M_{12} depicted in Figure 5.17.

5.6.3 A Refinement to the Chomsky Hierarchy in the Case of CFLs

The existence of *properly non-deterministic CFLs* suggests a slight refinement to the Chomsky hierarchy with respect to the context free and regular languages, as represented in Figure 5.18.

5.7 The Equivalence of the CFLs and the PDRs

The situation for FSRs and regular languages, i.e. that any regular grammar is represented by an *equivalent* FSR, and every FSR is represented by an equivalent regular grammar, is mirrored in the relationship between PDRs and CFGs. In this chapter, we have seen that for any CFG we can construct an equivalent NPDR. It turns out that for any NPDR we can construct an equivalent CFG. This will not be proved here, as the proof is rather complex, and it was considered more important to investigate the situation pertaining to the PDR as the *recogniser* for the CFLs, rather than the converse result.

One way of showing that every NPDR has an equivalent CFG is to show that from any NPDR we can create CFG productions that generate exactly the set of strings

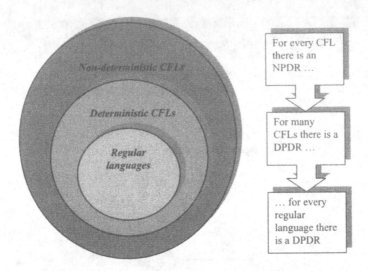

Figure 5.18. The CFLs: non-deterministic, deterministic and regular

that enable the NPDR to reach its halt state, leaving its stack empty. The details are not important, but the result is powerful, as it shows us that any language that cannot be generated by a CFG (we see two such languages in the next chapter) needs a machine that is more powerful than a PDR to recognise it.

Consider the "multiplication language",

$$\{a^i b^j c^{i \times j} \colon i, j \geqslant 1\},$$

introduced in Chapter 2. This is shown to be not a CFL at the end of Chapter 6, a result that means that the language cannot be recognised by a PDR. This leads us to speculate that we need a more powerful machine than the PDR to model many computational tasks. We return to this theme in the second part of the book, when we consider abstract machines as *computers*, rather than language recognisers, and we see the close relationship between *computation* and language recognition.

5.8 Observations on CFGs and CFLs

The lack of general equivalence between deterministic and non-deterministic CFLs is problematic for the design of programming languages. Efficient parsers can be designed from DPDRs. As argued in Chapter 4, non-determinism is undesirable, as it is computationally expensive. Moreover, *unnecessary* non-determinism is simply wasteful in terms of resources. It turns out that any deterministic CFL can be parsed by a pushdown machine that never has to look ahead more than one symbol in the input to determine the choice of the next transition.

Consider, for example, sentences of the language $\{a^i b^i \colon i \geqslant 1\}$. If we look at the next symbol in the input, it will either be an a or a b, and, like our deterministic machine M_3, above, we know exactly what to do in either case. Now consider strings of arbitrary *palindromes*, with no distinct central "marker" symbol. In this case, for any machine that could look ahead a fixed number of symbols, we could simply present that machine with a palindromic string long enough to ensure that the machine could not tell when it moved from the first to the second "half" of the input. Languages that can be parsed by a PDR (and are therefore context free) that never has to look ahead more than a fixed number, say k, of symbols in the input are called *LR(k) languages* (*LR* stands for *Look ahead Right*). The *LR(k)* languages are the deterministic CFLs. In fact, the deterministic languages are all *LR(1) languages*. Languages of arbitrary length palindromic strings with no distinguishable central marker are clearly not *LR(k)* languages, for any fixed k.

The preceding chapter closed with the observation that the language $\{a^i b^i : i \geqslant 1\}$ is not regular. It was promised that this would be proved in Chapter 6. Now we make a similar observation with respect to the CFLs. We argue that there are languages that are not context free, and cannot therefore be generated by a CFG, or recognised by an NPDR, and therefore must be either *type* 1 (*context sensitive*) or *type* 0 (*unrestricted*) languages (or perhaps neither).

In Chapter 6, we will see that both the "multiplication language" and the language:

$$\{a^i b^i c^i : i \geqslant 1\},$$

i.e. the set of all strings of the form *as* followed by *bs* followed by *cs*, where the number of *as* = the number of *bs* = the number of *cs*, are not context free. Given that the language $\{a^i b^i : i \geqslant 1\}$ *is* context free, it may at first seem strange that $\{a^i b^i c^i : i \geqslant 1\}$ is not. However, this is indeed the case. Part of the next chapter is concerned with demonstrating this.

5.9 Exercises

For exercises marked †, solutions, partial solutions, or hints to get you started appear in "Solutions to Selected Exercises" at the rear of the book.

5.1. Given the following grammar:

$S \rightarrow ABCA$

$A \rightarrow aA \mid XY$

$B \rightarrow bbB \mid C$

$X \rightarrow cX \mid \varepsilon$

$Y \rightarrow dY \mid \varepsilon$

$C \rightarrow cCd \mid \varepsilon.$

Produce an equivalent CNF grammar.

Note: remember that you first need to apply the rules for removing the empty string.

5.2†. $\{a^i b^j : i \geqslant 1, j = i \text{ or } j = 0\}$ is a *deterministic CFL*. However, as mentioned earlier in this chapter, it can only be accepted by a DPDR that does not always clear its *stack*. Provide a DPDR to accept the language, and justify the preceding statement.

5.3. Argue that we can always ensure that an NPDR clears its stack, and that NPDRs need have only one *halt state*.

5.4. Amend M_3^d, from earlier in the chapter (Figure 5.8), to produce *DPDRs* for the languages $\{a^i b^i c^j: i, j \geqslant 1\}$ and $\{a^i b^j c^j: i, j \geqslant 1\}$.

Note: in Chapter 6 it is assumed that the above two languages are deterministic.

5.5[†]. Justify by intuitive argument that $\{a^i b^j c^k: i, j \geqslant 1, i = j$ or $j = k\}$ (from Exercise 3.1 in Chapter 3) is a *non-deterministic CFL*.

5.6[†]. Design a program to simulate the behaviour of a DPDR. Investigate what amendments would have to be made to your program so that it could model the behaviour of an NPDR.

5.7. Design *DPDRs* to accept the languages $\{a^i b^j c^{i+j}: i, j \geqslant 1\}$ and $\{a^i b^j c^{i-j}: i, j \geqslant 1\}$.

Note: consider the implications of these machines in the light of the remarks about the computational power of the PDR, above.

Chapter 6

Important Features of Regular and Context Free Languages

6.1 Overview

In this chapter we investigate the notion of *closure*, which in terms of languages essentially means seeing if operations (such as *union*, or *intersection*) applied to languages of a given class (for example, *regular*) always result in a language of the same class. We look at some closure properties of the *regular* and *context free* languages (CFLs).

We next discover that the *Chomsky hierarchy* is a "proper" hierarchy, by introducing two theorems:

- the *repeat state theorem* for *finite state recognisers* (FSRs)
- the *uvwxy theorem* for CFLs.

We then use these theorems to show that:

- there are CFLs that are not *regular*
- there are *context sensitive* and *unrestricted* languages that are not *context free*.

6.2 Closure Properties of Languages

In terms of the *Chomsky hierarchy*, the notion of *closure* applies as follows:

1. choose a given type from the Chomsky hierarchy;
2. choose some operation that can be applied to any language (or pairs of languages) of that type, and which yields a language as a result;
3. if it can be demonstrated that when applied to each and every language, or pair of languages, of the chosen type, the operation yields a resulting language which is also of the chosen type, then the languages of the chosen type are said to be *closed under the operation*.

For example, take the set *union* operator, $\cup$, and the *regular* languages. If for *every* pair L_1, L_2, of regular languages the set $L_1 \cup L_2$ is also a regular language, then the regular languages would be *closed under union* (in fact they are, as we see, shortly).

In the following two sections, we investigate various closure properties of the regular languages and CFLs, with respect to the operations *complement*, *union*, *intersection*, and *concatenation*. As usual, we will be content with intuitive argument.

6.3 Closure Properties of the Regular Languages

6.3.1 Complement

As we discovered early on in the book, given some finite non-empty alphabet A, A^* denotes the *infinite* set of all possible strings that can be formed using symbols of that alphabet. Given this basis, a *formal language* was defined to be any subset of such a set. Take, for example, the language

$$R = \{a^i b^j c^k : i, j, k \geq 1\},$$

which is a subset of $\{a, b, c\}^*$.

Now, $\{a, b, c\}^* - R$ denotes the set of all strings that are in $\{a, b, c\}^*$ but *not in R* ("$-$" is called the *set difference* operator), i.e.

- all possible strings of *a*s and/or *b*s and/or *c*s except for those of the form "one or more *a*s followed by one or more *b*s followed by one or more *c*s".

$\{a, b, c\}^* - R$ is called the *complement* of R, and is, of course, a formal language, as it is a subset of $\{a, b, c\}^*$.

In general, then, if L is a language in A^*, $A^* - L$ is the *complement* of L. We are now going to see that

- if L is a regular language, then so is $A^* - L$.

In other words, we will see that the regular languages are *closed under* complement. This is best demonstrated in terms of *deterministic FSRs* (DFSRs). Our argument will be illustrated here by using the DFSR for R, the language described above, as an example.

We start with a DFSR that *accepts R*, as shown in Figure 6.1.

To the machine in Figure 6.1 we add a *null state* to make the machine completely deterministic. The resulting machine can be seen in Figure 6.2.

We take the machine in Figure 6.2 and make all its non-halt states into *halt states*, and all its halt states into non-halt states, achieving the machine shown in Figure 6.3.

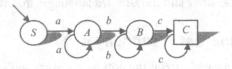

Figure 6.1. A DFSR for the language $R = \{a^i b^j c^k : i, j, k \geq 1\}$

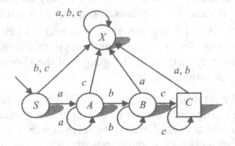

Figure 6.2. The DFSR from Figure 6.1 with a "null" state added to make it fully deterministic

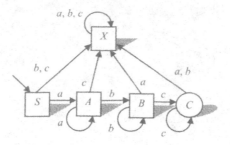

Figure 6.3. The "complement" machine of the DFSR in Figures 6.1 and 6.2

The FSR in Figure 6.3 will reach a halt state only for strings in $\{a, b, c\}^*$ for which the original machine (Figure 6.1) would not. It thus accepts the *complement* of the language accepted by the original machine. Note that as the machine's start state is also a *halt* state, the final machine also accepts as a valid input the empty string (ε), as it should, since the original machine did not.

It should be clear that the same method could be applied to *any* DFSR. As we saw in Chapter 4, there is a general *equivalence* between DFSRs and regular languages. What applies to all DFSRs, then, applies to all regular languages.

6.3.2 Union

We now see that

- if L_1 and L_2 are regular languages, then so is $L_1 \cup L_2$.

Consider again the DFSR for R, in Figure 6.1, above, and the DFSR for the language

$$\{a^i b^{2j}: i \geq 0, j \geq 1\},$$

which is shown in Figure 6.4.

Figure 6.5 shows a machine that is an amalgam of the machines from Figures 6.1 and 6.4. As can be seen in Figure 6.5, we have introduced an additional start state, labelled I, and for each outgoing arc from a start state of either original machine we draw an arc from state I to the destination state of that arc.

The machine in Figure 6.5 has three start states, but if we use our *subset algorithm* from Chapter 4 (Table 4.6) to make it deterministic, we obtain the FSR in Figure 6.6.

The machine in Figure 6.6 accepts the language:

$$\{a^i b^j c^k: i, j, k \geq 1\} \cup \{a^i b^{2j}: i \geq 0, j \geq 1\},$$

i.e. all strings accepted by either, or both, of the original machines.

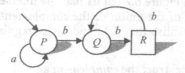

Figure 6.4. A DFSR for the language $\{a^i b^{2j}: i \geq 0, j \geq 1\}$

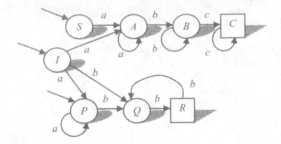

Figure 6.5. The non-DFSR that is the union machine for the FSRs in Figures 6.1 and 6.4. Note: the machine has three start states (*S, I,* and *P*)

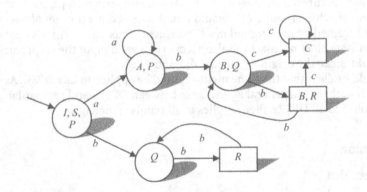

Figure 6.6. A deterministic version of the machine in Figure 6.5

You should again satisfy yourself that the suggested method could be applied to any pair of DFSRs.

6.3.3 Intersection

Now we establish that

- if L_1 and L_2 are regular languages, then so is $L_1 \cap L_2$

(the *intersection* of two sets is the set of elements that occur in both of the sets).

A basic law of set theory is one of *de Morgan's* laws, which is stated in Figure 6.7. Figure 6.8 shows an illustrative example to help to convince you that the rule is valid. The example is especially interesting in the light of discussions in part 3 of this book, when we consider de Morgan's laws applied to logical reasoning, rather than sets.

The law represented in Figure 6.7 tell us that the intersection of two sets is the same as the complement of the union of the complements of the two sets. Now, we have to do little to establish that if L_1 and L_2 are regular languages, then so is $L_1 \cap L_2$, since:

- we saw above how to construct the *complement* and
- we also saw how to create the *union*.

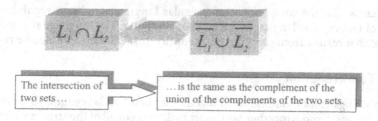

Figure 6.7. de Morgan's law for the intersection of two languages (sets)

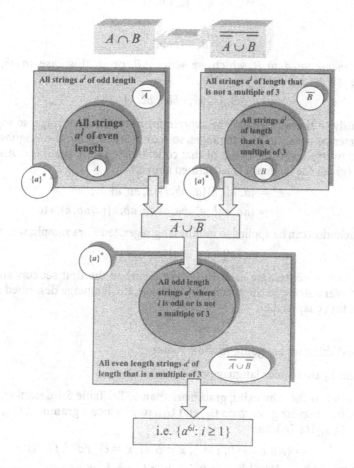

Figure 6.8. An illustration of de Morgan's law applied the intersection of the two languages $A = \{a^{2i}: i \geq 0\}$ and $B = \{a^{3i}: i \geq 1\}$

We can therefore take the two DFSRs for L_1 and L_2, say M_1 and M_2, and do the following: we create the

(a) "complement machines" for each of M_1 and M_2

(b) "union machine" from the results of (a)

(c) "complement machine" from the result of (b).

We know that the complement of a regular language is always regular, and the union of two regular languages is also always regular. The language recognised by the FSR that results from carrying out steps (a)–(c) will therefore also be regular.

6.3.4 Concatenation

As two strings can be *concatenated* together, so can two languages. To concatenate two languages A and B together, we simply make a set of all of the strings which result from concatenating each string in A in turn, with each string in B. So, for example, if

$$A = \{a, ab, \varepsilon\}$$

and

$$B = \{bc, c\},$$

then A *concatenated to* B, which we write AB, or in this case $\{a, ab, \varepsilon\}\{bc, c\}$ would be

$$\{abc, ac, abbc, bc, c\}.$$

Analogously to the way we use the *superscript* notation for strings, so we can use the superscript notation for *languages*, so that A^3 represents the language A concatenated to itself, and the result of that concatenated to A again (i.e. AAA), or in this case (given the language A as defined above):

$$A^3 = \{a, ab, \varepsilon\}\{a, ab, \varepsilon\}\{a, ab, \varepsilon\}$$
$$= \{aa, aab, a, aba, abab, ab, \varepsilon\}\{a, ab, \varepsilon\}, \text{etc.}$$

Concatenation can be applied to infinite languages, too. For example, we could write

$$\{a^i b^j : i, j \geqslant 1\}\{c^i : i \geqslant 1\},$$

to mean the set containing each and every string in the first set concatenated to each and every string in the second. In this case, the language described is R from earlier in this chapter, i.e.

$$\{a^i b^j c^k : i, j, k \geqslant 1\}.$$

The next thing we are going to show is that

- if L_1 and L_2 are regular languages, then so is $L_1 L_2$.

This is easier to show by using grammars than FSRs. Table 6.1 describes the combining of two regular grammars G_{13} and G_{14}, to produce a grammar, G_{15}, such that $L(G_{15}) = L(G_{13})L(G_{14})$, i.e.

$$= \{a^i b^j x : i \geqslant 0, j \geqslant 1, x = b \text{ or } x = c\}\{a^i d^j : i, j \geqslant 1\}$$
$$= \{a^h b^i x a^j d^k : h \geqslant 0, i, j, k \geqslant 1, x = b \text{ or } x = c\}.$$

It should be clear that the method in Table 6.1 would work for any pair of regular grammars. Note that the method could also be used repeatedly, beginning with a grammar G, to create a grammar that generated $L(G)^n$, for any finite n.

6.4 Closure Properties of the Context Free Languages

In this section, we investigate the closure properties of the CFLs in general. We find that the situation is not so straightforward for the CFLs as for the regular languages.

Table 6.1. Combining two regular grammars to create a regular grammar that generates the concatenated language

Two example grammars	
G_{13} $\quad S \to aS \mid bB$ $\quad B \to b \mid c \mid bB$ so that: $L(G_{13}) = \{a^i b^j x : i \geqslant 0, j \geqslant 1, x = b \text{ or } x = c\}$	G_{14} $\quad S \to aS \mid aA$ $\quad A \to dA \mid d$ $L(G_{14}) = \{a^i d^j : i, j \geqslant 1\}$

To produce a composite grammar that generates $L(G_{13})L(G_{14})$:

Step	Description	Result
1	Rename the non-terminals in the second grammar so that the two grammars have no non-terminals in common	G_{14} is now: $\quad X \to aX \mid aA$ $\quad A \to dA \mid d$ with X as the start symbol
2	In the first grammar, replace each production of the form $x \to y$, where y is a single terminal, by a production $x \to y\,Z$, where Z is the start symbol of the second grammar	G_{13} is now: $\quad S \to aS \mid bB$ $\quad B \to bX \mid cX \mid bB$
3	Put all the productions together. They now constitute one grammar, the start symbol of which is the start symbol of the first grammar. The resulting grammar generates the language that is the language generated by the first grammar concatenated to the language generated by the second	Composite grammar, G_{15}: $\quad S \to aS \mid bB$ $\quad B \to bX \mid cX \mid bB$ $\quad X \to aX \mid aA$ $\quad A \to dA \mid d$

We also find that there are differences between the closure properties of the CFLs in general, and the *deterministic CFLs*, the latter being described in Chapter 5 as those CFLs recognised by *deterministic pushdown recognisers* (DPDRs). For the CFLs to be closed under an operation means that the language resulting from the application of the operation to *any* CFL (or pair of CFLs, if the operation applies to two CFLs) is also a CFL. Moreover, for the *deterministic* CFLs to be closed under an operation means that the result must always be a *deterministic* CFL.

To support the arguments, we consider the following languages:

$$X = \{a^i b^i c^j : i, j \geqslant 1\}$$

and

$$Y = \{a^i b^j c^j : i, j \geqslant 1\}.$$

First of all, let us observe that the two languages are context free. In fact, X and Y are both *deterministic* CFLs, as we could design DPDRs to accept each of them, as you were asked to do in the exercises at the end of Chapter 5.

In any case, they are certainly context free, as we can show by providing CFGs for each of them. The two grammars, G_x and G_y, are shown in Table 6.2.

To simplify our constructions, the two grammars in Table 6.2 have no *non-terminals* in common.

6.4.1 Union

The first result is that the CFLs are closed under *union*, i.e.

- if L_a and L_b are CFLs, then so is $L_a \cup L_b$.

Table 6.2. Two example grammars, G_x and G_y, to generate the languages $X = \{a^i b^i c^i : i, j \geq 1\}$ and $Y = \{a^i b^j c^j : i, j \geq 1\}$, respectively

$G_x(L(G_x) = X)$		$G_y(L(G_y) = Y)$	
$A \to BC$	{start symbol is A}	$D \to EF$	{start symbol is D}
$B \to aBb \mid ab$		$E \to aE \mid a$	
$C \to cC \mid c$		$F \to bFc \mid bc$	

Suppose that for L_a and L_b there are CFGs, G_a and G_b, respectively. Then to produce a grammar that generates the union of L_a and L_b, we carry out the process described in Table 6.3 (the example on the right of Table 6.3 uses the two grammars G_x and G_y from Table 6.2).

Referring to Table 6.3, you should be able to see that the grammar resulting from step 3 generates all of the strings that could be generated by one grammar or the other. Step 1 is necessary so that any derivation uses only productions from one of the original grammars. If G_a and G_b are the original grammars, the resulting grammar generates $L(G_a) \cup L(G_b)$. You can also see that if the two starting grammars are context free, the resulting grammar will also be context free.

The CFLs are therefore *closed under union*.

Table 6.3. Constructing a CFG to generate the union of two CFLs. The example column features grammars G_x and G_y from Table 6.2

Step	Description	Example
1	Arrange it so the two grammars have no non-terminals in common	G_x and G_y are already like this
2	Collect all the productions of both grammars together	
3	Add the two productions $S \to P \mid Q$, where S is a new non-terminal and P and Q are the start symbols of the two grammars S is now the start symbol of the composite grammar	

Now we focus on our particular example. The new grammar that we obtained from G_x and G_y generates $X \cup Y$, where X and Y are the languages above. This set, which will be called Z, is

$$Z = X \cup Y = \{a^i b^j c^k : i, j, k \geqslant 1, i = j \text{ or } j = k\}.$$

Now, while this is a CFL, and represents the union of two *deterministic* CFLs, it is not itself a deterministic CFL. To appreciate this, consider a PDR to accept it. Such a machine could not be deterministic, as when it was reading the as in the input it would have to assume that it was processing a string in which the number of as was equal to the number of bs, which may not be the case (it may be processing a string in which the number of bs was equal to the number of cs). There would therefore be a need for *non-determinism*. For a class of languages to be closed under an operation, the result of every application of the operation must also be a language of that class. So our example (and there are many more) shows us that

- the *deterministic CFLs* are not *closed under union*.

However, as we showed above, in general, the CFLs *are* closed under union.

6.4.2 Concatenation

In this section, we see that

- if L_a and L_b are CFLs, then so is $L_a L_b$.

Remember that, if X and Y are languages, then XY denotes the language resulting from the *concatenation* of each and every string in X with each and every string in Y. Table 6.4 shows a simple method for taking two CFGs, G_a and G_b, and producing from them a new CFG, G, which generates $L(G_a)L(G_b)$. As before, we use our grammars G_x and G_y, from Table 6.2, to illustrate.

The new grammar created in Table 6.4 ensures that any *terminal string* that could be derived by the first grammar is immediately followed by any terminal string that the second grammar could produce. Therefore, the new grammar generates the language generated by the first grammar concatenated to the language generated by the second grammar. The resulting grammar in Table 6.4 generates the set XY, such that

$$XY = \{a^i b^i c^j a^s b^t c^t : i, j, s, t \geqslant 1\}.$$

The method of construction specified in Table 6.4 could be applied to any pair of context free grammars (CFGs), and so *the CFLs are closed under concatenation*.

However, the *deterministic* CFLs are not closed under concatenation. This is not demonstrated by our example, because XY happens to be deterministic in this case. The two *deterministic* CFLs

$$\{a^i b^j : i \geqslant 1, j = 0 \text{ or } j = i\}$$

and

$$\{b^i c^j : i = 0 \text{ or } i = j, j \geqslant 1\}$$

concatenated together result in a language which is a *non*-deterministic CFL. You are asked in the exercises to argue that this is indeed the case.

Table 6.4. Constructing a CFG to generate the concatenation of two CFLs. The example features grammars G_x and G_y from Table 6.2

Step	Description	Example
1	Arrange it so the two grammars have no non-terminals in common	G_x and G_y are already like this
2	For each production in the first grammar with S (the start symbol of the grammar) on its left-hand side, put R (the start symbol of the second grammar) at the end of the right-hand side of that production	New production created: $A \rightarrow BC\underline{D}$ (the only one created, since G_x has only one production with A, its start symbol, on the left-hand side)
3	Add the rest of the first grammar's productions, and all of the second grammar's productions to the productions created in 2 The start symbol of the first grammar is the start symbol of the new grammar	New grammar:

$A \rightarrow BCD$ — *modified productions*

$B \rightarrow aBb \mid ab$
$C \rightarrow cC \mid c$ — *rest of G_x*

$D \rightarrow EF$
$E \rightarrow aE \mid a$ — *all of G_y*
$F \rightarrow bFc \mid bc$

Our concatenation result for the deterministic CFLs is thus:

- the *deterministic CFLs* are not *closed under concatenation*.

6.4.3 Intersection

Thus far, we have seen that the closure situation is the same for the CFLs as for the regular languages, with respect to both union and concatenation. However, in terms of *intersection*, the CFLs differ from the regular languages, in that the CFLs are not closed under intersection. To appreciate this, consider the two sets we have been using in this section, i.e.

$$X = \{a^i b^i c^j : i, j \geqslant 1\}$$

and

$$Y = \{a^i b^j c^j : i, j \geqslant 1\}.$$

What is the intersection $(X \cap Y)$ of our two sets? A little thought reveals it to be the set of strings of as followed by bs followed by cs, in which the number of as equals the number of bs *and* the number of bs equals the number of cs, i.e.

$$X \cap Y = \{a^i b^i c^i : i \geqslant 1\}.$$

This language was stated to be not context free at the end of the last chapter, and this will be demonstrated later in this chapter. Moreover, both X and Y are *deterministic* CFLs, so this counter example shows us that:

- both the *deterministic* and the *non-deterministic CFLs* are not *closed under intersection*.

6.4.4 Complement

We can argue that it is also the case that the CFLs are not closed under *complement* (i.e. that the complement of a CFL is not necessarily itself a CFL). You may recall that above we used de Morgan's law (Figure 6.7) to show that the regular languages were closed under intersection.

We can use the same law to show that the CFLs cannot be closed under complement. If the CFLs *were* closed under complement, the above law would tell us that they were closed under intersection (as we showed that they were closed under *union* earlier on). However, we have already shown that the CFLs are not closed under intersection, so to show that this is the case is absurd. We must therefore reject any assumption that the CFLs are closed under complement.

The above argument is our first example of proof by a technique called *reductio ad absurdum*. We simply assume the logical negation of the statement we are trying to prove, and on the basis of this assumption follow a number of logically sound arguments that eventually lead to a non-sensical conclusion. Since the steps in our argument are logical, it must be our initial assumption that is false. Our initial assumption is false, but it is actually the negation of the statement we are trying to prove. Therefore, the statement we are trying to prove is indeed true.

In this case, what our *reductio ad absurdum* proof shows us is that

- the *CFLs* are not *closed under complement*.

This result is very important, at least with respect to the *PDR*. Imagine you are writing a *parser* for a programming language that you have designed. Suppose you implement your parser to be equivalent in power to a non-deterministic PDR (NPDR). It turns out, unknown to you, that your language is one for which the complement is not a CFL. Now, this is not likely to cause problems for your parser if its input consists entirely of *syntactically correct* programs. This is highly unlikely, I'm sure you will agree, if you have ever done any programming! However, there will be one (or more) syntactically incorrect "programs" for which your parser will not be able to make any decision at all (it will probably enter an infinite loop). If, for every string we input to the parser, the parser (which remember is no more powerful than a PDR) could output an indication of whether *or not* the program is syntactically legal, the same parser could be used to accept the complement of our original language (i.e. we simply interpret the original parser's "yes" as a "no", and vice versa). However, the fact that the CFLs are not closed under complement tells us that for some languages a parser equivalent in power to a PDR will be unable to make the "no" statement. I suggest you re-examine the *acceptance conditions* for NPDRs in the previous chapter (Table 5.5). Now you can probably see why there is no condition that dictates that the machine must *always* halt, only that it must always halt in the case of *sentences* of the language.

The issue of whether or not *abstract machines* eventually stop at some time during a computational process is central to computer science. However, we defer consideration of such issues until Part 2 of this book.

All is not lost, however, for two reasons. Firstly, if your programming language is a *deterministic* CFL, its complement will also be a deterministic CFL. Remember, at the end of Chapter 5 we saw that a deterministic CFL was as a language for which a pushdown machine did not have to look more than one symbol ahead in the input. We saw that deterministic CFLs are called *LR*(1) *languages* for that reason. Well, in such cases there are only a finite number of symbols the machine expects to encounter when it looks ahead to the next one. They form two sets at each stage: the symbols the machine expects to see next, and the symbols it does not expect to see next. If at any stage, the machine looks ahead and finds a symbol which at that stage is *not* expected, the machine simply rejects the input as invalid (i.e. moves to a halt state that denotes *rejection*, as opposed to a halt state that denotes *acceptance*). We can make sure, then, that our machine always halts with "yes" or "no". We can then simply exchange the acceptance/rejection status of our machine's halt states, and we have a machine that accepts the complement of the original language. So,

- the *deterministic CFLs* are *closed under complement*.

6.5 Chomsky's Hierarchy Is Indeed a Proper Hierarchy

In this section, we sketch out two fundamental theorems of formal languages. One is for the *regular languages*, though it actually applies to the FSR, and is called the *repeat state theorem*. The other, for the CFLs, is in some ways similar to this, and is called the *uvwxy theorem*. The usefulness of these theorems is that they apply to every single language in their class, and can therefore be used to show that a given language is not in the class. To appreciate this point, think about it like this: if every element in a set must possess a particular property, and we come across an object that does not have that property, then the object cannot belong to our set. If it is true that every bird has feathers, and we come across a creature that is featherless, then that creature is certainly not a bird.

If we use the *repeat state theorem* to show that a certain language cannot be regular, but we can show that the same language is context free, then we know that the regular languages are a *proper subset* of the CFLs. We already know that every regular language is context free, but then we will have also shown that there are CFLs that are not regular. Moreover, if we use the *uvwxy theorem* (considered later in this chapter) to show that a given language cannot be context free, and we then show that it is *type* 0, we then also know that the CFLs are a proper subset of the type 0 languages.[1] We have then established that *Chomsky's hierarchy* is a real hierarchy.

Now, some students experience problems in thinking in terms of proper subsets. This is because when you say to them "all *x*s are *y*s", they automatically assume that you also mean "all *y*s are *x*s". To demonstrate the silliness of this inference, consider *parrots* and *birds*. If I said to you "all parrots are birds", you would not then think "all birds are parrots". The set of all regular languages corresponds to the set of all parrots, and the set of all CFLs corresponds to the set of all birds. One set is *properly* contained in the other.

6.5.1 The "Repeat State Theorem"

Any FSR that has a *loop* on some path linking its start and halt states recognises an infinite language.

Consider the example FSR, *M*, in Figure 6.9.

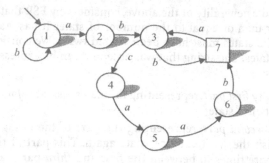

Figure 6.9. An FSR, *M*, that contains a *loop* of states

M, in Figure 6.9, has 7 states. Consider any string acceptable to *M* that is greater than 6 symbols in length. That string must have resulted in the machine passing through one of its states more than once. After all, *M* has only 7 states, and it would need 8 states to accept a string of length 7, if no state was to be visited more than once. Consider, as a suitable example, the string *abbacaab*. This string is 8 symbols in length. The sequence of states visited during the acceptance of this string is indicated in Figure 6.10.

Let us choose any state that *M* visited more than once in accepting our string. Choose state 3. State 3, as the diagram shows, was the third and fifth state visited. Consider the substring of *abbacaab* that caused *M* to visit state 3 twice, i.e. *ba*. Where *ba* occurs in *abbacaab*, we could copy it *any* number of times, and we would still obtain a string that *M* would accept. As an example, we will copy it four times: *abbabababacaab*. Furthermore, we could omit *ba*, giving *abcaab*, a string that is also accepted by our machine.

The other state that was visited twice in the acceptance of *abbacaab* was state 7 (the halt state). We now choose this as our example repeated state. The substring of *abbacaab* that caused state 7 to be visited twice was *acaab*. As before, our machine *M* would accept the original string with no copies of *acaab* in it: *abb*. It would also accept the original string with any number of copies inserted after the first occurrence, for example, three copies: *abbacaabacaabacaab*.

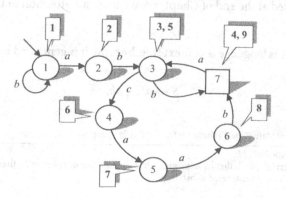

Figure 6.10. The order in which the states of *M* are visited when the input string is *abbacaab*

To appreciate the generality of the above, consider any FSR that accepts a string of length greater than or equal to the number of states it has. As we said before, one or more of its states must have been visited twice. We consider any of those "visited-twice" states. The string that was accepted can be expressed as three concatenated strings:

- A possibly empty *first part*, representing the states visited *before* the machine got to its "visited-twice" state.
- A non-empty *second part*, representing the part of the string that caused the machine to visit the "visited-twice" state again. This part of the string can be copied 0 or more times in between the *first* and *third* parts, and the resulting string will also be acceptable to the machine.
- A possibly empty *third part*, representing the states visited *after* the machine has left the "visited-twice" state loop.

In our examples for the string *abbacaab*, above, when *state* 3 was the chosen "*visited-twice*" *state*:

- *first part* was *ab*
- *second part* was *ba*
- *third part* was *caab*.

Alternatively, when *state* 7 was the chosen "*visited-twice*" *state*:

- *first part* was *abb*
- *second part* was *acaab*
- *third part* was empty.

In many cases, *first part* may be empty, if the loop began at the start state. In other cases, both *first* and *third parts* may be empty, if the loop encompasses every state in the machine.

The repeat state theorem is expressed formally in Table 6.5.

As was established in Chapter 4, any language accepted by a DFSR is a regular language. Therefore, our theorem applies to any *infinite* regular language.

6.5.2 A Language that Is Context Free but Not Regular

Now, as promised at the end of Chapter 4, we turn our attention to the language

- $\{a^i b^i: i \geq 1\}$.

We know that this language is context free, because it is generated by the CFG G_3, of Chapter 2:

$$S \to aSb \mid ab.$$

Table 6.5. The repeat state ("*vwx*") theorem for regular languages

---

For any deterministic FSR, M:

If M accepts a string z, such that $|z| \geq n$, where n is the number of states in M, then z can be represented as three concatenated substrings

$z = vwx$

in such a way that all strings $vw^i x$, for all $i \geq 0$ are also accepted by M

The question is, could it also be a regular language? We shall now use our new theorem to show that it cannot be a regular language, using the technique of *reductio ad absurdum* introduced earlier in this chapter.

Our assumption is that $\{a^i b^i: i \geq 1\}$ *is* a regular language. If it is regular it is accepted by some FSR, M. M can be assumed to have k states. Now, k must be a definite number, and clearly our language is infinite, so we can select a sentence from our language of the form $a^j b^j$, where $j \geq k$. The repeat state theorem then applies to this string. The repeat state theorem applied to our language tells us that our sentence $a^j b^j$ can be split into three parts, so that the middle part can be repeated any number of times within the other two parts, and that this will still yield a string of as followed by an equal number of bs. But how can this be true? Consider the following:

- the "*repeatable substring*" cannot consist of as followed by bs, as when we repeated it we would get as followed by bs followed by as ... Our machine would accept strings that are not in the language $\{a^i b^i: i \geq 1\}$.

This means that the repeatable substring must consist of as alone or bs alone, However:

- the repeatable substring cannot consist of as alone, or when we repeated it more than once, the as in our string would outnumber the bs, and our machine would accept strings not in the language $\{a^i b^i: i \geq 1\}$.

A similar argument obviously applies to a repeatable substring consisting entirely of bs.

Something is clearly amiss. Every statement we have made follows logically from our assumption that $\{a^i b^i: i \geq 1\}$ is a regular language. We are therefore forced to go right back and reject the assumption itself. $\{a^i b^i: i \geq 1\}$ is not a regular language.

Our result tells us that an FSR could not accept certain strings in a language like $\{a^i b^i: i \geq 1\}$ without accepting many more that were not supposed to be acceptable. We simply could not constrain the FSR to accept only those strings in the language. That is why an FSR can accept a language such as $\{a^i b^j: i, j \geq 1\}$, as you will find out if you design an FSR to do so. However, the language $\{a^i b^i: i \geq 1\}$, which is a *proper subset* of $\{a^i b^j: i, j \geq 1\}$ is beyond the computational abilities of the FSR.

One interesting outcome of discovering that $\{a^i b^i: i \geq 1\}$ is not regular is that we can use that fact to argue that a language such as Pascal is not regular. In a sense, the set of Pascal <compound statement>s is of the form:

$\{$begin^i x end^i : x is a sequence of Pascal <statement>s$\}$.

The repeat state theorem tells us that many strings in this part of the Pascal language cannot be accepted by an FSR unless the FSR accepts many other strings in which the number of *begin*s does not equal the number of *end*s. By extension of this argument, the *syntax* of Pascal cannot be represented as a regular grammar. However, as mentioned in Chapter 3, the syntax of Pascal can be expressed as a CFG. The same applies to most programming languages, including, for example, LISP, with its incorporation of strings of arbitrary length "balanced" parentheses.

You are asked in the exercises to provide similar justifications that certain other CFLs cannot be regular. It is thus clear that the regular languages are properly contained in the class of CFLs. In fact, we have shown *two* things about Chomsky's hierarchy:

1. the *regular languages* are a *proper subset* of the CFLs in general, and, more specifically,
2. the *regular languages* are a *proper subset* of the *deterministic* CFLs.

Point 2 is demonstrated by considering that $\{a^i b^i : i \geq 1\}$ is not regular, but is a CFL, and is also deterministic (as we saw in Chapter 5). So there is at least one language that is a deterministic CFL that is not regular. In fact, none of the deterministic CFLs we considered when establishing the *closure* results for CFLs in this chapter are regular. You might like to convince yourself of this, by using the repeat state theorem.

6.5.3 The "*uvwxy*" Theorem for CFLs

The theorem described in this section is a CFL counterpart of the *repeat state theorem* described above. However, we establish our new theorem by using properties of CFGs and *derivations* rather than the corresponding abstract machines (*PDRs*). You may find the argument difficult to follow. It is more important that you appreciate the implications of the theorem, rather than understand it in great detail.

Let us assume that we have an ε free CFG that is in *Chomsky normal form* (CNF). That we can assume that any CFG is so represented was justified in Chapter 5.[2] As an example, we will use a CNF version of G_3 ($S \rightarrow aSb \mid ab$), which we will simply call G:

$$S \rightarrow AX \mid AB$$
$$X \rightarrow SB$$
$$A \rightarrow a$$
$$B \rightarrow b.$$

G has four *non-terminal* symbols. This means that if we construct a derivation tree for a sentence in which there is a path from S with more than four non-terminals on it, one of the non-terminals will appear more than once.

Now, consider the *derivation tree* for the sentence $a^4 b^4$, shown in Figure 6.11 and labelled *DT*1.

In particular, we focus on the path in *DT*1 proceeding from the top of the tree to the non-terminal marked "*low X*", in *DT*1. Sure enough, this path features a repeated non-terminal. In fact, both S – three times – and X – also three times – are repeated on this path. We can choose any pair of the repeated non terminals, so we will choose the two instances of the non terminal X that have been labelled "*low X*" and "*high X*". Five *substrings*, that *concatenated* together make up the sentence, have been identified and labelled in Figure 6.11 as follows:

u   this is the string derived in the part of the tree from S to the *left* of *high X*, and is the string *aa*

v   this is the string derived from *high X* to the *left* of *low X*, and is the string *a*

w   this is the string derived from *low X*, i.e. *abb*

x   this is the string derived from *high X* to the *right* of *low X*, i.e. *b*

y   this is the string derived from S to the *right* of *high X*, and is *b*.

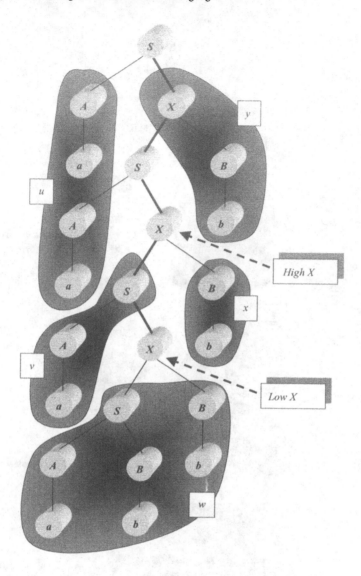

Figure 6.11. Derivation tree $DT1$ for the string a^4b^4, showing the characteristic context free $uvwxy$ form

Concatenating all the above strings in order, i.e. $uvwxy$, gives us our sentence a^4b^4.

Careful thought may now be required. The derivation from *high X* in $DT1$ is simply the application of a sequence of productions, the first having X as its left-hand side. G is a CFG, so the derivation we did from *high X* in $DT1$ could be repeated from *low X*. We would still have a sentence, as the whole derivation started with S, and a terminal string is derived from *high X* in $DT1$.

Doing the above gives us $DT2$, as shown in Figure 6.12. $DT2$ is the derivation tree for the sentence a^5b^5. As $DT2$ shows, the way we have obtained the new derivation ensures that we now have *two copies* of the substrings v and x, where there was one

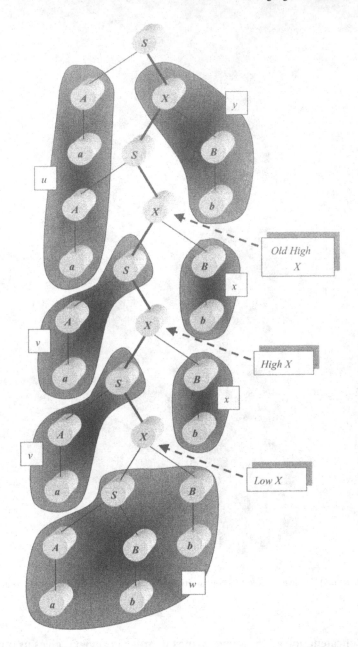

Figure 6.12. Derivation tree $DT2$ for the string a^5b^5, showing the uv^2wx^2y form of $DT1$ in Figure 6.11

copy of each, respectively, in the $DT1$ sentence. Expressed in an alternative way, we have:

$$uv^2wx^2y \ (= a^5b^5).$$

What we have just done could be repeated, i.e. we could do the *high X* derivation we did in $DT1$ from the new *low X* near the bottom of $DT2$, in Figure 6.12. Then we

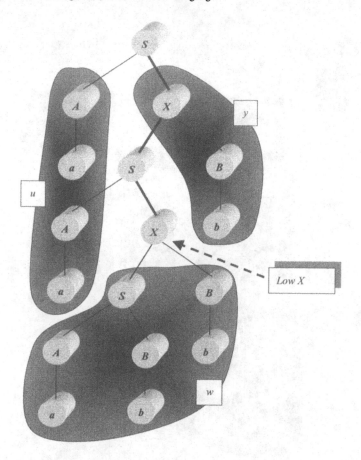

Figure 6.13. Derivation tree *DT*3 represents the *uwy* form of *DT*1 (Figure 6.11)

would have uv^3wx^3y. Moreover, we could repeat this process indefinitely, i.e. we can obtain uv^iwx^iy, for all $i \geqslant 1$.

Finally, referring back to *DT*1 (Figure 6.11), the derivation originally carried out from *low X* could have been done from *high X* instead. If we do this, we obtain *DT*3, as shown in Figure 6.13. *DT*3 is the derivation tree for the sentence a^3b^3. As can be seen from Figure 6.13, this time there are only three of our substrings, namely *u*, *w*, and *y*. We have $uwy = a^3b^3$.

As stated in Chapter 2, for any string, *s*, $s^0 = \varepsilon$, the *empty string*, and for any string *z*, $z\varepsilon = z$, and $\varepsilon z = z$. *uwy* can therefore be represented as uv^0wx^0y.

The constructions used to obtain the three trees, *DT*1–3 (Figures 6.11–6.13), and our accompanying discussion tell us that, for the language $L(G)$, i.e. $\{a^ib^i : i \geqslant 1\}$, we have found a string, a^4b^4, which is such that:

a^4b^4 can be represented as concatenated substrings *u*, *v*, *w*, *x*, and *y*, such that $uvwxy = a^4b^4$, and for all $i \geqslant 0$, uv^iwx^iy is in $L(G)$.

We initially chose the two *X*s called *high X* and *low X* in *DT*1. However, any other equal pair of non-terminals on the path could have been chosen, and a similar

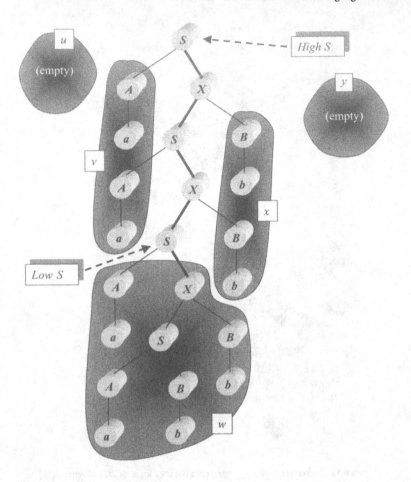

Figure 6.14. Derivation tree *DT1a*, an alternative *uvwxy* form for *DT1* (Figure 6.11)

argument could have been made. For example, we could have chosen two of the repeated *S* nodes. A possible arrangement in this case is shown in *DT1a*, for which see Figure 6.14.

In Figure 6.14 there is no string *u* to the left of the top *S* down to *high S* (as *high S is* the top *S*), and similarly there is no string *y* to its *right*, both *u* and *y* are empty, i.e. $u = y = \varepsilon$. So we now have:

$$u = \varepsilon, \quad v = aa, \quad w = aabb, \quad x = bb, \quad y = \varepsilon.$$

This time, if we duplicated the original *high S* derivation from *low S* in Figure 6.14, we would have uv^2wx^2y, i.e. ε concatenated to *aa* concatenated to *aa* concatenated to *aabb* concatenated to *bb* concatenated to *bb* concatenated to ε, which is a^6b^6. The next time we repeated it we would obtain uv^3wx^3y, i.e. a^8b^8, and so on. If we duplicated the *low S* derivation from *high S*, we would obtain uv^0wx^0y. This, since all of the substrings except for *w* is empty, simply leaves us with *w*, i.e. *aabb*. So again, our statement above:

> a^4b^4 can be represented as concatenated substrings u, v, w, x, and y, such that $uvwxy = a^4b^4$, and for all $i \geq 0$, uv^iwx^iy is in $L(G)$,

is true for this case also, as long as we permit u and y to be empty.

How general is the $uvwxy$ type construction? It can be argued to apply to *any* CFG that generates a sentence long enough so that its derivation tree contains a path from S that contains two (or more) occurrences of one non-terminal symbol. Any CFG that generates an *infinite* language *must* be able to generate a derivation tree that contains a path like this. Such a CFG, as we showed in Chapter 5, can have its ε free part converted into CNF. If there are n non-terminals in a CNF grammar, by the time we generate a sentence that is greater in length than 2^{n-1} (*numerically speaking*), we know we have a path of the type we need. To appreciate this, suppose that a CNF grammar has 3 non-terminals, i.e. $n = 3$. The shallowest tree for a sentence of length *equal* to 2^{n-1} (4) is shown in Figure 6.15.

Notice that the longest paths in the tree in Figure 6.15 have length, or number of *arcs*, 3, and, since the last *node* on the path would be a terminal, such paths have only 3 non-terminals on them. Such paths are not long enough to ensure that they feature repeated non-terminals. Our hypothetical grammar has 3 non-terminals, so to be absolutely sure that our tree contains a path of length greater than 3 we must generate a sentence of length *greater than* 4. You may need to think about this, perhaps drawing some possible extensions to the above tree will convince you. Remember, the grammar is assumed to be in CNF. You should also convince yourself that the 2^{n-1} argument applies to any CNF grammar.

It is not always necessary to generate a long sentence to obtain our "repeated non-terminal path". Our grammar G can be used to create derivation trees for sentences of length less than 8 (2^3, since G has four non-terminals) that have paths of the type we require. However, here we are interested in defining conditions that apply in general, which requires us to be *sure* that we can obtain a path of the type required.

Let G stand for *any* ε free CFG. We now know that if we generate a sentence of sufficient length the derivation tree for that sentence will have a path, starting from the S at the root of the tree, upon which the same non-terminal, say X, appears twice or more. We choose two occurrences of X on that path, and call one *high* X and one *low* X. The sentence can be split into five concatenated substrings as

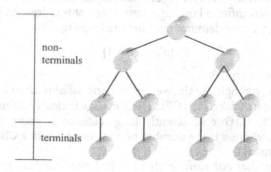

non-
terminals

terminals

Figure 6.15. The shallowest derivation tree for a sentence that is four symbols long produced by CNF productions

Table 6.6. The "*uvwxy*" theorem for CFLs

For any infinite CFL, L:
   Any sufficiently long sentence, z, of L can be represented as five concatenated substrings
   u, v, w, x, and y,
   such that $z = uvwxy$, and uv^iwx^iy is also in L, for all $i \geq 0$

before (use the derivation trees $DT1$–3 and $DT1a$ in Figures 6.11–6.14 for illustration, if you wish):

u the string derived from S at the root and to the left of *high X*. This string may be empty, either because $S = high\ X$ (as in $DT1a$ above), or because our path proceeds directly down the left of the tree.

v the string derived from *high X* to the *left* of *low X*. This string may be empty if the path from *high X* to *low X* is on the left. However, if this string is empty, x (see below) will not be. As we assume a CNF grammar, *high X* must expand initially to *two* non-terminals. Either (a) one of these is *low X*, or (b) must eventually expand into two non-terminals of which one is *low X*. The other one expands into part of v or x, depending on the particular tree.

w the string derived from *low X*. This string will not be empty, as the grammar is ε free.

x as for v but derived from *high X* to the *right* of low X. Same argument as for v applies here. In particular, cannot be empty if v is.

y as for u but to the *right* of *high X*. Also may be empty, for analogous reasons to those given above, for u.

As was shown by our example derivation trees $DT1$–3, above, we can always repeat the "deriving what was derived from the high X in the original tree from low X" routine as many times as we wish, thus obtaining the characteristic uv^iwx^iy form, for all $i \geq 1$. Furthermore, we can derive what was derived from low X in the original tree from high X, giving us the uv^0wx^0y case.

Enough has been said to demonstrate the truth of the theorem that is formally stated in Table 6.6.

6.5.4 $\{a^ib^ic^i: i \geq 1\}$ Is Not Context Free

As for the *repeat state theorem* for regular languages, we can use our *uvwxy theorem* to show that a given infinite language *cannot* be context free. As promised at the end of Chapter 5, we now demonstrate that the language

$$\{a^ib^ic^i: i \geq 1\}$$

is not context free.

The proof is very simple. Again, we use *reductio ad absurdum* as our technique. We assume that $\{a^ib^ic^i: i \geq 1\}$ *is* a CFL. The language is clearly infinite, so the *uvwxy theorem* applies to it. Our sufficiently long sentence is going to be greater in length than 2^{n-1}, where n is the number of non-terminals in a CNF grammar that generates our language.

We can assume that our sufficiently long sentence is $a^kb^kc^k$, for some k, where $k > 2^{n-1}$. The theorem tells us that this sentence can be represented as five concatenated substrings $uvwxy$, such that uv^iwx^iy are also in $\{a^ib^ic^i: i \geq 1\}$ for all $i \geq 1$. However, how are substrings of $a^k b^k c^k$ to be allocated to u, v, w, x, and y in this way?

- neither v nor x can be strings of as and bs, or bs and cs, or as, bs and cs, as this would result in symbols being in the wrong relative order when v and x were repeated.

So this means that the repeatable strings v and x must consist of strings which are either all as, all bs or all cs. However,

- this would also be inappropriate, as even if, say, v was aaa and x was bbb, when v and x were repeated, though the number of as would still equal the number of bs, the number of each would exceed the number of cs. A similar argument can be used to show that no other allocation of substrings to v and x would work.

We must again reject our original assumption. $\{a^i b^i c^i : i \geq 1\}$ is not a CFL.

In fact, $\{a^i b^i c^i : i \geq 1\}$ is a *context sensitive* language, since it is generated by the following *context sensitive grammar*, G_{16}:

$$S \rightarrow aSBC \mid aBC$$
$$CB \rightarrow BC$$
$$aB \rightarrow ab$$
$$bB \rightarrow bb$$
$$bC \rightarrow bc$$
$$cC \rightarrow cc,$$

as you might like to justify for yourself.

The arguments in the following section may require some effort to appreciate. The section can be omitted, if required, though the result, as stated in the title of the section, should be appreciated.

6.5.5 The "Multiplication Language" Is Not Context Free

Now we consider one final refinement to our *uvwxy theorem*. We focus on a derivation tree for a sentence of length greater than 2^{n-1}, as described above. We choose the longest path (or one of them, if there are several). Now, as our *high X* and *low X*, we pick the *lowest two repeated nodes on this path*, a constraint that was not enforced in our earlier examples. The string derived from *high X* (i.e. the one we call *vwx*). Is derived by a tree with *high X* as its root in which the path that passes through *low X* is the longest path. We selected the two Xs because they were the lowest repeated nodes on our longest path, therefore *high X* and *low X* are the *only* repeated nodes on our "sub-path". Therefore, the path must have maximum length $n + 1$, with the last node on the path being a terminal. Furthermore, the longest string that could be generated from *high X* (the one we call *vwx*) must be such that its length is less than or equal to 2^n (formally, $|vwx| \leq 2^n$).

We have established that for any sentence, z, such that $|z| > 2^{n-1}$, where there are n non-terminal symbols in a CNF grammar that generated z, we can define the *uvwxy* substrings so that $|vwx| \leq 2^n$. To see why this can be useful, we consider a language introduced at the end of Chapter 2, i.e.

$$\{a^i b^j c^{i \times j} : i, j \geq 1\},$$

which was generated by our *type* 0 *grammar*, G_4 (towards the end of Chapter 2). It was stated at the close of Chapter 2 that the above language is not context free. We now use the *uvwxy* theorem, along with our newly established result obtained immediately above, to demonstrate this.

Again, we use *reductio ad absurdum*. We assume that $\{a^i b^j c^{i \times j} : i, j \geqslant 1\}$ *is* a CFL. This means that the *uvwxy* theorem applies to a sufficiently long string. Consider a string, z from our language. $z = a^{2n} b^{2n} c^{(2n)^2}$, where n is the number of non-terminals in a CNF grammar for the language. Obviously, the length of z is greater than 2^{n-1}. We can then, as shown in the preceding paragraph, establish a *uvwxy* form for this sentence, such that $|vwx| \leqslant 2^n$ (remember, n is the assumed number of non-terminals). Now, by a similar argument to that applied in the case of the language $\{a^i b^j c^i : i \geqslant 1\}$ above, we can certainly justify that v and x *cannot be mixtures of different terminals*. This means that v must consist either entirely of *a*s or entirely of *b*s, and x must be *c*s only (otherwise when we increase the number of *a*s or *b*s for v^i, the *c*s would not increase accordingly). Now,

- suppose v is a string of *a*s. Then, since x must consist of *c*s only, w must include all of the *b*s. This cannot be so, as both v and x are non-empty, so $|vwx|$ exceeds 2^n (since there are 2^n *b*s in w alone).

So v cannot be a string of *a*s then, which means that:

- v must be a string of *b*s. Now, suppose $v = b^k$, for some $k \geqslant 1$. Then x must be $c^{2n \times k}$. This is because each time $uv^i wx^i y$ inserts a new copy of v (b^k) into the *b*s, it inserts a new copy of x into the *c*s. To ensure that the number of *c*s is still the same as the number of *a*s times the number of *b*s, we must insert 2^n *c*s (2^n being the number of *a*s) for each of the k *b*s. However, again this cannot be so, because if $|x| = 2^n \times k$, then $|vwx|$ is certainly $> 2^n$.

If the language had been context free, we could, for our sufficiently long sentence, have found a *uvwxy* form such that $|vwx| \leqslant 2^n$. We assumed the existence of such a form, and it led us to a contradiction. We are therefore forced, once more, to reject our original assumption, and conclude that $\{a^i b^j c^{i \times j} : i, j \geqslant 1\}$ is not a CFL.

The "length constraint", i.e. that the *vwx* in the *uvwxy* form is such that $|vwx| \leqslant 2^n$, where n is the number of non-terminals in a CNF grammar, makes the *uvwxy* theorem especially powerful. It would have been difficult to establish that $\{a^i b^j c^{i \times j} : i, j \geqslant 1\}$ is not a CFL by arguing simply in terms of *uvwxy* patterns alone. You may wish to try this to appreciate the point.

6.6 Preliminary Observations on the Scope of the Chomsky Hierarchy

The fact that $\{a^i b^j c^{i \times j} : i, j \geqslant 1\}$ is not context free suggests that we need more powerful machines than *PDRs* to perform *computations* such as multiplication. This is indeed the case. However, the abstract machine that is able to multiply arbitrary length numbers is not that far removed from the pushdown machine. The study of the machine, the *Turing machine* (TM), will be a major feature of the second part of this book. It also turns out that the TM is the recogniser for the context sensitive (type 1) and unrestricted (type 0) languages, which is how we introduce it in the next chapter.

In the preceding sections we have seen that there is indeed a hierarchy of *formal languages*, with each class being properly contained within the class immediately above it. The question arises: are the type 0 languages a *proper subset* of the set of all formal languages? As we said earlier, a formal language is simply any set of strings. Much later in this book (Chapter 11), we will see that there are indeed formal languages that are outside the class of type 0 languages, and thus cannot be generated by any grammar specified by the Chomsky hierarchy.

6.7 Exercises

For exercises marked †, solutions, partial solutions, or hints to get you started appear in "Solutions to Selected Exercises" at the rear of the book.

6.1. Produce a *DFSR* to accept the language $\{a^{4i}: i \geqslant 1\}$, then use the *complement* construction described earlier in the chapter to produce a machine that accepts the language $\{a^{4i+j}: i \geqslant 0, 1 \leqslant j \leqslant 3\}$.

6.2. As mentioned in the text above, both $L_1 = \{a^i b^j: i \geqslant 1, j = 0 \text{ or } j = i\}$ and $L_2 = \{b^i c^j: i = 0 \text{ or } i = j, j \geqslant 1\}$, are *deterministic CFLs*, but the language $L_1 L_2$ is not. Justify this.

6.3†. As expressed above, the first sentence of the section on the *repeat state theorem* is not generally true. Rewrite the sentence so that it is.

6.4†. Use the *repeat state theorem* to show that $\{a^i b^j c^{i+j}: i, j \geqslant 1\}$ and $\{a^i b^j c^{i-j}: i, j \geqslant 1\}$, both shown to be *deterministic* by Exercise 7 of Chapter 5, are not *regular* languages.

6.5†. Use the *uvwxy theorem* to show that $\{xx: x \in \{a, b\}^*\}$ is *not context free*. (In $\{xx: x \in \{a, b\}^*\}$ each string consists of an arbitrary mixture of *a*s and/or *b*s of even length (or empty) and the second half of the string is an exact copy of the first, for example, *aaabbbaaaabbba*.)

Notes

1. We omit the *context sensitive* (type 1) languages from this discussion. Only the ε free CFGs are context sensitive, as context sensitive grammars are not allowed to have any ε productions.
2. We leave out the single ε production, if it exists, and concentrate only on the ε free part of the grammar.

Chapter 7
Phrase Structure Languages and Turing Machines

7.1 Overview

This chapter introduces the *Turing machine* (TM), an abstract machine for language recognition and computation.

We examine TMs as language recognisers. You will appreciate that:

- TMs can do anything that *finite state recognisers* (FSRs) and *pushdown recognisers* (PDRs) can do
- TMs are more powerful than *non-deterministic PDRs* (NPDRs)
- TMs are the recognisers for the *phrase structure* (*type* 0) languages in general.

7.2 The Architecture of the Turing Machine

The *abstract machine* we consider in this chapter, and for much of Part 2 of this book, is called the *TM*. Alan Turing was the name of the English mathematician who first described such a machine in 1936. A TM is essentially a finite state machine, like a *FSR*, but with some very simple additional facilities for input and output, and a slightly different way of representing the input sequences.

7.2.1 "Tapes" and the "Read/Write Head"

The input to a TM is regarded as being contained on a "*tape*". The tape is divided into "*tape squares*", so that each symbol of the input sequence occupies one square of the tape. We assume that the tape is not restricted in length, so that we can place an input sequence of any length we like on the tape (we assume that any unused part of the tape consists of *blank* tape squares). For example, the input sequence *aabbaa* may occupy part of a tape such as that depicted in Figure 7.1.

Hitherto, our abstract machines have been assumed to *read* their input sequences from left to right, one symbol at a time. However, a TM can move along its tape in either direction, examining, or "reading" one tape square at a time.

Let us now consider the output of a TM. The TM simply uses its tape for output as well as input. On reading an input symbol from a particular tape square in a particular state, the TM can replace the input symbol on that particular square with any designated symbol. As a simple example, consider a tape containing a sequence

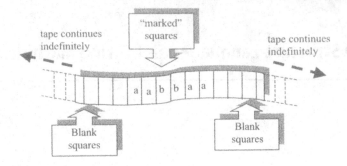

Figure 7.1. The tape of a TM, containing the string *aabbaa*

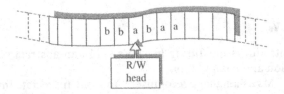

Figure 7.2. The *read/write head* located on a square of the tape

of *a*s and/or *b*s, where we require as output a sequence representing the input sequence, but with all *b*s replaced by *a*s, and vice versa.

An example tape is shown in Figure 7.2, as the hypothetical TM described immediately above has completed its processing of the first three symbols (assume that the tape is being processed from left to right).

In order to clarify the notion of a TM "examining" one square of a tape, and placing a symbol on a tape, we say the TM possesses a *read/write head* (we sometimes simply say *R/W head* , or just *head*). The read/write head is at any time located on one of the tape squares, and is assumed to be able to "sense" which symbol of some finite alphabet occupies that square. The head also has the ability to "write" a symbol from the same finite alphabet onto a tape square. When this happens, the symbol written replaces the symbol that previously occupied the given tape square. The read/write head can be moved along the tape, one square at a time, in ways we discuss later.

7.2.2 Blank Squares

In Figure 7.2, we examined the state of the tape of an imaginary TM that replaced *b*s by *a*s (and vice versa), when the machine had dealt with only some of the input symbols on its tape. Let us now consider the example tape for Figure 7.2 when the machine has processed the last symbol (the *a* at the right-hand end) of the marked portion of the tape. At this point, since the machine is replacing *a*s by *b*s, and vice versa, the tape will be as shown in Figure 7.3.

When our machine reaches the situation shown in Figure 7.3, it will attempt to read the symbol on the next tape square to the right, which of course is blank. So that our machine can terminate its processing appropriately, we allow it to make

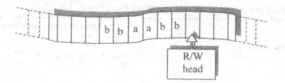

Figure 7.3. The *read/write head* moves past the right-hand end of the marked portion of a tape, and reaches a blank square

a transition on reading a blank square. In other words, we assume that the read/write head of a TM can sense that a tape square is blank. Moreover, the TM can write a symbol onto a blank square and even "blank out" an occupied tape square.

We denote the blank symbol by an empty box (suggesting that it is represented as a blank tape square), thus:

Note that in some books, the blank is represented as a 0, or a β (Greek letter beta).

7.2.3 TM "Instructions"

Like *FSRs* (Chapter 4) and *PDRs* (Chapter 5), TMs can be represented as drawings, featuring *states* and labelled *arcs*. An arc in a TM is, like the arc between two states of a PDR, labelled with *three* symbols (though what the three symbols mean is different, as we will now see).

By way of illustration, Figure 7.4 shows part of a TM. Assume that the machine is in state *A*, and that the tape and position of the read/write head are as shown on the right of Figure 7.4.

The TM represented in Figure 7.4 would carry out the activities described in Table 7.1, assuming that it starts in state *A*.

We now use the conceptual framework provided above as the basis of a more rigorous definition of TMs.

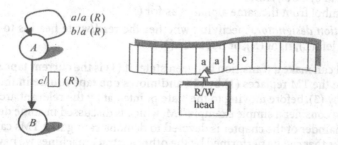

Figure 7.4. A fragment of a TM, along with an appropriate tape set up

Table 7.1. A trace of the behaviour of the TM fragment from Figure 7.4

Action	State of tape immediately following action
Read an *a*, write an *a*, move right (*R*), stay in state *A*	
Read an *a*, write an *a*, move right (*R*), stay in state *A*	
Read a *b*, write an *a*, move right (*R*), stay in state *A*	
Read a *c*, write a *blank*, move right (*R*), move to state *B*	

7.2.4 TMs Defined

A TM has essentially the same pictorial formulation as the FSR or PDR, except that each arc of a TM is labelled with *three* instructions:

1. *an input symbol* which is either
   or a symbol from a finite alphabet
2. an *output* symbol, either
   or a symbol from the same alphabet as for (1),
3. a *direction designator*, specifying whether the read/write head is to move one square left (*L*), right (*R*), or not at all (*N*).

The TM can make a transition between states if (1) is the current tape symbol, in which case the TM replaces (1) by (2), and moves one tape square in the direction specified by (3), before moving to the state pointed at by the relevant arc.

Next, we consider a simple example TM, which is discussed in some detail. Much of the remainder of the chapter is devoted to demonstrating that TMs can perform all the tasks that can be performed by the other abstract machines we have hitherto considered, and many tasks that they could not.

7.3 The Behaviour of a TM

Figure 7.5 shows our first full TM, T_1.

T_1 processes input tapes of the form shown in Figure 7.6.

As shown in Figure 7.6, T_1 (Figure 7.5) expects the marked portion of its input tapes to consist of an *a* followed by zero or more blank squares, followed by either *b* or *c*. T_1's head is initially positioned at the *a*, and the output is a tape of the form depicted in Figure 7.7.

As Figure 7.7 shows, the *b* or *c* of the input tape (Figure 7.6) has been "erased" from its initial position – replaced by a blank – and placed in the tape square immediately to the right of the *a*, with T_1's head finally positioned at that square. We will call this moving of a symbol from one part of a tape to another "*shuffling*" the symbol. We can talk of *right shuffling* and *left shuffling*. T_1 does a *left shuffle* of the symbol denoted by *x* in Figure 7.6.

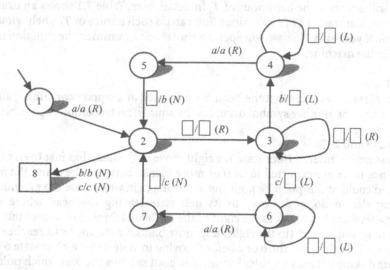

Figure 7.5. The TM, T_1

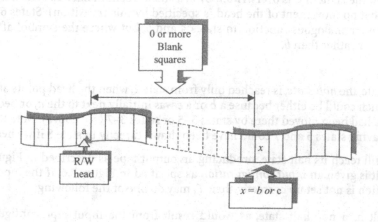

Figure 7.6. The input tape set up for T_1 of Figure 7.5

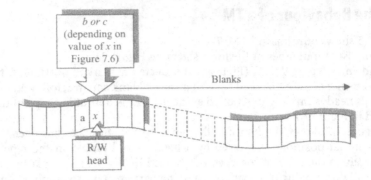

Figure 7.7. The tape when T_1 of Figure 7.5 halts, having started on a tape as specified in Figure 7.5

We will not trace the behaviour of T_1 in detail, here. Table 7.1 shows an example of how we can trace a TM's behaviour. You can do such a trace of T_1's behaviour for yourself, if you wish. Now we will spend a short time examining the function of the states in the machine:

State 1
The *start state*. Assumes that the head is positioned at a square containing an *a*. If there is no *a*, or another symbol occupies the square, no transition is possible.

States 2–5 and 2–7
When state 2 is entered from state 1, a right move of the head has just been carried out. Since there are expected to be 0 or more blanks between the *a* and the *b* or *c*, the head could at this point be pointing at a *b* or a *c* (in which case the machine has nothing else to do, and moves to its halt state leaving the head where it is). Otherwise, there is a blank to the right of the *a*. The machine then moves into state 3, where it "skips over" (to the right) any more blanks, until the head reaches the *b* or *c*. State 3 also "erases" the *b* or *c* before moving to state 4 (for a *b*) or state 6 (for a *c*). State 4 skips left over any blanks until the head reaches the *a*, at which point the head moves back to the blank square to the right of *a*. State 5 then ensures that the blank to the right of *a* is overwritten by *b*, as the transition back to state 2 is made (note that no movement of the head is specified by this transition). States 6 and 7 perform an analogous function to states 4 and 5, but where the symbol after the blanks is *c* rather than *b*.

State 8
This state, the *halt state*, is reached only from state 2, when the head points at a *b* or a *c* (which could be either because a *b* or a *c* was initially next to the *a*, or because a *b* or a *c* had been moved there by states 3–5 or states 3–7 (see above). Note that the arcs leaving states 5 and 7 could have been drawn directly to state 8 if wished.

T_1 will reach its halt state, producing an output tape as described in Figure 7.7, when it is given an *input configuration* as specified in Figure 7.6. If the input configuration is not set up correctly, then T_1 may do one of the following:

(a) halt in a non-halt state, as would result from the input tape configuration shown in Figure 7.8.

(b) never halt, which would be the result of the input tape configuration shown in Figure 7.9.

You may like to satisfy yourself that (a) and (b) are true for the input configurations of Figures 7.8 and 7.9, respectively. You should also observe that a third general possibility for a TM's halting situation, i.e.

(c) reaches halt state, but produces output not conforming to specification,

is not possible in the case of T_1, if only blanks lie to the right of the b or c. The latter comment relates to the fact that the output specification requires a totally blank tape to the right of the head when T_1 halts.

The immediately preceding discussion indicates the need for care in describing a TM's expected *input configuration* (i.e. the marked portion of its tape, and initial location of the read/write head). If we carefully specify the input conditions, then we need not consider what happens when an input situation that does not meet our specification is presented to the machine. Of course, we should also specify the ways in which the *output configuration* (the state in which the machine halts, the contents of its tape, and the final location of its read/write head) is to represent the "result".

The remainder of this chapter will be devoted to demonstrating that for language recognition, TMs are indeed more powerful than the other abstract machines we have studied thus far.

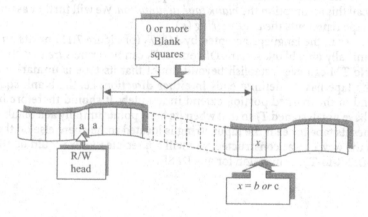

Figure 7.8. A tape that would cause problems for T_1 of Figure 7.5. There is an unexpected a to the right of the head, which would cause T_1 to get stuck in state 2

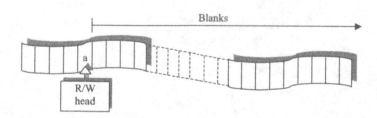

Figure 7.9. A tape that would cause even bigger problems for T_1 of Figure 7.5. There is no b or c present, so there is nothing to stop T_1 going along the tape forever (in state 3)

7.4 TMs as Language Recognisers

7.4.1 Regular Languages

We do not have to expend much effort to demonstrate that TMs are at least as powerful as FSRs. Recall from Chapter 4 that for each *regular language*, there exists a *deterministic FSR* (DFSR) to recognise that language. A DFSR is simply a TM that always moves its read/write head to the right, replacing each symbol it reads by that same symbol, on each transition.

Thus, the DFSR, M_5^d, from Chapter 4, reproduced here as Figure 7.10, becomes the TM, T_2, shown in Figure 7.11.

We have taken advantage of the TM's ability to write, as well as read, symbols, and its ability to detect the end of the input string when reaching a blank square, to ensure that T_2 has two useful features:

1. it prints T for any valid string, and F for any invalid string, and

2. it has only one halt state.

We henceforth assume that all of our TMs have only one halt state. An exercise asks you to convince yourself that this is a reasonable assumption.

An important point should be noted. T_2 prints F and goes directly to its halt state if the head is initially located on a blank square. In a sense, this reflects an assumption that if the head is initially at a blank square then the tape is entirely blank. We will call this assumption the *blank tape assumption*. We will further assume that a blank tape represents the *empty string* (ε).

As ε is not in the language accepted by M_5^d, T_2 (of Figure 7.11) prints an F if its head is initially on a blank square. Of course, T_2 can never be *sure* that the tape is empty. No TM can ever establish beyond doubt that its tape is unmarked. As we know, the tape has no definite ends in either direction, i.e. the blank squares at either end of the marked portion extend indefinitely. It should therefore be clear why we have not designed T_2 so that when its head points initially at a blank square, it searches its tape to see if the input string is located somewhere else on the tape.

From the way T_2, was constructed, you will appreciate that we could do the same type of DFSR-to-TM conversion for any DFSR.

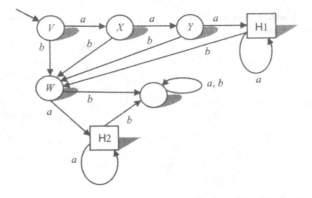

Figure 7.10. The deterministic finite state recogniser, M_5^d, of Chapter 4

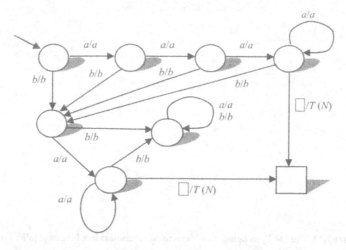

Figure 7.11. TM T_2, which does the same job as DFSR M_5^d (Figure 7.10). Since all instructions but three involve a rightward move, the move designator (R) has been omitted. Every state apart from those that write T is assumed to write F and move to the halt state if a blank square is encountered

7.4.2 Context Free Languages

Recall from Chapter 5 that the abstract machine associated with the *context free languages* (CFLs) is the *NPDR*. Recall also that some of the CFLs are *deterministic* (can be recognised by a *DPDR*) while some are not (can be recognised by an NPDR, but not a DPDR).

It can be shown that any DPDR can be converted into a TM. We will not consider this here, but merely point out that the basic method is for the TM to use part of its tape to represent the stack of the PDR, thus working with two parts of the tape, one representing the input string, the other representing the stack. However, one usually finds that a TM to do the job done by a DPDR is better designed from scratch. Here, we consider T_3, which is equivalent to M_3^d of Chapter 5 (Figure 5.8), the DPDR that accepts the language

$$\{a^i b^i : i \geq 1\},$$

which we know from Chapter 6 to be *properly* context free, i.e. context free but not regular. T_3 is shown in Figure 7.12.

T_3 expects on its tape an arbitrary string of as and/or bs (using the notation of Chapter 2, a string in $\{a, b\}^*$), finally printing T if the string is in the language $\{a^i b^i : i \geq 1\}$, and F if it is not. Note that, like T_2 (Figure 7.11), T_3's read/write head begins on the leftmost symbol of the input. Moreover, T_3 embodies the same *blank tape assumption* as did T_2. T_3 uses the symbols X and Y to "tick off" the as and corresponding bs, respectively. You should trace T_2's behaviour on example sentences and non-sentences. An example of how a trace can be done is shown in Table 7.1.

Now for the *non-deterministic* CFLs. Recall from Chapter 5 that some CFLs can be accepted by an *NPDR*, but not a DPDR. Later, in Chapter 11, we briefly consider the conversion of NPDRs into TMs. However, to demonstrate that TMs are more powerful than NPDRs, we see in Figure 7.13 a TM, T_4, that recognises the language of possibly empty *palindromic* strings of as and bs. Recall that in Chapter 5 such

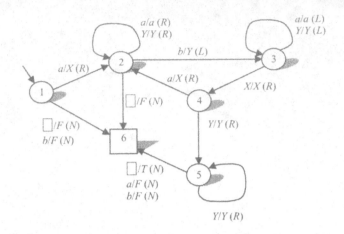

Figure 7.12. The TM T_3 recognises the deterministic context free language $\{a^i b^i: i \geq 1\}$

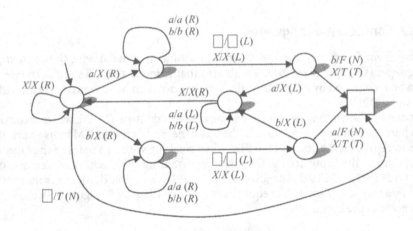

Figure 7.13. The Turing machine T_4 deterministically recognises a language of palindromic strings that requires a non-deterministic push-down recogniser

languages were shown to be non-deterministic CFLs, and therefore beyond the processing power of the DPDR.

T_4 accepts only those strings in $\{a, b\}^*$ that are palindromic, and, what is more, does the job *deterministically*.

T_4 uses only one marker, X, and works from the outside in, "ticking off" matching symbols at the extreme right and left ends of the string. This time, the blank tape assumption results in a T being printed if the read/write head of T_4 starts on a blank, as the empty string is in $\{a, b\}^*$ and is palindromic. Again, you should make sure you understand how the machine works, since more complex TMs follow.

7.4.3 TMs Are More Powerful than PDRs

The above discussion implies that TMs are as powerful, in terms of language processing, as FSRs and (deterministic and non-deterministic) PDRs. This claim is justified later in this chapter. However, by using a single example we can show that TMs are actually *more* powerful than PDRs. Consider the TM, T_5, shown in Figure 7.14, that processes the language

$$\{a^i b^i c^i : i \geq 1\},$$

which was demonstrated in Chapter 6 to be properly *context sensitive* (i.e. context sensitive but not context free).

T_5 requires an input configuration of the form shown in Figure 7.15.

It reaches its halt state for valid or invalid strings. For *valid* strings (i.e. strings of one or more *a*s followed by the same number of *b*s followed by the same numbers of *c*s), the output configuration is as specified in Figure 7.16.

The input sequence of Figure 7.15 is altered by T_5 to produce a tape of the form described in Figure 7.16 (the symbols X, Y, and Z are used as auxiliary markers). Again, the output of a T indicates that the input sequence represents a sentence of the language in question.

For *invalid* strings (i.e. any string in $\{a, b, c\}^*$ which is not in $\{a^i b^i c^i : i \geq 1\}$), however, when the halt state is reached, the configuration will be as shown in Figure 7.17.

In the case of invalid strings then, the read head will be left pointing at an *F*, as shown in Figure 7.17.

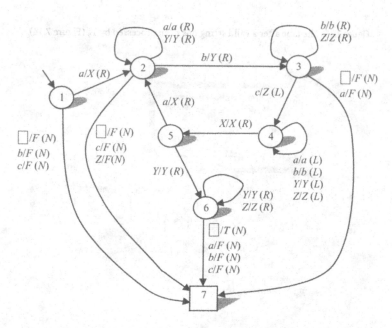

Figure 7.14. The TM T_5 recognises the context sensitive language $\{a^i b^i c^i : i \geq 1\}$

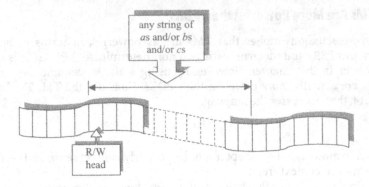

Figure 7.15. The input tape set up for T_5 of Figure 7.14

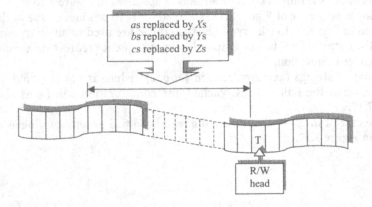

Figure 7.16. The tape after a valid string has been processed by T_5 (Figure 7.14)

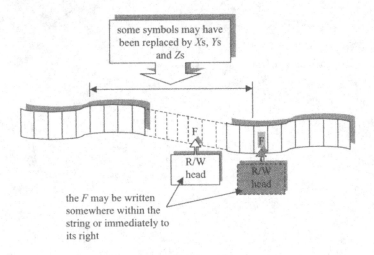

Figure 7.17. The tape after an invalid string has been processed by T_5 (Figure 7.14)

When students design language recognition machines such as T_5, their machines sometimes omit to check for the possible occurrence of extra symbols at the end of an otherwise valid string. For the language $\{a^i b^i c^i : i \geq 1\}$, such machines will indicate that a valid string immediately followed by another string in $\{a, b, c\}^+$ is valid, when, of course, it is not.[1]

However, students are not the only guilty parties when it comes to failing to check for invalid symbols at the end of an otherwise valid input. I have encountered a Pascal compiler that simply ignores all of the source file text that occurs after an end followed by a full stop (in Pascal, the only end terminated by a full stop is the last one in the program, and corresponds to a begin that announces the start of the program block). This caused severe difficulties to a novice who had accidentally put a full stop after an end halfway down their program, and happened to have an extra begin near the beginning. The program compiled perfectly, but when run only carried out the first part of what was expected by the programmer! The Pascal compiler in question was accepting an invalid input string, in this case with potentially serious consequences.

7.5 Introduction to (TM) Computable Languages

We have seen that TMs are more powerful language recognisers than FSRs and PDRs. Regular languages are associated with FSRs in the same way as CFLs are associated with PDRs. It seems reasonable to ask, then, if there is a class of languages that are associated with TMs in a similar way. In other words, is there a class of *grammar* in the *Chomsky hierarchy*, so that for any language generated by a grammar in that class, there is some TM that can recognise that language? Conversely, for any TM that recognises a language, can we be sure that the language is in the same class? In fact, there is such a class of languages, and it turns out that it is exactly that defined by the most general Chomsky classification, i.e. the general *phrase structure languages* (PSLs) themselves, these being those generated by the type 0, or *unrestricted*, grammars. Type 0 grammars were defined in Chapter 2 as being those with productions of the form

$$x \to y, \quad x \in (N \cup T)^+, \ y \in (N \cup T)^*,$$

i.e. a non-empty arbitrary string of terminals and non-terminals on the left-hand side of each production, and a possibly empty arbitrary string of terminals and non-terminals on the right-hand sides.

Let us call the languages that can be recognised by a TM the *TM Computable languages*. For convenience, and with some justification, as we will see later in the book, we will refer to them as *computable languages*. We eventually discover that the computable languages include, *but are not restricted to*, all of the regular languages, all of the CFLs, and all of the context sensitive languages.

In Chapter 4, the correspondence between regular languages and FSRs was specified. We saw how we can take any regular grammar and construct an FSR to accept exactly the same set of strings as that generated by the grammar. We also showed that we could take any FSR and represent it by a regular grammar that generated the same set of strings as that accepted by the FSR (we did a similar thing with respect to NPDRs and the CFLs in Chapter 5). Analogously, to show the correspondence between TMs and unrestricted languages we would have to show how to create a type 0 grammar from any TM, and vice versa. However, this is rather more complicated than is the case for the more restricted grammars and machines, so the result will merely be sketched here.

We first consider the context sensitive languages, as their defining grammars, the *context sensitive grammars*, cannot have the empty string on the right-hand side of productions. We then generalise our method to deal with the type 0 grammars in general. The definition of context sensitive productions is as for type 0 productions given above, with the additional restriction that the right-hand sides of context sensitive productions are not allowed to be shorter in length than the left-hand sides.

7.6 The TM as the Recogniser for the Context Sensitive Languages

The machine T_5, shown in Figure 7.14, recognises the language

$$\{a^i b^i c^i : i \geq 1\}.$$

In Chapter 6 we established that this language is not context free. We also saw that it is, in fact, a context sensitive language, as it is generated by the following grammar, G_{16}:

$$S \rightarrow aSBC \mid aBC$$
$$CB \rightarrow BC$$
$$aB \rightarrow ab$$
$$bB \rightarrow bb$$
$$bC \rightarrow bc$$
$$cC \rightarrow cc.$$

We wish to establish that for any context sensitive language we can provide an equivalent TM. T_5 itself is not enough to convince us of this, as it was *designed* for a specific language, and not constructed through using a method for creating TMs from the productions of the corresponding grammar. We therefore require a method for constructing TMs for context sensitive grammars that we could apply to any such grammar. Here, we will consider the application of such a method to the grammar G_{16}. I then leave it to you to convince yourself that the method could be applied to any context sensitive grammar.

7.6.1 Constructing a Non-deterministic TM for Reduction Parsing of a Context Sensitive Language

The TM, T_6, that has been constructed from the productions of G_{16}, is partly depicted in Figure 7.18.

You will note that T_6 is in three sections, namely sections A, B, and C, all of which have *state* 1 in common. Sections A and B are shown in full, while section C is only partly specified. The machine is far more complex than T_5 (Figure 7.14.), a TM that was designed to recognise the same language. T_6 was constructed from the productions of a grammar (G_{16}), and its purpose is merely to demonstrate a general method for constructing such TMs.

The first thing to note is that T_6 is *non-deterministic*. Its overall process is to model the *reduction* method of parsing introduced in Chapter 3. Beginning with a

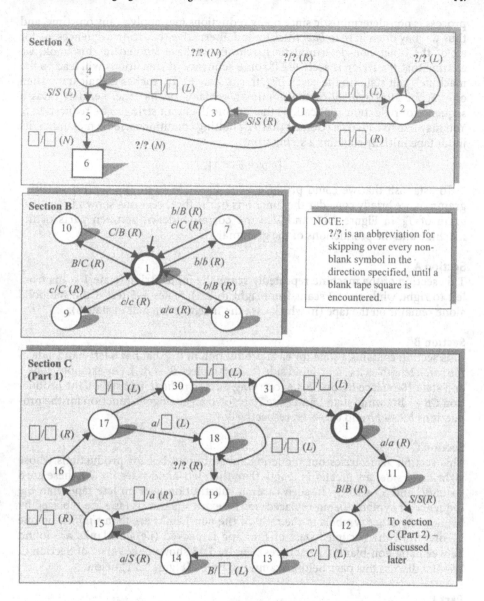

Figure 7.18. The *non-deterministic* TM T_6 recognises the context sensitive language $\{a^i b^i c^i : i \geqslant 1\}$

terminal string, if the right-hand side of any of our productions is a substring of that string, we can replace it by the left-hand side of such a production. The process is complete if we manage to reduce the string to the start symbol S alone. For a TM, this means we finish up in a halt state with the tape that initially contained the input string containing only S.

T_6 carries out reduction parsing using the grammar G_{16}, by effectively "scanning" the input string repeatedly from left to right, then right to left, attempting to find a sequence of reductions that will leave only the symbol S on the tape. The process is non-deterministic since many reductions may apply at any one time, and thus we may often have to *backtrack* to find alternative reduction sequences. As for all of the other non-deterministic machines we have studied in this book, we assume that the string is accepted if some sequence of transitions will lead to the machine's halt state (in the case of T_6, this is *state* 6). T_6 reaches its halt state when only S, the *start symbol* of the grammar, remains on its tape, i.e. if it finds a sequence of reduction applications that enable the input string to be reduced to S. You may like to convince yourself that T_6's halting conditions will be met only if its input tape initially contains a string from

$$\{a^i b^i c^i : i \geq 1\}.$$

To establish that we could produce a "T_6-like" TM from any context sensitive grammar, we briefly consider the functions of the three sections shown in the diagram of T_6, in Figure 7.18, and also the correspondence between parts of the machine and the productions of the grammar.

Section A
This section of the machine repeatedly scans the input string (state 1 scans from left to right, while state 2 scans from right to left). States 3–5 test if the symbol S alone remains on the tape (in which case the machine can halt in state 6).

Section B
This section contains a state for each production in G_{16} that has a left-hand side of the same length as its right-hand side (for example, $CB \rightarrow BC$). For example, state 1 and state 10 replace BC by CB, i.e. carrying out the *reduction* based on the production $CB \rightarrow BC$, while state 1 and state 7 perform an analogous function for the productions $bB \rightarrow bb$ and $bC \rightarrow bc$, respectively.

Section C
This section also carries out reduction substitutions, but for productions whose right-hand sides are greater in length than their left-hand sides (in G_{16}, these are $S \rightarrow aSBC$ and $S \rightarrow aBC$). In such cases, a substitution uses up less tape than the sequence of symbols being replaced, so the extra squares of tape are replaced by blanks. Any other symbols to the right of the new blanks are then *shuffled* up one by one until the occupied part of the tape is packed (i.e. no blanks are found between any non-blank tape squares). Figure 7.18 shows only Part 1 of Section C. We first discuss this part, before examining Part 2 (Figure 7.21) below.

Part 1
This section of T_6 deals with the production $S \rightarrow aBC$, i.e. "reduces" aBC to S. To illustrate, Figure 7.19 shows what happens when we are in state 11 with the tape configuration as shown.

Note that, for simplicity, Figure 7.18 shows only the loop of states (states 17, 18, 19, 15, and 16) that shuffle an a leftwards so that it goes immediately to the right of the next non-blank symbol to its left. We have omitted the 10 states that would be needed to shuffle any other symbols of the alphabet (i.e. b, c, S, B, C) that may be required. To consider why two states are needed to deal with each symbol, consider the part of T_6 that does the shuffling job for a, depicted in Figure 7.20.

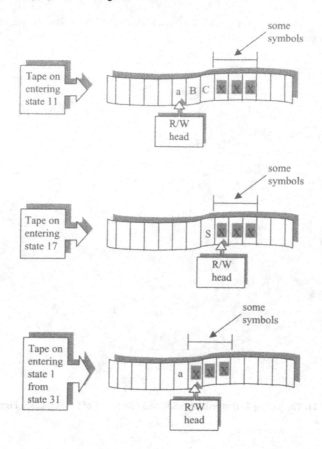

Figure 7.19. Stages in the reduction of *aBC* to *S* by the TM T_6 (the state numbers refer to those in Figure 7.18)

So, states 15, 16, and 17 are common to the shuffling loop for each symbol in the alphabet. We need, say, states 20 and 21 to shuffle any *b*s, states 22 and 23 to shuffle any *c*s, states 24 and 25 to shuffle any *S*s, states 26 and 27 to shuffle any *B*s, and states 28 and 29 to shuffle any *S*s. That is why the state in Figure 7.20 to the right of state 17 is named state 30.

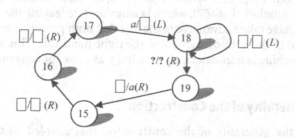

Figure 7.20. The fragment of T_6 (Figure 7.18) that "shuffles" an *a* leftwards over a blank portion of the tape so it is placed immediately to the right of the next non-blank tape square

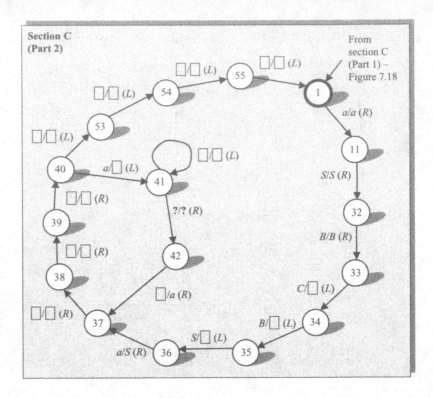

Figure 7.21. The part of T_6 that reduces $aSBC$ to S (for Part 1 of Section C, see Figure 7.18)

Part 2

This second part of section C of T_6, depicted in Figure 7.21, deals with the production $S \rightarrow aSBC$, i.e. "reduces" $aSBC$ to S. It is very similar to part 1 (see Figure 7.18), that deals with $S \rightarrow aBC$, and was discussed immediately above.

As for *section C part 1*, we have only included the pair of states (state 41 and state 42) between states 40 and 38 that shuffle the a to the left (note that in this case, shuffling involves *three* blanks, as the right-hand side of $S \rightarrow aSBC$ has four symbols). As for Part 1, we assume that there is a similar pair of states linking states 40 and 38 for each of the other alphabet symbols (b, c, S, B, C). Thus we need 10 more states (assumed to be numbered 46–57), which is why the state leaving the loop has been labelled 53. We have taken advantage of the fact that both productions dealt with by *section C* have a as the leftmost symbol of the right-hand side. This was not necessary (as our machine is non-deterministic), but was done for convenience.

7.6.2 The Generality of the Construction

To appreciate the generality of the construction that enabled us to produce T_6 from G_{16}, consider the following. The definition of context sensitive productions tells us that the left-hand side of each production does not exceed, in length, the right-hand side (i.e. productions are of the form $x \to y$, where $|x| \le |y|$). Therefore we can represent any context sensitive grammar by a TM with:

A *Section A* part

To scan back and forth over the input between the start state and another state, enabling the halt state to be reached only when the start symbol, S, appears alone on the tape.

A *Section B* part

That consists of a distinct path for each production in the grammar with equal length left- and right-hand sides, each path leaving and re-entering the start state. The path for each production performs the reduction associated with that production, i.e. when the right-hand side of a production appears anywhere on the input tape, it can be replaced by the left-hand side of that production.

A *Section C* part

That consists of a distinct path for each production in the grammar with left-hand side shorter in length than the right-hand side, each path leaving and re-entering the start state. As in section B, the path for each production performs the reduction associated with that production. However, in section C cases the newly placed symbols occupy less tape space than the symbols replaced, so each path has a "shuffling routine" to ensure that the tape remains in packed form. In G_{16}, only the productions $S \to aSBC$ and $S \to aBC$ are of section C type, and both represent reductions where the replacement string is a single symbol. In general, of course, a context sensitive production could have *several* symbols on the left-hand side. This represents no real problems however, as demonstrated by the sketch of part of a TM that does the reduction for the context sensitive production $XY \to PQR$, shown in Figure 7.22.

As you can see from Figure 7.22, in this example we need only to skip over one tape square to the right of the newly written XY. This is because the right-hand side of the production $XY \to PQR$ is one symbol longer than the left. We thus replace the PQ with XY, and the R with a blank. We then need to move over that blank to see if there is a symbol there to be shuffled up, and so on.

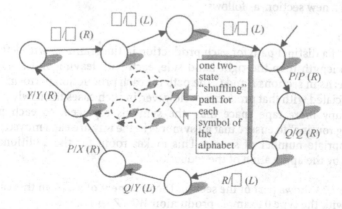

Figure 7.22. A Turing machine fragment to replace PQR by XY, representing a reduction involving the type 1 production $XY \to PQR$

In Chapter 5, the NPDRs that we created from the productions of context free grammars (CFGs) recognised the language generated by the grammar by being able to model any derivation that the grammar could carry out. Our *non-deterministic* TMs operate by modelling every possible sequence of reductions for the corresponding context sensitive grammar. Such TMs reach their halt state only if the string on the input tape can be reduced to S. The construction sketched out above ensures that this would happen only for inputs that were sentences of the language generated by the original grammar.

Enough has been said to establish the following:

- The recogniser for the context sensitive languages is the TM. Context sensitive languages are computable languages.

You may be wondering why the statement immediately above does not say "*non-deterministic* TM". In Chapter 10 we will see that any non-deterministic TM can be simulated by a deterministic TM. Hence, the non-deterministic TMs of this chapter could be replaced by equivalent deterministic TMs.

7.7 The TM as the Recogniser for the Type 0 Languages

Now, the sole difference between context sensitive productions and *unrestricted*, or *type* 0, productions is that the latter are not restricted to having right-hand sides that are greater than or equal in length to the left-hand sides. This means that, say, a production such as $WXYZ \rightarrow Pa$, which is not a valid context sensitive production, is a perfectly reasonable type 0 production. Moreover, type 0 grammars can contain ε productions (see Table 5.1, Chapter 5). Productions such as $WXYZ \rightarrow Pa$ do not present a severe problem, as we simply add another section to our TMs to deal with these types of production, as discussed next (we will return to ε productions shortly).

7.7.1 Amending the Reduction Parsing TM to Deal with Type 0 Productions

Assuming that we already have sections A, B, and C as defined for our TMs derived from context sensitive grammars (for example, Figure 7.18), we add to such a construction a new section, as follows:

Section D
Consists of a distinct path for each production in the grammar with left-hand side longer in length than the right-hand side, each path leaving and re-entering the start state. As in sections B and C, the path for each production performs the reduction associated with that production. However, in such cases the newly placed symbols occupy *more* tape space than the symbols replaced, so each path has a "shuffling routine" to ensure that all symbols in the way are each moved to the right an appropriate number of spaces. This makes room for the additional symbols required by the application of the reduction.

Figure 7.23 shows part of the section D component of a TM, in this case the part dealing with the type 0 example production $WXYZ \rightarrow Pa$.

For our example production, there are two extra symbols to insert when performing the associated *reduction*, as the right-hand side of our production – Pa – has

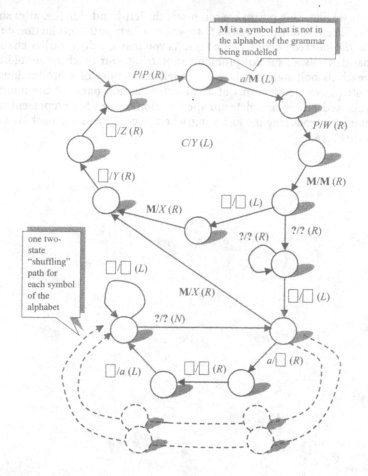

Figure 7.23. A TM fragment to replace *Pa* by *WXYZ*, representing a reduction involving the type 0 production *WXYZ* → *Pa*

two fewer symbols than the left-hand side – *WXYZ*. A machine of the type illustrated in Figure 7.23 therefore has to shuffle all symbols (if any) that were to the right of the *Pa* in question two tape squares to the right, leaving two blanks immediately to the right of *Pa* (which is eventually changed to *WX*). Then *YZ* can be inserted. The symbol *M* is used as a marker, so that the machine can detect when to stop shuffling. When the *M* has a blank next to it, either there are no symbols to its right, or we have shuffled the symbol next to it to the right, and so no more symbols need shuffling. You should convince yourself that a *section D*-type construction of the type shown in the example could be carried out for any production where the right-hand side is non-empty and of smaller length than the left-hand side.

7.7.2 Dealing with the Empty String

As promised above, we now consider ε productions. In fact, we can treat these as a special case of the *section D*-type construction. We simply allow for the fact that on either side of any non-blank symbol on the tape, we can imagine there to be as many εs as we like. This means we can insert the left-hand side (i.e. after shuffling to make sufficient room, in a *section D*-type way) of a type 0 ε production *anywhere we like on the marked portion of the tape*. As you may recall, in earlier chapters we noted that this was a cause of problems in parsing. Our machine should only be able to reach its halt state if it inserts the left-hand sides of ε productions in the appropriate places, i.e. in terms of a correct reduction parse of the input string being processed. So, we complete our specification of the TM to represent any type 0 grammar by introducing the following, where necessary, to the machine sketched out in Figure 7.18.

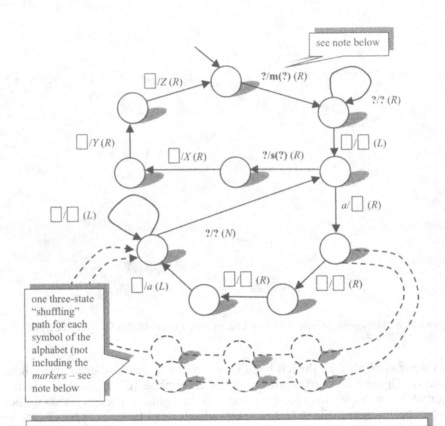

NOTE:
- We allocate a unique "marker" symbol to each of the symbols in the grammar
- Let $m(x)$ denote the marker for symbol x
- Let $s(m)$ denote the symbol associated with marker m
  Then
  ?/m(?) means "replace the symbol currently at the r/w head by its marker"
  ?/s(?) means "replace the marker currently at the r/w head by its associated symbol"
- Example: suppose the alphabet is a, b, X, with markers T, U, V, respectively, then
  - **?/m(?) (R)** means "$a/T(R)$, $b/U(R)$, $X/V(R)$", and
  - **?/s(?) (R)** means "$T/a(R)$, $U/b(R)$, $V/X(R)$"

Figure 7.24. A TM fragment to insert XYZ in the input tape, thus modelling the reduction based on the type 0 production $XYZ \rightarrow \varepsilon$

Section E

Consists of a distinct path for each production in the grammar with ε on the right-hand side, each path leaving and re-entering the *start state*. As in sections B, C, and D, the path for each production performs the reduction associated with that production. Figure 7.24 shows a sketch of a *section E* construction for the type 0 production $XYZ \to \varepsilon$.

The *section E* construction, as illustrated in Figure 7.24, inserts the left-hand side of an ε production *after* any symbol on the tape. There are two problems remaining. Firstly, we must deal with an insertion *before* the leftmost symbol. Secondly, we need to cater for an insertion on a blank tape. Considering the second case first, if ε is in the language generated by the grammar, the corresponding machine should be able to reach its halt state appropriately. Our *blank tape assumption* dictates that we represent ε as the input string by setting up our machine so that its head points initially to a blank square. We can solve both problems with very similar constructions to *section E*, though there are a few differences. I leave you to sketch the construction yourself.

7.7.3 The TM as the Recogniser for All Types in the Chomsky Hierarchy

Enough has now been said to support the following claim:

• The recogniser for the PSLs in general is the TM. PSLs are computable languages.

(For similar reasons to those discussed above, we again refer to the "TM", not the "*non-deterministic* TM".)

We argued above that any *DFSR* could be converted into an equivalent TM. It was then implied that TMs could recognise any of the CFLs. However, by showing that any type 0 grammar can be represented by a TM that accepts the same language, we have also shown that the same applies to the other types of language in Chomsky's hierarchy. This is because *any production of any grammar of type 0, type 1, type 2, and type 3 is catered for in one of the sections of our construction.*

However, if we were to implement the non-deterministic TMs that are produced by our construction, we would have very inefficient parsers indeed! That is the whole point of defining restricted types of machine to deal with restricted types of language.

7.8 Decidability: A Preliminary Discussion

We close this chapter by briefly discussing one of the most important concepts related to language processing, i.e. *decidability*. In doing this, we anticipate a more thorough treatment of the topic in Chapter 11.

7.8.1 *Deciding* a Language

Consider the TMs T_2, T_3, T_4, and T_5 in this chapter (Figures 7.11–7.14, respectively). Each accepts an input tape containing a terminal string from the appropriate alphabet. If each of these machines is started with a valid input configuration, it will halt in a halt state, and as it reaches its halt state it will:

• print T if the input sequence was a sentence of the language, otherwise print F.

A language-recognising TM that conforms to the above behaviour is said to *decide* the corresponding language (because given any string of the appropriate terminals it can say "yes" if the string is a sentence *and* "no" if the string is not a sentence).

For example, consider some of the decisions made by T_5 (of Figure 7.14), that decides the language $\{a^i b^i c^i : i \geq 1\}$, as shown in Table 7.2.

The formal definition of *deciding* a language is given in Table 7.3.

Table 7.2. Various input sequences that are accepted or rejected by T_5 of Figure 7.14

Input string	T_5 input and output
aabbcc (valid string)	Initial configuration:
	Final configuration:
aaabbcc (invalid string – too many s)	initial configuration:
	Final configuration:
caaaa (invalid string – starts with *c*)	Initial configuration:
	Final configuration:

Table 7.3. Deciding a language

Given some language L, with alphabet A, a TM *decides* that language if for any string, x, such that $x \in A^*$:
    if $x \in L$
        the machine indicates its acceptance,
    and
    if $x \notin L$
        the machine indicates its rejection of x

A language for which there is a TM that *decides* that language is called a *decidable* language

7.8.2 *Accepting* a Language

Suppose we amend T_5 (of Figure 7.14) to produce the machine T_7, depicted in Figure 7.25.

We have simply removed from T_5 all arcs entering the halt state which result in writing the symbol F, and removed the instructions that write an F from the arc between *states* 6 and 7. Like T_5, T_7 will now reach its halt state and write a T for valid input strings. However, unlike T_5, for invalid strings T_7 will eventually stop in one of its non-halt states, and make no F indication at all. For example, given the input *aaabbcc*, (the second example used for T_5 in Table 7.2), T_7 would eventually come to a stop in *state* 2, being unable to make any move from that state in the way that T_5 could.

A machine that can output T for each and every valid string of a language (and never outputs T for any *invalid* strings) is said to *accept* that language. Compare this carefully with the definition of *deciding*, from Table 7.3. Clearly, a machine that

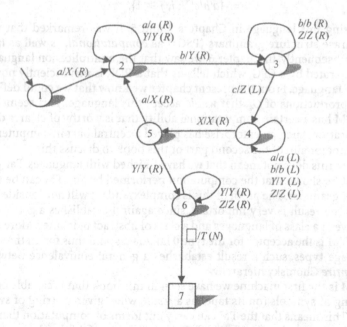

Figure 7.25. The TM T_7 (based on T_5 of Figure 7.14) also recognises the context sensitive language $\{a^i b^i c^i : i \geq 1\}$. Unlike T_5, T_7 does not print F for invalid strings

Table 7.4. Accepting a language

Given some language L, with alphabet A, a TM *accepts* that language if for any string, x, such that $x \in A^*$:
  if $x \in L$
    the machine indicates its acceptance of x.

A language for which there is a TM that *accepts* that language is called an *acceptable* language.

---

decides a language also accepts it, but the converse is not necessarily true. T_5, T_6, and T_7 all *accept* the language $\{a^i b^i c^i : i \geq 1\}$, but only T_5 *decides* it.

The formal definition of accepting a language is given in Table 7.4.

7.9 End of Part 1

In this chapter, we have encountered *computable languages, decidable languages,* and now *acceptable languages*. From the way in which the construction of TMs from type 0 grammars was developed, it can be seen that such TMs would *accept*, rather than decide, the language generated by the grammar from which the machine was constructed. In other words, *computable languages are acceptable languages*. However, we have also seen that some computable languages are decidable (since TMs T_2, T_3, T_4, and T_5 all *decide* their respective languages). A question addressed in Chapter 11 asks whether all computable languages are also *decidable*. We find that in general they are not, though the more restricted languages of the Chomsky hierarchy are.

In Chapters 2 and 6, we saw the type 0 grammar, G_4, which generates what we call the *multiplication language*:

$$\{a^i b^j c^{i \times j} : i, j \geq 1\}.$$

In discussing this language in Chapters 2 and 5, it was remarked that we could regard phrase structure grammars (PSGs) as *computational*, as well as *linguistic*, devices. Subsequently, in Chapter 6, we saw that the multiplication language could not be generated by a CFG, which tells us that a PDR is insufficiently powerful to accept the language. From the present chapter we know that we could define a TM, from the productions of G_4, that *would* accept this language. This seems to imply that the TM has a certain computational ability that is worthy of clearer definition and exploration. Such an enterprise has played a central part in computer science, and it is a major aim of the second part of this book to discuss this.

However, this does not mean that we have finished with languages. Far from it. It can in fact be shown that the computation performed by any TM can be modelled by a type 0 grammar. The proof of this is complex, and we will not consider it in this book, but the result is very important, since again it establishes a general equivalence between a class of languages and a class of abstract machines. More than this, since the TM is the acceptor for the type 0 languages, and thus for all the subordinate language types, such a result establishes a general equivalence between TMs and the *entire* Chomsky hierarchy.

The TM is the first machine we have seen in this book that is capable of producing a string of symbols (on its tape) as a *result*, when given a string of symbols as its *input*. This means that the TM can carry out forms of computation that can provide richer answers than "yes" or "no". The second part of this book investigates the nature of the computational power of the TM. In the process of this, we discover that the TM is more powerful than any digital computer. In fact, the TM is the most powerful computational model we can define. Eventually, our investigations bring us full circle, and enable us to precisely specify the relationship between *formal languages, computable languages,* and *abstract machines.* In doing so, we are able to answer a question that was asked at the end of Chapter 6: are there *formal languages* that are a proper *superset* of the *type* 0 languages?

7.10 Exercises

For exercises marked †, solutions, partial solutions, or hints to get you started appear in "Solutions to Selected Exercises" at the rear of the book.

7.1[†]. Sketch out a general argument to demonstrate that any TM needs only one halt state, and also that the single halt state requires no outgoing arcs.

7.2. Design TMs to decide the following languages:

(a) $\{a^{2i+1}c^jb^{2i+1}: i, j \geqslant 1\}$

(b) $\{xx: x \in \{a, b\}^*\}$

(c) $\{a^ib^jc^{i+j}: i, j \geqslant 1\}$

(d) $\{a^ib^jc^{i-j}: i, j \geqslant 1\}$

(e) $\{a^ib^jc^{i\times j}: i, j \geqslant 1\}$

(f) $\{a^ib^jc^{i \text{ div } j}: i, j \geqslant 1\}$ (div is *integer division*).

Hint: in all cases, first design the machine so that it deals only with acceptable strings, then add additional arcs so that it deals appropriately with non-acceptable strings.

7.3. The following is a recursive definition of "well-formed parenthesis" strings (WPEs):

( ) is a *WPE*

if X is a *WPE*, then so is (*X*)

if X and *Y* are *WPEs*, then so is *XY*.

(a) Define a TM to decide the language of all WPEs.

(b) The language of all WPEs is context free. Show this by defining a CFG to generate it.

Hint: the recursive definition above converts almost directly into a CFG.

Notes

1. Recall from Chapter 2 that $\{a, b, c\}^+$ is the set of all non-empty strings of *as* and/or *bs* and/or *cs*.

Part 2

Machines and Computation

Chapter 8
Finite State Transducers

8.1 Overview

In this chapter, we consider the types of computation that can be carried out by *finite state transducers* (FSTs), which are the *finite state recognisers* (FSRs) of Chapter 4, with output.

We see that FSTs can do:

- various limited tasks involving memory of input sequences
- *addition* and *subtraction* of arbitrary length numbers
- restricted (constant) *multiplication* and *modular division*.

You then learn that FSTs:

- cannot do multiplication of arbitrary length numbers
- are not a very useful model of real computers (despite some superficial similarities with them).

8.2 Finite State Transducers

An *FST* is a *deterministic FSR* (DFSR) with the ability to output symbols. Alternatively, it can be viewed as a *Turing machine* (TM) with two tapes (one for input, one for output) that

- always moves its *head* to the *right*
- cannot recognise *blank squares* on its *tape*.

As the machine makes a transition between two states (according to the input symbol it is reading, of course), it outputs one symbol. In addition to performing recognition tasks, such a machine can also carry out computations. The purpose of this chapter is to investigate the scope of the FST's computational abilities. We shall see that we need a more powerful type of machine to carry out all of the computations we usually expect machines (i.e. computers) to be able to do. As we will see, this more powerful machine is the one we encountered in Chapter 7. It is the TM.

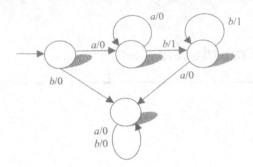

Figure 8.1. An FST that recognises the regular language $\{a^i b^j : i \geqslant 1\}$. The last symbol it outputs indicates if it accepts (1) or rejects (0) the input string

8.3 FSTs and Language Recognition

First of all, you should appreciate that the FSRs used for *regular language* recognition in Chapter 4 are simply a special case of FSTs. We could have defined FSRs so that on each transition they produce an output symbol, and we could have defined these output symbols to be significant in some way. In our first example machine, we assume that, when the input is exhausted, the *last* symbol the machine outputs is a 1, then the input string is *accepted*, and if the last symbol output is a 0, then the string is *rejected*. Figure 8.1 shows such a machine.

The FST in Figure 8.1 accepts the language $\{a^i b^j : i, j \geqslant 1\}$. It indicates acceptance or rejection of an input string by the last symbol it outputs as the last symbol of the input is read (1 = *accept*, 0 = *reject*). Some examples of the machine's output are shown in Table 8.1. The significant symbol of the output (the last) is underlined in each case.

As you can see, the machine in Figure 8.1 has no state designated as a *halt state* or *accept state*. This is because the *output* is now used to tell us whether or not a string is accepted.

Table 8.1. Example input strings, and corresponding output strings, for the FST of Figure 8.1. The last symbol output indicates the machine's decision

Input string	Output string	Accept or reject
aaab	0001	Accept
aa	00	Reject
aabba	00110	Reject
aabbb	00111	Accept

8.4 FSTs and Memory

An interesting application of FSTs is as "memory" devices. For example, FST_1, in Figure 8.2, processes inputs that are strings of binary digits. As output, the machine produces strings of binary digits, representing the input string "shifted" two digits to the right.

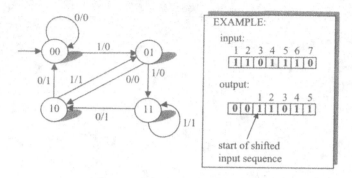

Figure 8.2. FST_1, a binary two-digit "shift" machine

You may think that the names of FST_1's states look strange. Of course, the state names do not affect the operation of the machine, but they have been chosen for symbolic reasons, which should become apparent as the chapter continues.

The next machine, FST_2, does a similar task, but with a right shift of *three* digits. FST_2 is shown in Figure 8.3.

Now consider FST_3, shown in Figure 8.4, that does a "shift two to the right" operation for inputs consisting of strings in $\{0, 1, 2\}^+$ (remember from Chapter 2

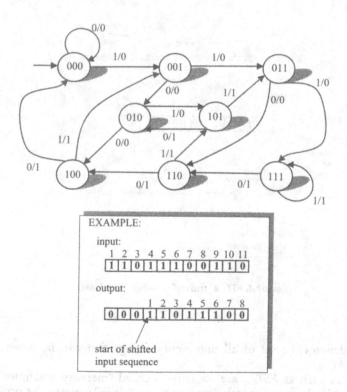

Figure 8.3. FST_2, a binary three-digit "shift" machine

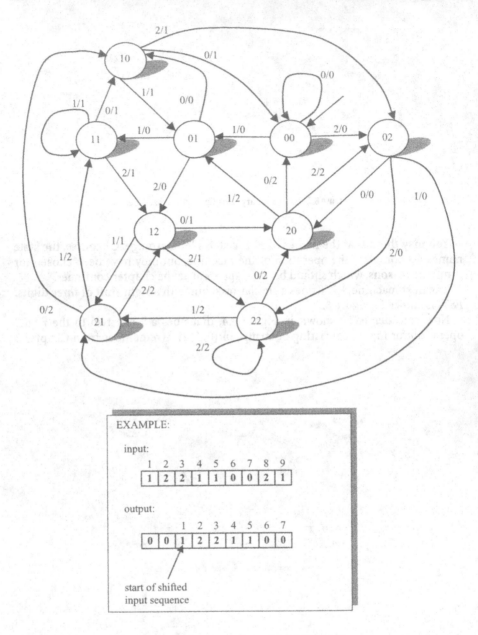

EXAMPLE:

input:

1	2	3	4	5	6	7	8	9
1	2	2	1	1	0	0	2	1

output:

	1	2	3	4	5	6	7	
0	0	1	2	2	1	1	0	0

start of shifted
input sequence

Figure 8.4. FST_3, a "trinary" two-digit "shift" machine

that A^+ denotes the set of all non-empty strings that can be taken from the alphabet A).

Machines such as FST_{1-3} are sometimes called "memory machines", as they perform a simple memory task. Imagine that a "clock" governs the processing of the input, a concept we will return to later in the book. At each "tick" of the clock, the next symbol is read, and the appropriate symbol is output. Then:

FST_1 stores a symbol of a two symbol alphabet for two "ticks" and then outputs it

FST_2 stores a symbol of a two symbol alphabet for *three* "ticks" and then outputs it

FST_3 stores a symbol of a *three* symbol alphabet for *two* "ticks" and then outputs it.

You should be able to see that we could design memory machines for an alphabet of any number, *s*, of symbols, and for any number, *t*, of clock ticks, *providing that we know in advance the values of s and t.* You should also be able to see that we need to increase the number of states as the values of *s* and *t* increase.

In fact, and this is where the names of the states in our three machines are significant, there is a simple numeric relationship between the number of symbols, *s*, in the alphabet, the number of ticks, *t*, that each input symbol has to be "remembered" for, and the minimum number of states required to do the job. There are s^t (numerically speaking) distinct sequences of length *t* when the alphabet has *s* symbols, and the only way the FST can remember which of the s^t sequences it has just read is to have a state for each one, so the FST needs a minimum of s^t states. As you can see, the state names in the machines in this section represent, for each machine, the s^t sequences of length *t* we can have for the *s*-symbol alphabet in question.

So,

FST_1 where $t = 2$ and $s = 2$, has $s^t = 2^2 = 4$ states

FST_2 where $t = 3$ and $s = 2$, has $s^t = 2^3 = 8$ states

FST_3 where $t = 2$ and $s = 3$, has $s^t = 3^2 = 9$ states.

8.5 FSTs and Computation

In this section, we shall see that FSTs possess reasonably sophisticated computational abilities, so long as we are fairly creative in defining how the input and/or output sequences are to be interpreted. As an example of such "creativity", we will first look in a new way at the tasks carried out by our machines FST_{1-3} (Figures 8.2–8.4), in the preceding section.

It should be pointed out that there is no trickery in the way we are going to interpret the input and output of our machines. We are simply defining the type of input expected by a particular machine, and the form that the output will take. This, of course, is a crucial part of problem solving, whether we do it with formal machines or real programs.

8.5.1 Simple Multiplication

Beginning with FST_1, the machine that shifts a binary sequence two digits to the right, let us consider that the sequence of 1s and 0s presented to the machine is first subjected to the following (using 1011 as an example):

1. The string 00 is appended to the rightmost end (giving 101100, for the example).
2. The resulting sequence is interpreted as a binary number *in reverse* (for the example we would then have 001101, which is 13 in *decimal* (base 10).

The output sequence is then also interpreted as a binary number in reverse.

For the example sequence 101100, FST_1 would produce as output 001011, which in reverse (110100) and interpreted as a binary number is decimal 52.

You can convince yourselves that under the conditions defined above, FST_1 performs *multiplication by* 4.

For FST_2, we apply similar conditions to the input, except that for condition (1) – see above – we place *three* 0s at the rightmost end of the input sequence. FST_2 then performs *multiplication by* 8.

For FST_3, the same conditions apply as those for FST_1, except that the input and output number are interpreted as being in *base* 3, in which case FST_3 performs *multiplication by* 9. Thus, given the input 121100 (i.e. 001121, or 43 decimal), FST_3 would produce as output 001211 (i.e. 112100), which, according to our specification, represents 387 (i.e. 43 × 9).

We have seen that FSTs can carry out certain restricted multiplication tasks. Later in this chapter we will see that there are limitations to the abilities of FSTs with respect to multiplication in general.

8.5.2 Addition and Subtraction

FSTs are capable of adding or subtracting binary numbers of any size. However, before we see how this is done, we need to define the form of our input and output sequences.

The input

When *we* add or subtract two numbers (assuming that the numbers are too long for us to mentally perform the calculation), we usually work from right to left (least significant digits first). It seems reasonable, then, in view of the need to deal with any "carry" situations that may arise, that our machines can do likewise. In the previous section we came across the notion of the input being in reverse. Clearly, this method can also be used here, and will then represent the machine processing the pairs of corresponding digits in the way that we usually do.

Apart from certain minor problems that will be addressed shortly, the remaining major problem is that we have *two* numbers to add or subtract, whereas our machines accept only one sequence as input. Our solution to this problem is to present the two binary numbers (to be added or subtracted) as *one sequence*, in which the digits of one of the numbers are interleaved with the digits of the other. This, of course, means that the numbers must be of the same length. However, we can ensure that this is the case by placing sufficient 0s at the leftmost end of the shorter of the two numbers before we reverse them and present them to our machines.

Table 8.2 summarises the preparation of input for the FSTs for addition and subtraction which we will define shortly.

We now turn to the output.

The output

For our addition and subtraction FSTs to model our behaviour, they must process the two numbers by examining pairs of corresponding digits, in least to most significant digit order of the original numbers. However, as an FST can only read one

Table 8.2. Data preparation for binary addition and subtraction FSTs

To prepare two binary numbers for input to the addition or subtraction FSTs:

1. Ensure numbers are the same length by padding the shorter number with leading 0s

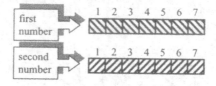

1a. (For addition only) Place an extra leading 0 on each number, to accommodate possible final carry

2. Reverse both numbers

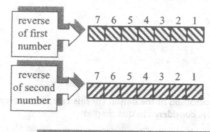

3. Make one sequence by interleaving the corresponding digits of each reversed number

   (Note: for subtraction FST, begin with second number)

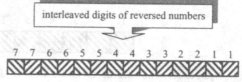

---

digit at a time, it will not be able to output a 0 or 1 representing a digit of the answer until it has examined the second digit of each pair of corresponding digits. Moreover, our FSTs must output a symbol for each input symbol encountered. For the first digit in each corresponding pair, then, our FSTs will output the symbol "*". Note that there is no special significance to this symbol, it is simply being used as a "marker".

Thus, given input of the form indicated by the diagram that accompanies step 3 in Table 8.2, our FSTs will produce output as represented in Figure 8.5.

The output shown in Figure 8.5 represents the digits of the answer *in reverse*, with each digit preceded by *. The *s are not part of the answer, of course, and so we read the answer as specified in Figure 8.6.

The machines

We are now ready to present the two machines, FST_4, the binary adding FST (Figure 8.7), and FST_5, the binary subtractor (Figure 8.8). Each machine is accompanied by an example showing a particular input and the resulting output. I leave it to you to convince yourself that the machines operate as claimed.

It is important to note that for both machines there is no limit to the length of the input numbers. Pairs of binary numbers of any length could be processed. It is clear, therefore, that the FST is capable of adding and subtracting arbitrary length binary numbers.

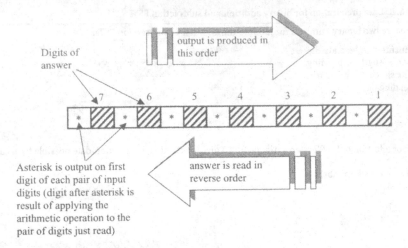

Figure 8.5. Specification of the output (in this case, for a seven-digit number) for the arithmetic finite state transducers considered in this chapter

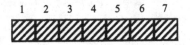

Figure 8.6. How we construct the answer from the output specified in Figure 8.5

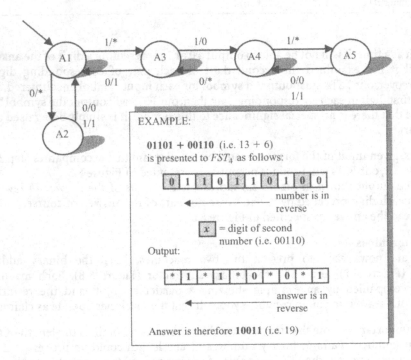

Figure 8.7. A binary adder FST, FST_4

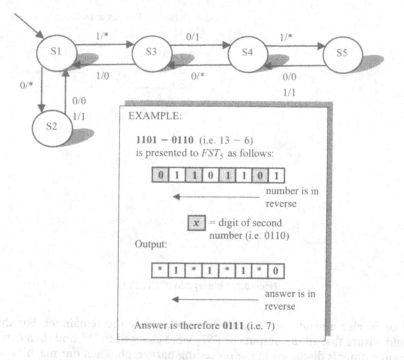

Figure 8.8. A binary subtractor FST, FST_5. If the result of subtracting the numbers would be negative, the result is undefined

8.5.3 Simple Division and Modular Arithmetic

Certain restricted forms of division are also within the capabilities of the FST. For example, FST_6, in Figure 8.9, performs integer division of a binary number by 3, returning the result as a binary number. In some programming languages, integer division is represented by the infix operator *div*, so FST_6 actually does x *div* 3, where x is the input number.[1] However, FST_6 does not require the input sequence to be in reverse. FST_6 carries out integer division from the *most* significant end of the number, as we usually do. The machine also outputs the result in the correct order. Figure 8.9 includes an example input sequence, and its corresponding output sequence.

We can quite easily use FST_6 as the basis of a machine that computes the quotient *and* the remainder, by making the following observations about states of FST_6:

- $D0$ represents a remainder of 0 (i.e. no carry)
- $D1$ represents a remainder of 1 (i.e. a carry of 1)
- $D2$ represents a remainder of 2 (i.e. a carry of 2).

Now, the remainder resulting from a division by 3 requires a maximum of *two* binary digits, as it can be 0, 1, or 10 (i.e. 0, 1 or 2). We could make sufficient space for the remainder, by placing "***" (three asterisks) after the input sequence. I leave it to you to decide how to amend the machine to ensure that the quotient is followed by the binary representation of the remainder. The first of the three asterisks should be left unchanged by the machine, to provide a separator between quotient and remainder.

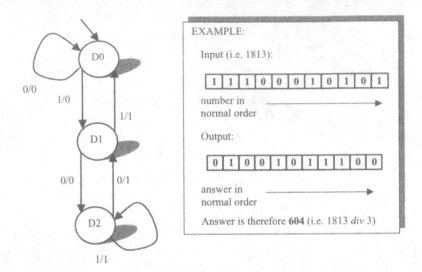

Figure 8.9. A binary "*div* 3" FST, *FST*₆

We could also amend our machine to provide only the remainder. For this we would ensure that all the outputs of FST_6 were changed to "*", and then carry out the amendments discussed in the preceding paragraph. Then our machine would carry out an operation similar to that called *mod* in programming languages, except it would be restricted to *mod* 3.

8.6 The Limitations of the FST

We have seen that FSTs can be defined to do restricted forms of multiplication and division, and also that FSTs are capable of unrestricted addition and subtraction. You will note that the preceding sentence does not include the word "binary". This is because we could define adders and subtractors for any number base (including base 10), since there are only a finite number of "carry" cases for addition and subtraction in *any* given number system. Similarly, we can define FSTs to do multiplication or division by any fixed multiplicand or divisor, in any number base. However, the binary machine is much simpler than would be the corresponding FST for, say, operations on base 10 numbers. Compare the complexity of FST_1 (Figure 8.2) and FST_3 (Figure 8.4), for example. I will leave it to you to consider the design of FSTs for number systems from base three upwards.

In fact, our intuition tells us that the binary number system is sufficient; after all, we deal every day with a machine that performs *all* of its operations using representations that are essentially binary numbers. We never use that as a basis for arguing that the machine is insufficiently powerful! We return to the issue of the universality of binary representations several times in subsequent chapters.

For now, our purpose is to establish that FSTs are not sufficiently powerful to model all of the computational tasks that we may wish to carry out. We have seen that the FST can perform unrestricted addition and subtraction. The question arises; can we design an FST to do unrestricted multiplication? We shall eventually demonstrate that the answer to this question is "no". This is obviously a critical

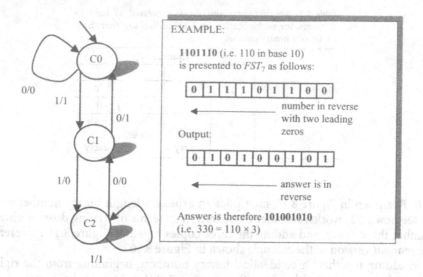

Figure 8.10. A binary "×3" FST, FST_7

limitation, since the ability to perform multiplication is surely one of the basic requirements of useful computing systems. First, however, we show that there are *some* multiplication operations we *can* perform with FSTs.

8.6.1 Restricted FST Multiplication

Earlier in this chapter, we saw that FSTs are capable of multiplying a base n number by some constant c, where c is a power of n (for example, FST_1 was used to multiply a binary – base 2 – number by 4, i.e. 2^2). Suppose we wish to multiply a number, say a binary number, by a number that cannot be expressed as an integral power of 2, say 3? One way to achieve such tasks is to effectively "hardwire" the multiplicand 3 into the machine, in a similar way that the divisor 3 was represented in the *div* 3 machine, FST_6 (Figure 8.9).

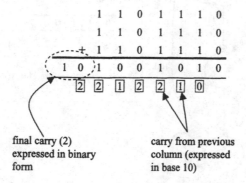

Figure 8.11. Multiplying a binary number by 3 by adding the number three times. The calculation is the same as that represented by the example in Figure 8.10. See also Table 8.3

Table 8.3. Possible carry situations, expressed as base 10 numbers, for multiplication by 3 by adding together three copies of a binary number

		Column digit			
		1		0	
Carry	0	1	1	0	0
	1	0	2	1	0
	2	1 result	2 new carry	0 result	1 new carry

FST_7, shown in Figure 8.10, multiplies an arbitrary length binary number by 3. To see how FST_7 works, consider its operation in terms of writing down a binary number three times, and adding the three copies together. Figure 8.11 represents the *manual* version of the example shown in Figure 8.10.

In adding together three identical binary numbers, beginning from the right-most column, we can either have three 1s (i.e. $1 + 1 + 1$) = 11, i.e. 1 carry 1, or three 0s (0 carry 0), i.e. a carry of 1 or 0. If there is a carry of 0, we continue as in the previous sentence. Otherwise, we have a carry of 1 and the next column could be three 1s (i.e. $1 + 1 + 1 + 1$ – adding in the carry) so we write down a 0 and carry 2 (decimal). If the *next* column is three 0s, we have $0 + 0 + 0 + 10 = 10$, i.e. 0 carry 1 – back to carry 1, which has been discussed already. Otherwise, we have three 1s and a carry of 10, i.e. $1 + 1 + 1 + 10 = 101 = 1$ carry 2 – back to carry 2, which, again, we have already considered. The whole situation is summarised in Table 8.3.

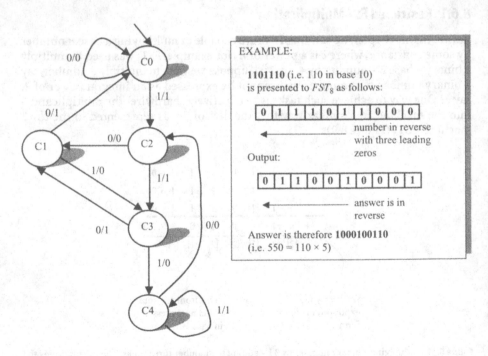

EXAMPLE:

1101110 (i.e. 110 in base 10) is presented to FST_8 as follows:

0	1	1	1	0	1	1	0	0	0

number in reverse with three leading zeros

Output:

0	1	1	0	0	1	0	0	0	1

answer is in reverse

Answer is therefore **1000100110** (i.e. $550 = 110 \times 5$)

Figure 8.12. FST_8 multiplies any binary number by 5

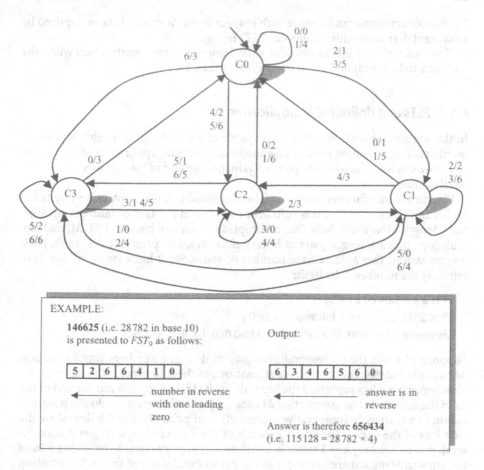

EXAMPLE:

146625 (i.e. 28 782 in base 10)
is presented to FST_9 as follows: Output:

| 5 | 2 | 6 | 6 | 4 | 1 | 0 |

← number in reverse
with one leading
zero

| 6 | 3 | 4 | 6 | 5 | 6 | 0 |

← answer is in
reverse

Answer is therefore **656434**
(i.e. 115 128 = 28 782 × 4)

Figure 8.13. FST_9 multiplies any base 7 number by 4

The states in Figure 8.10 were labelled C0–2 to represent the three carry situations found in Table 8.3. From Table 8.3, you should be able to see why only three states are required for FST_7. Moreover, you should also convince yourself that we could not define an FST of fewer than three states to carry out the same task.

A similar method can be used to create a machine, FST_8 that multiplies a binary number by 5. The machine is shown in Figure 8.12, and as you can see, requires 5 states.

Finally, we consider in Figure 8.13 the rather unusual FST_9, that multiplies a base 7 number by 4 (!).

I leave it to you to convince yourself that the above machines, FST_{7-9}, operate as claimed.

The above multipliers, FST_{7-9}, reflect a construction suggesting that a machine to multiply a number in any base by a constant c, needs no more states than c (i.e. one for each of the possible base 10 carry values). As we saw above:

FST_7 where $c = 3$, had 3 states
FST_8 where $c = 5$, had 5 states
FST_9 where $c = 4$, had 4 states.

Similar observations can be made with respect to the number of states required by a constant divisor machine (consider FST_6 in Figure 8.9).

However, you may like to consider what happens to our construction when the numbers to be multiplied are in bases greater than 10.

8.6.2 FSTs and Unlimited Multiplication

In this section we see that FSTs are incapable of multiplication of arbitrary length numbers. The argument used is essentially an informal application of the *repeat state theorem* introduced in Chapter 6. Again, the method of "proof" used is *reductio ad absurdum*.

We can assume, for reasons that will be discussed in later chapters, and hinted at earlier in this chapter, that it is sufficient to focus on the task of multiplying arbitrary length *binary* numbers. Our assumption is that we have an FST, M, that can multiply arbitrary length pairs of binary numbers, and print out the result. Now, we can assume that M has a fixed number of states. Say it has k states. We ask M to multiply the number 2^k by itself:

2^k is a 1 followed by k 0s

$2^k \times 2^k$ (i.e. 2^{2k}) is a 1 followed by $2k$ 0s

(example: $2^3 = 1000, 2^3 \times 2^3 = 2^6 = 1\,000\,000$, i.e. $8 \times 8 = 64$).

Suppose M reads the corresponding digits of its input numbers simultaneously. M has only k states, and it must use at least one of them to process the input digits. However, after it has processed the input digits it still has to print out the rest of the 0s in the answer. Let us assume that M uses only one of its states to do the multiplication (an extremely conservative estimate!), and print out the 1 followed by the first k 0s of the answer (one digit for each of the k pairs of input digits it encounters). It then has only $k - 1$ states left with which to print out the remaining k 0s of the answer. M must therefore *loop* when it prints out the rest of the 0s. It is printing out the 0s without any further input of digits, so there is no way of controlling the number of 0s that M prints. The only way to make an FST generate a fixed number, k, of symbols without input is to have a sequence of $k + 1$ states to do it. This notion was discussed in the context of FSRs in Chapter 4, and formed the basis of the *repeat state theorem* of Chapter 6.

We are therefore forced to reject our initial assumption: the machine M cannot exist. Multiplication of arbitrary length binary numbers is beyond the power of the FST. By extension of this argument, multiplication of arbitrary length numbers in *any* number base is also beyond its power.

8.7 FSTs as Unsuitable Models for Real Computers

It is unfortunate that FSTs are not powerful enough to perform all of the operations that we expect computers to do. As you can probably guess from considering the decision programs that represent FSRs (Chapter 4), FSTs also lend themselves to conversion into very simple programs. However, at the very least, we expect any reasonable model of computation to be able to deal with arbitrary multiplication and division. As we have seen, this basic requirement is beyond the capabilities of FSTs.

You may think you detect a flaw in an argument that assumes that real computers can perform operations on arbitrarily large data. Most of us who have ever written any non-trivial programs have experienced having to deal with the fact that the computer we use is limit*ed*, not limit*less*, in terms of its storage and memory capacity. Since a computer is, in terms of available storage at any given point in time, finite, it could be regarded as a (huge) FST. As such, for any given computer, there is, even for basic operations such as multiplication, a limit to the size of the numbers that can be dealt with. In this sense, then, a computer is an FST.

It turns out, however, that treating a digital computer as if it were an FST is not, in general, a useful position to adopt, either theoretically or practically.[2] Let us look at the practical issues first. When a deterministic FST enters the same state (say q) twice during a computation, it is looping. Moreover, once the machine reaches state q, repeating the input sequence that took it from q back to q (perhaps via several other states) will result in the same behaviour being exhibited again. Once again, this was the basis of the repeat state theorem for FSRs in Chapter 6.

Suppose a program is performing some calculations, subsequently to print some output at the terminal, and receives no output from external sources once it has begun (i.e. it operates on data stored internally). The *state* of the computer running the program is its entire internal configuration, including the contents of its memory, its registers, its variables, its program counter, and so on. Viewed in this way, if such a state of the computer is ever repeated during the execution of the program, the program will repeat the whole cycle of events that lead from the first to the second occurrence of that configuration over and over again, *ad infinitum*. If this happens we have obviously made a (fairly common) mistake in our coding, resulting in what is often called an *infinite loop*.

It would be useful if, before we marketed our programs, we could test them to make sure that such infinite loops could never occur. Let us suppose that we have some way of examining the internal state of the machine each time it changes, and some way of deciding that we had reached a state that had occurred before. In theory, this is possible because, as stated above, a computer at a given point in time is a finite state machine.

Suppose that our computer possesses 32 KB of total storage, including data, registers, primary memory, and so on. Each individual bit of the 32 K can be either 1 or 0. Thus we have a total of 262 144 bits, each of which can be either 1 or 0. This means that, at any point our machine is in one of $2^{262\,144}$ states. $2^{262\,144}$ is about $10^{79\,000}$. This is an *incredibly* large number. In chapter 10 we come across a *really* fast (over three million million instructions per second) hypothetical machine that would take 10^{80} *years* (a 1 followed by *eighty* 0s) to solve a problem involving passing through 10^{100} states. Our computer may thus have to go through a huge number of internal states before one of these states is repeated. Waiting for even a tiny computer, like our 32K model here, to repeat a state because it happens to be an FST is thus a rather silly thing to do!

In the remainder of this book, we study the deceptively simple abstract machine that was introduced in Chapter 7, i.e. the *TM*. The TM is essentially an FST with the ability to move left and right along a single input *tape*. This means that the TM can overwrite symbols on its tape with other symbols, and can thus take action on the basis of symbols placed there earlier in its activities. In subsequent chapters, we find that this simple machine is actually *more* powerful than any digital computer. In fact, in some senses the TM is the most powerful computational device of all. However, we also discover that there is something about the problem of whether or

not a certain computation will eventually stop, that was discussed in terms of the
FST, above, that puts its general solution beyond *any* computational device.

8.8 Exercises

*For exercises marked †, solutions, partial solutions, or hints to get you started appear
in "Solutions to Selected Exercises" at the rear of the book.*

8.1. FSTs can be used to represent Boolean logic circuits (such circuits are dis-
      cussed in Chapter 13). Design FSTs to perform "bitwise" manipulation of
      their binary number inputs. For machines needing two input values,
      assume an input set-up similar to that defined for FST_4 and FST_5, the addi-
      tion and subtraction machines of Figures 8.7 and 8.8, respectively. The
      operations you may wish to consider are:

      and
      or†
      not
      nand
      nor.

8.2†. Design an FST to convert a binary number into an *octal* (base 8) number.

      *Hint: each sequence of three digits of the binary number represents one digit
      (0…7) of the corresponding octal number.*

Notes

1. An infix operator is written in between its operands. The arithmetic *addition* operator, "+", is thus
    an example of an infix operator.
2. This is not altogether true. The FST is used, for example, to model aspects of computer networks
    (see Tanenbaum, 1988, in the "Further Reading" section). Here, I am referring to its ability to model
    arbitrary computations.

Chapter 9
Turing Machines as Computers

9.1 Overview

In this chapter, we begin our investigation into the computational power of *Turing machines* (TMs), that we first encountered in Chapter 7.

We find that TMs are capable of performing computational tasks that *Finite state transducers* (FSTs) cannot do, such as

- multiplication
- division

of arbitrary length binary numbers.

We also see a sketch of how TMs could perform any of the operations that computers can, such as:

- logical operations
- memory accessing operations
- control operations.

9.2 Turing Machines and Computation

In Chapter 7, *TMs* for language recognition were introduced. We saw that the TM is the recogniser for all of the languages in the *Chomsky hierarchy*. Chapter 8 demonstrated that the *FST*, which is essentially a TM that can move only right on its *tape*, is capable of some fairly sophisticated computational tasks, including certain restricted types of multiplication and division. We also observed that any given digital computer is essentially finite in terms of storage capacity at any moment in time. All this seems to suggest that viewing the computer as an FST might be a useful perspective to take, especially since an FST can be expressed as a very simple program. However, we saw that the huge number of internal states possible in even a small computer meant that notions typically applied to FSTs, such as the *repeat state theorem*, were of limited practical use.

In this chapter, we see that, in terms of *computation*, the TM is a much better counterpart of the computer than is the FST. Simple as it is, and resulting as it does from addition of a few limited facilities to the FST, the TM actually exceeds the computational power of any single computer, of any size. In this sense, the TM is an abstraction over all computers, and the theoretical limitations to the power of the TM are often the *practical* limitations to the digital computer, not only at the present time, but also at any time in the future. However, we shall descend from these heady philosophical heights until the next chapter. For now, we concentrate on demonstrating that the TM is capable of the basic computational tasks we expect from any *real* computer.

9.3 TMs and Arbitrary Binary Multiplication

Chapter 8 established that FSTs are not capable of multiplication of arbitrary length numbers, in any number base. In this section, we sketch out a TM to perform multiplication of two binary numbers of any length. I will leave it to you to realise that similar TMs could be defined for any number base (cf. the treatment of different number bases in Chapter 8). However, here we choose the binary number system, for obvious reasons.

9.3.1 Some Basic TM Operations

A side effect of defining a TM for multiplication is that we will appreciate a notion actually formulated by Turing himself in conjunction with his machines, and one which occurs in some form in various programming paradigms. Essentially, we shall build up our binary multiplier by creating a "library" of TM "subroutines". These subroutines, like subroutines in typical programming languages ("procedures" and "functions", in Pascal), represent operations that we wish to carry out many times, and thus do not want to create over and over again. However, as there are no straightforward mechanisms for parameter passing in TMs, our subroutines are more of an analogy to *macros*, i.e. we have to imagine that the "sub-machine" that constitutes the macro is actually inserted into the appropriate place in the overall machine. This means that we have to "call" our sub-machine by ensuring that the "calling" machine configures the tape and R/W head appropriately.

Some of the operations required by our multiplier TM are now specified. Our multiplication routine will work by repeated addition, based on the common "shift and add" long multiplication method (we will examine this in more detail in the next section). Clearly, then, we need an *ADD* routine. Our *ADD* routine will make use of two sub-machines, *WRITE-A-ZERO* and *WRITE-A-ONE*, as specified in Figure 9.1.

Referring to Figure 9.1, recall from Chapter 7 that we use "?/?" to mean "replace any *non-blank* symbol of the alphabet for the machine by the same symbol".

Our *ADD* routine adds two arbitrary length binary numbers, adding pairs of digits in least to most significant order (as we usually do). The machine assumes that the two numbers are initially represented on a tape as specified in Figure 9.2.

So, *ADD* expects on its tape a "*" followed by the first number, followed by a "*" followed by the second number. Its head is initially located on the least significant digit of the second number. *ADD* will not be concerned with what is to the left of the relevant part of the input tape, but will assume a blank immediately to the right of the second number. The left asterisk will be used by *ADD* to indicate the current carry situation during the computation ("*" = no carry, C = carry). The machine will write the result of the addition to the left of the currently occupied part of the

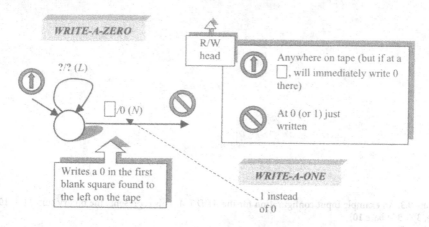

Figure 9.1. TM fragments "*WRITE-A-ZERO*" and "*WRITE-A-ONE*", as used by TMs later in the chapter

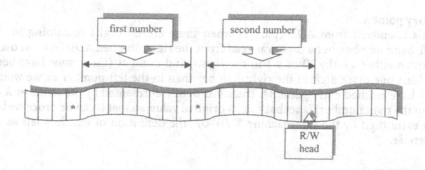

Figure 9.2. The input tape set up for the *ADD* TM

input tape (and thus assumes that the first blank found to the left of the left "∗" indicates the extreme right of the unoccupied left-hand side of the tape).

The description of *ADD* will be supported by an example input tape, and we will examine the effects of each major part of *ADD* on the contents of this tape. The example input tape we will use is shown in Figure 9.3.

9.3.2 The "ADD" TM

Figure 9.4 shows part one of the *ADD* TM.

Part 1 of *ADD*, shown in Figure 9.4, deals with the adding of corresponding digits of the two numbers until the digits of one or both of the numbers have been used up. Note that *X* and *Y* are used as markers for 0 and 1, respectively, so that *ADD* can detect how far along each number it has reached, at any stage, and thus avoid processing the same digit more than once. When the final digit of one or both of the numbers has been reached, part 2 of *ADD* (shown in Figure 9.6) takes over.

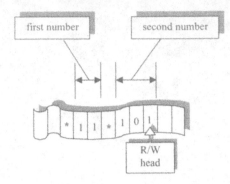

Figure 9.3. An example input configuration for the *ADD* TM. This represents the binary sum 11 + 101 (i.e. 3 + 5 in base 10)

At the time we are ready to enter *ADD* part 2, our example tape is configured as shown in Figure 9.5.

As can be seen in Figure 9.6, there are two entry points *A* and *B*, to *ADD* part 2.

Entry point A

This is entered from *ADD* (part 1) when there are no digits remaining in the left-hand number, but a digit was read from the right number. *ADD* has just over-written either a 0 (by *X*) or a 1 in the right-hand number (there must have been at least one more digit in the right number than in the left number or we would not have reached entry point *A*). Thus, this digit (represented by the leftmost *X* or *Y* in the right number) is set back to its original value, so that it can be processed as an extra digit by the sub-machine *TIDY-UP*, the definition of which is left as an exercise.

Entry point B

This is entered from *ADD* (part 1) when there are no digits remaining in the right-hand number. This case is simpler than that for *entry point A*, as no extra digits have been marked off (*ADD* always looks at the digit in the *right* number first, before it attempts to find the corresponding digit in the left number).

Entry points *A* and *B* both ensure that the head is located at the square of the tape representing the carry situation. Remember that this is the square immediately to the left of the left number, and is occupied with the symbol "*" for no carry, and *C* for carry.

Our example tape (Figure 9.3) would have resulted in entering add at *entry point A*, as the end of the right-hand number was reached first. Immediately before the machine enters the component labelled *TIDY-UP*, which you are invited to complete as an exercise, our example tape from Figure 9.3 would be as described in Figure 9.7.

ADD should now complete its computation, the result being the binary number at the left of the occupied portion of the tape. For our example tape, Figure 9.8 shows an appropriate outcome from *TIDY-UP*, and thus an appropriate representation of the result of *ADD*, given the input tape of Figure 9.3.

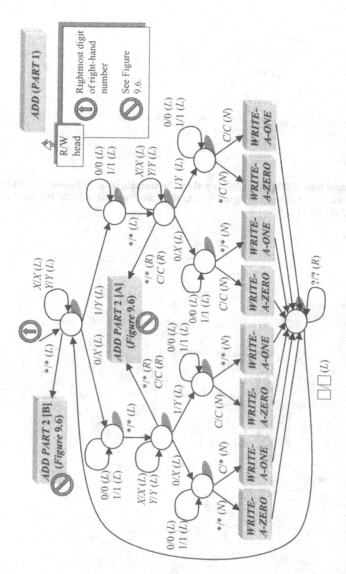

Figure 9.4. The *ADD* Turing machine (part 1). *ADD* adds together two binary numbers, beginning with a tape as specified in Figure 9.2. The answer is written to the left of the occupied portion of the initial tape

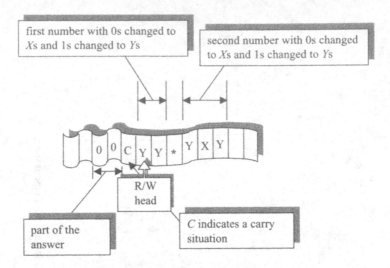

Figure 9.5. The output from *ADD* part 1 (Figure 9.4) for the example input of Figure 9.3. This is as the tape would be on entry to *ADD* part 2 (Figure 9.6). As there are fewer digits in the left-hand number, entry will be to *ADD* part 2 [A]

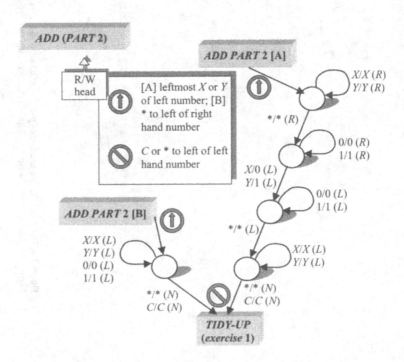

Figure 9.6. The *ADD* TM (part 2 – for part 1 see Figure 9.4). This part of *ADD* deals with any difference in the length of the two input numbers

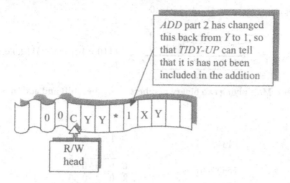

Figure 9.7. The output from *ADD* part 2 (Figure 9.6) for the example input of Figure 9.5. This is as the tape would be on entry to *TIDY-UP*, which you are asked to construct in exercise 1 at the end of this chapter

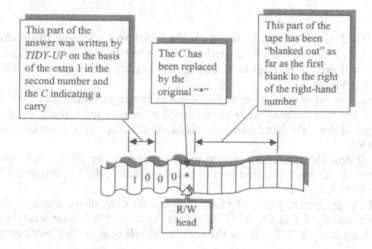

Figure 9.8. The expected output from *TIDY-UP* after processing the tape of Figure 9.7

9.3.3 The "MULT" TM

We are now ready to sketch out the description of a TM, *MULT*, to multiply two arbitrary length binary numbers. As mentioned earlier, our machine uses the "shift and add" method. For example, suppose we wish to compute 1110×101. This would be carried out as shown in Figure 9.9.

An example of the same method being employed for the multiplication of two base 10 numbers may provide a more familiar illustration. Figure 9.10 shows the method being used for the base 10 computation 1345×263.

The "shift and add" process is simpler for binary numbers, since the same rule applies for placing zeros at the end of each stage, but we only ever have to multiply at intermediate stages by either 1 or 0. Multiplying by 1 involves simply copying the number out exactly, while multiplying by zero can, of course, be ignored. An algorithm to do "shift and add" binary multiplication is given in Table 9.1.

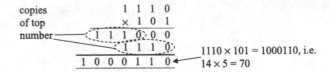

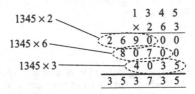

Figure 9.9. Multiplying two binary numbers using the "shift and add" method

Figure 9.10. Multiplying two decimal numbers using the "shift and add" method

Our *MULT* machine will be based on the "shift and add" algorithm shown in Table 9.1. The detailed implementation will be left to the exercises. The *MULT* machine will expect its input to be configured exactly as that for the *ADD* TM, as specified in Figure 9.2.

An example *input configuration* for *MULT* is shown in Figure 9.11. Note that we have decided that the read/write head should be positioned initially on the rightmost digit of the right-hand number, this number being called n_2 in the algorithm of Table 9.1.

We will now sketch out the actions to be carried out by *MULT*, in terms of the algorithm of Table 9.1. The first part of the algorithm is the "if" statement (shown as "part 1" in Table 9.1).

The head is currently over the first (rightmost) digit of n_2, which in the algorithm is called d_1. If d_1 is a 0, *MULT* initialises the answer to 0, otherwise the answer is initialised to n_1. *MULT* will use the area to the left of n_1 as the "work area" of its

Table 9.1. The "shift and add" method for multiplying two binary numbers

$\{n_1$ and n_2 are the two numbers to be multiplied and d_i represents the ith digit from the right of $n_2\}$

```
begin
    if d_i = 1 then
        ans := n_1
    else
        ans := 0
    endif

    for each digit d_i, where i ≥ 2, in n_2 do
        if d_i ≠ 0 then
            ans := ans + (n_1 with i − 1 0s appended to its right)
        endif
    endfor
end
```

P
a
r
t
1

P
a
r
t
2

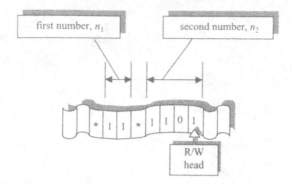

Figure 9.11. An example input configuration for the *MULT* TM. This represents the binary calculation 11 × 1101 (i.e. 3 × 13 in base 10)

tape, and leave the answer there when it terminates, separated by a single blank from the "*" preceding n_1. "$ans := n_1$" thus means "copy n_1 into the answer area". For this we need a copy routine, which we will call *COPY-L*. *COPY-L* is defined in Figure 9.12.

COPY-L needs little explanation. The symbols X and Y are again used as temporary markers for 0 and 1, respectively. The machine leaves the copied sequence exactly as it finds it. Clearly, the machine is designed to expect an alphabet of (at least) $\{*, X, Y, 0, 1\}$, which is due to the requirements of the multiplier.[1] The machine

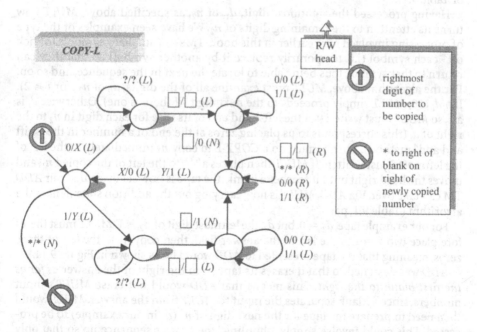

Figure 9.12. The *COPY-L* Turing machine. Copies a binary number preceded by "*" from the position of the read/write head to left end of marked portion of tape. Leaves a blank square between copied sequence and existing occupied tape

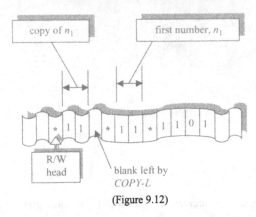

(Figure 9.12)

Figure 9.13. The state of the input tape from Figure 9.11, following MULT's execution of the statement $ans := n_1$ (Table 9.1). The first number, n_1, has been copied to the left of n_1

could easily be amended to deal with other alphabets. It would also be a simple matter to create a corresponding *COPY-R* ("copy right") machine, if required.

If $d_1 = 0$, then *MULT* simply writes a 0 in the answer area. In our example here, $d_1 = 1$, so *MULT* must move its head to the rightmost digit of n_1, then enter *COPY-L*. At this point, *MULT* could also place a "*" at the left-hand end of the newly written answer sequence.

The example tape would then be as shown in Figure 9.13.

MULT will now deal with the next part of the algorithm, i.e. the "for loop" (part 2 of Table 9.1).

Having processed the rightmost digit, d_1, of n_2, as specified above, *MULT* now turns its attention to the remaining digits of n_2. We have seen examples of the type of processing involved here earlier in this book. TMs use auxiliary symbols to "tick off" each symbol (i.e. temporarily replace it by another symbol) so that they can return to that symbol, thus being able to locate the next in the sequence, and so on. For the algorithm above, *MULT* must examine all of the digits, d_i, of n_2 (for $i \geqslant 2$). If a d_i is 0, *MULT* simply proceeds to the next digit (if there is one). Otherwise d_i is a 1, so *MULT* must write a 0 at the left-hand end of its tape for each digit in n_2 to the right of d_i (this corresponds to us placing zeros at the end of a number in the "shift and add" method). Then it must do a *COPY-L* to copy n_1 immediately to the left of the leftmost 0 just written. If *MULT* then places a "*" to the left of the copied n_1, and moves the head right until it reaches a blank, the tape is now set up so that our *ADD* TM can take over. Via *ADD*, *MULT* is now carrying out the addition statement in the algorithm (Table 9.1, part 2).

For our example tape, $d_2 = 0$, but d_3, the leftmost digit of n_2, is 1. *MULT* must therefore place two zeros to the left of the answer field, then copy n_1 to the left of those zeros, meaning that the tape presented to *ADD* would be as shown in Figure 9.14.

ADD was designed so that it erases its tape from the right of the answer *as far as the first blank to the right*. This means that *ADD* would not erase *MULT*'s input numbers, since a blank separates the input to *MULT* from the answer. *MULT* would then need to prepare the tape for the next digit of n_2 (d_4 in our example) to be processed. This could involve simply "shuffling" the answer sequence up so that only one blank separates it from *MULT*'s input sequence. Our example tape would now be as displayed in Figure 9.15 (subject to some slight repositioning of the "*").

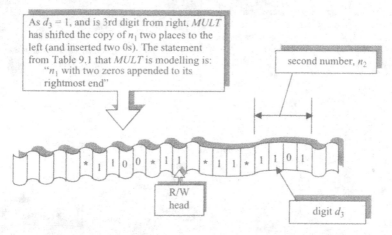

Figure 9.14. The state of the input tape from Figure 9.11 immediately before *ADD* is used by *MULT* to add 1100 to 11. *MULT* is processing digit d_3 of n_2

MULT would now deal with d_4 of n_2 in an analogous way to its treatment of d_3, described above. This time, *three* zeros would be needed, so the tape on input to *ADD* would be as shown in Figure 9.16.

Figure 9.17 shows a suitably tidied up form of the final result .

Sufficient detail has been presented to demonstrate that such a *MULT* machine could be implemented. The exercises ask you to do this. It is clear that such a *MULT* *machine* could process binary numbers of any length. We established in the preceding chapter that multiplication of arbitrary length numbers is beyond the capabilities of the *FST*. We now know, therefore, that the TM possesses greater computational power than does the FST.

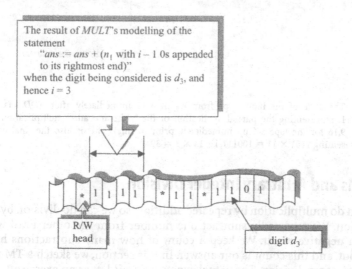

Figure 9.15. The state of the input tape from Figure 9.11 immediately after *ADD* has performed 1100 + 11, representing the partial application of the "shift and add" multiplication procedure (see Figure 9.14 for the tape set up immediately prior to this)

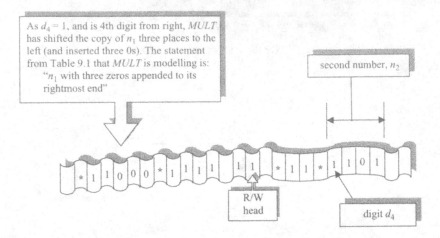

As $d_4 = 1$, and is 4th digit from right, *MULT* has shifted the copy of n_1 three places to the left (and inserted three 0s). The statement from Table 9.1 that *MULT* is modelling is: "n_1 with three zeros appended to its rightmost end"

second number, n_2

R/W head

digit d_4

Figure 9.16. The state of the input tape from Figure 9.11 immediately before *ADD* is used by *MULT* to add 11000 to 1111. *MULT* is processing digit d_4 of n_2

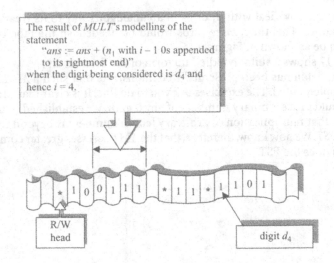

The result of *MULT*'s modelling of the statement
"*ans* := *ans* + (n_1 with i − 1 0s appended to its rightmost end)"
when the digit being considered is d_4 and hence $i = 4$.

R/W head

digit d_4

Figure 9.17. The state of the input tape from Figure 9.11 immediately after *ADD* has performed 11000 + 1111, representing the partial application of the "shift and add" multiplication procedure (see Figure 9.16 for the tape set up immediately prior to this). This is also the final output from *MULT*, representing 1101 × 11 = 100111, i.e. 13 × 3 = 39

9.4 TMs and Arbitrary Integer Division

As we can do multiplication by repeated addition, so we can do division by repeated subtraction. We repeatedly subtract one number from the other until we obtain zero or a negative result. We keep a count of how many subtractions have been carried out, and this count is our answer. In this section, we sketch a TM that does arbitrary integer division. The actual construction is left as an exercise.

Our machine, *DIV*, will perform division by repeated subtraction. We therefore require a *SUBTRACT* sub-machine for our *DIV* machine (as we needed an *ADD* machine for *MULT*).

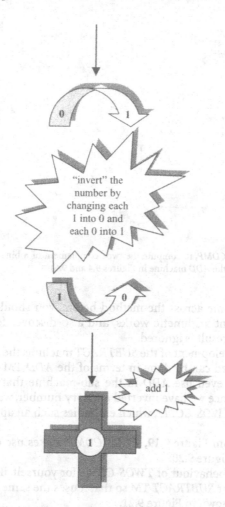

Figure 9.18. Obtaining the "twos complement" of a binary number

9.4.1 The "SUBTRACT" TM

It turns out that we can do subtraction in terms of *addition*, and you may be familiar with the approach considered here. For example, $8-5$ is really the same as saying $8 + (-5)$. If we have a way of representing the negation of the second number, we can then simply add that negation to the first number. The usefulness of this method for *binary* numbers relates to the fact that the negation of a number is simple to achieve, using a representation called the *twos complement* of a binary number. The process of obtaining the twos complement of a binary number is described in Figure 9.18.

So, to compute $x - y$, for binary numbers x and y, we compute the twos complement of y and add it to the first number. Referring to the example of $8 - 5$ mentioned above, we can represent this in binary as $1000 - 101$. The twos complement of 101 is 011 (the inverse is 010, and $010 + 1 = 011$). $1000 + 011 = 1011$. Now, this result appears to be 11 (eleven) in base 10. However, after the addition we ignore the leftmost digit of the result, and thus our answer is 011 in binary (i.e. 3 in base 10), which is the correct result.

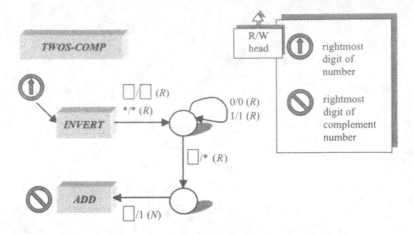

Figure 9.19. A TM, *TWOS-COMP*, to compute the twos complement of a binary number (the *INVERT* machine is in Figure 9.20, the *ADD* machine in Figures 9.4 and 9.6)

If you have not come across the method before, you should convince yourself that twos complement arithmetic works, and also discover for yourself why the leftmost digit of the result is ignored.

In terms of the development of the *SUBTRACT* machine, the subtraction method we have just described can be done in terms of the *ADD* TM described earlier in the chapter. We can even use *ADD* in the sub-machine that computes the twos complement, since once we have inverted a binary number, we then need to *add* 1 to the result. A TM, *TWOS-COMP*, which embodies such an approach, is described in Figure 9.19.

As you can see from Figure 9.19, *TWOS-COMP* makes use of the TM, *INVERT*, which is defined in Figure 9.20.

You can study the behaviour of *TWOS-COMP* for yourself, if you wish.

We could set up our *SUBTRACT* TM so that it uses the same input format as that used for *ADD*, and shown in Figure 9.21.

We assume that the "second number" of Figure 9.21 is to be subtracted from the "first number".

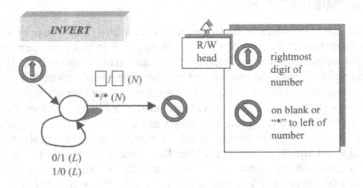

Figure 9.20. A TM, *INVERT*, that is used by *TWOS-COMP* (Figure 9.19). It inverts each digit of a binary number

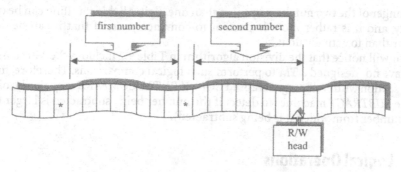

Figure 9.21. The input tape set up for the *SUBTRACT* TM

The exercises ask you to complete the *SUBTRACT* machine. This simply involves using the *ADD* machine to add the twos complement of one number to the other number. The main problem, as for the *MULT* machine, concerns the need to shuffle the various intermediate sequences around to ensure that essential information is not corrupted; but it is not too difficult to achieve this.

9.4.2 The "DIV" TM

We now sketch a TM for performing integer division by repeated subtraction. Again, its implementation is left as an exercise. An algorithm to carry out the task is specified in Table 9.2.

As for the *MULT* and *SUBTRACT* TMs, the main difficulty in implementing the *DIV* TM is in ensuring that the sub-machines are "called" properly, and do not accidentally access inappropriate parts of the marked tapes. This needs careful consideration of where each sub-machine expects asterisks or blanks as delimiters, and which parts of the tape a sub-machine erases after completing its computation. One way is to use different areas of the tape (separated by one or more blanks) for different purposes. For example, you might represent the *count* variable from the algorithm in one place, the original n_1 and n_2 in another place. Sometimes, you may have to *shuffle* up one marked area to make room for the result from another marked area. However, as the main operations that increase the tape squares used are copying and adding, your machine can always be sure to create enough room between marked areas to accommodate any results. For example, if we add two binary numbers we know that the length of the result cannot exceed the length of

Table 9.2. The "repeated subtraction" method for division, using the TMs *SUBTRACT* and *ADD*

{the calculation performed is $n_1 \div n_2$}
begin
    count := 0
    temp := n_1

    while $n_2 \leqslant$ temp do
      temp := *SUBTRACT*(temp, n_2)
      count := *ADD*(count, 1)
    endwhile

> $T(x, y)$ is a functional representation of the statement: "apply the TM, T, to the tape appropriately configured with the numbers x and y"

    {*count* is the result}
end

the longer of the two numbers by more than one digit. All this shuffling can be quite
tricky, and it is rather more important to convince yourself that it can be done,
rather than to actually do it.

You will notice that the division algorithm in Table 9.2 includes the ≤test. As yet
we have not designed a TM to perform such logical comparisons. Therefore, in the
next section we develop a generalised *COMPARE* TM. *COMPARE* may also be of use
to the *SUBTRACT* machine (to detect if the number being subtracted is larger than
the number from which it is being subtracted).

9.5 Logical Operations

The generalised comparison TM, *COMPARE*, tells us whether two arbitrary length
binary numbers are equal, or whether one number is greater than, or less than the
first number. *COMPARE*'s input configuration will follow that for *ADD*, *MULT*, and
SUBTRACT, and is shown in Figure 9.22.

COMPARE will write the symbol =, <, or > in place of the "*" between the two
numbers in Figure 9.22 according to whether the "second number" is equal to, less
than, or greater than the "first number", respectively. Thus, for example, given the
example input tape configuration shown in Figure 9.23, *COMPARE* will produce an
output configuration as shown in Figure 9.24.

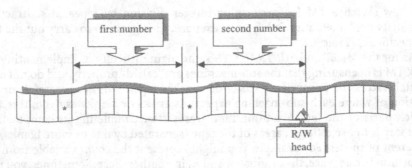

Figure 9.22. The input tape set up for the *COMPARE* TM (Figures 9.25 and 9.26), that tells us the
relationship between the first and second number

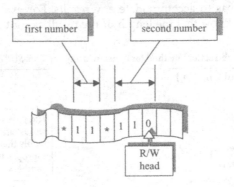

Figure 9.23. An example input configuration for the *COMPARE* TM (Figures 9.25 and 9.26). This
represents the question *"what is the relationship between 110 and 11?"*

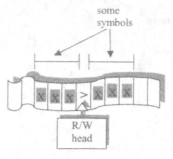

Figure 9.24. The output from the *COMPARE* TM (Figures 9.25 and 9.26) for the input specified in Figure 9.23. Given the original input, the output represents the statement "110 is *greater than* 11"

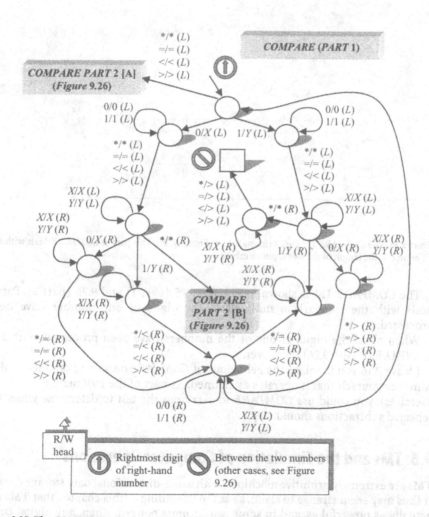

Figure 9.25. The *COMPARE* TM (part 1). Does a "bitwise" comparison of two binary numbers, beginning with a tape as specified in Figure 9.22. Answer is a symbol ("<", ">", or "=") written between the two numbers

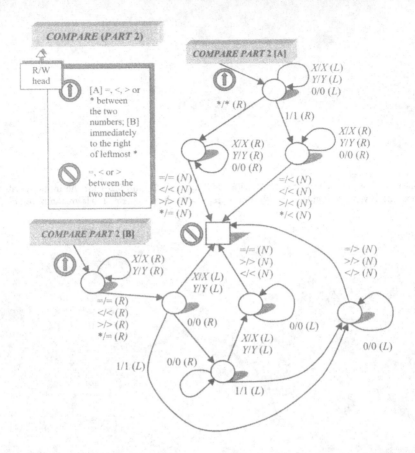

Figure 9.26. The *COMPARE* TM part 2 (for part 1 see Figure 9.25). This part of *COMPARE* deals with any difference in the length of the two input numbers

The *COMPARE* TM is shown in Figures 9.25 (part 1) and 9.26 (part 2). Part 1 deals with the comparison until all of the digits of one number have been processed.

When all of the digits of one of the numbers have been processed, part 2 of *COMPARE* (Figure 9.26) takes over.

I leave it to you to follow the behaviour of *COMPARE* on some examples, and to convince yourself that it operates as claimed. As part of the *DIV* machine (see the exercises), you could use *COMPARE* to carry out the test to determine when the repeated subtractions should stop.

9.6 TMs and the Simulation of Computer Operations

TMs are extremely primitive machines, with a one-dimensional data storage device. It thus may seem strange to claim, as at the beginning of this chapter, that TMs are actually as powerful as, and in some senses *more powerful* than, *any* digital computer can be. To provide evidence for this claim, we have seen TMs that do multiplication and division of arbitrary length binary numbers. Such operations are beyond the capabilities of any machine with a fixed amount of storage, such as any real

Table 9.3. TM simulation of arithmetic operations

Example operations	TM implementation
Division*	DIV
Multiplication*	MULT
Addition*	ADD
Subtraction*	SUBTRACT

Power (x^y)*	result := 1 {write a 1 in the result area of tape}
	while $y \neq 0$ {COMPARE$(y, 0)$ in another area of the tape}
	    result := result $\times x$ {MULT(result, x) in another area of tape}
	    $y := y - 1$ {SUB$(y, 1)$ in another area of tape}
	endwhile

* Computer limitations: size of operands/size of result.

computer at a given point in time. We have also seen how logical operations such as comparison of (again, arbitrarily large) numbers can be carried out by our TMs.

In this chapter, we considered only the binary number system. You should not see this as a limitation, however, as your experience with computers tells you that ultimately, everything that is dealt with by the computer, including the most complex program, and the most convoluted data structure, is represented as binary codes. Even the syntactic rules (i.e. the *grammars*) of our most extensive programming languages are also ultimately represented (in the compiler) as binary codes, as is the compiler itself.

In the following chapter, we take our first look at *Turing's thesis*, concerning the computational power of TMs. Before this, we consider more evidence supporting the claim that TMs are as powerful as any computers by briefly mapping out how many internal functions performed by, and many representations contained within, the digital computer, can be modelled by a TM. This is not an exhaustive map, but a sketch of the forms that the TM implementation of computer operations could take. I leave it to you to convince yourself of the general applicability of TMs to the modelling of any other operations found in the modern computer.

Tables 9.3–9.6 show examples for each specified internal operation. For each example, we suggest a TM that can model that operation (some of these are from

Table 9.4. TM simulation of logical operations

Example operations	TM implementation	Computer limitations
$<, > =$ $\leqslant, \geqslant, \neq$	COMPARE Example: "$\neq$" (TM NOT-EQ)	Length of operands Same as above

COMPARE's halt
state is a non-halt
state in NOT-EQ

=/F (N)
</T (N)
>/T (N)

bitwise *NOT*	INVERT	Same as above

Table 9.5. TM simulation of memory/array structure and access

Example operations	TM implementation	Computer limitations

One-dimensional array	$index$ * $item_1$ * $item_2\ldots$
	
	{$index$ is the number of the item to be accessed}
	1. To the left of $index$, write a 0 (to represent a *count* variable initialised to 0)
	2. *ADD*(*count*, 1)
	3. *COMPARE*(*count*, *index*)
	   if they are equal,
	     proceed right to the first "*" to the right of $index$. If a blank is encountered, HALT, otherwise copy the digits between that "*" and the next "*" (or blank) somewhere to the left of $index$, then go back rightwards along the tape changing all Xs back to *s, then HALT
	   else
	     proceed right to the first "*" to the right of $index$, change it to a X, and go to 2.
		Item length/address length/array size

Indirect addressing	As above, but in step 3 overwrite $index$ with copied number and go to step 1	Item length/address length/array size
Multi-dimensional arrays	EXAMPLE: $3 \times 2 \times 2$ three-dimensional array i * j * k * 1,1,1–1,2,2 **I** 2,1,1–2,2,2 **I** 3,1,1–3,2,2 **I** 2,1,1 **K** 2,1,2 **J** 2,2,1 **K** 2,2,2 i, j, k: indexes of item to be retrieved (i.e. item k of item j of item i) K: marker for increment of k index J: marker to set k item number to 1 and increment j item number I: marker to set both k and i item number to 1 and increment i item number	As above plus permissible number of dimensions

Table 9.6. TM simulation of program execution

Example operations	TM implementation	Computer limitations
Running a program	Use binary numbers to represent codes for operations such as *ADD*, *MULT*, memory access and so on	Size of program
	Set up data and memory appropriately on tape	Amount of data
	Use a portion of the tape as data, and a portion to represent sequences of operations (i.e. the *program*)	Number of operations

earlier in the chapter). In each case, certain limitations that would apply to any digital computer, at a given moment in time, but do not apply to TMs, are given.

Table 9.3 suggests how TMs can model arithmetic operations. In Table 9.4, logical operations are considered. Memory and array representation and accessing feature in Table 9.5, while Table 9.6 briefly considers the TM simulation of program execution (a subject which is given much more attention in Chapters 10 and 11).

It is worth focussing briefly on memory and array operations (Table 9.5), as these did not feature earlier in this chapter. In the examples given in Table 9.5 we have used memory structures where each element was a single binary number. This is essentially the memory organisation of the modern digital computer. I leave it to you to appreciate how complex data items (such as records, etc.) could also be incorporated into the scheme.

Careful consideration of the preceding discussion will reveal that we seem to be approaching a claim that TMs are capable of performing any of the tasks that real computers can do. Indeed, the remarks in the third columns of Tables 9.3–9.6 suggest that computers are essentially limited compared to TMs. These observations lead us on to *Turing's thesis*, which forms a major part of the next chapter.

9.7 Exercises

For exercises marked †, solutions, partial solutions, or hints to get you started appear in "Solutions to Selected Exercises" at the rear of the book.

9.1. Define the TM *TIDY-UP* for the *ADD* machine, described above.

9.2. Complete the definitions of the TMs *MULT* (using the *ADD* TM), *SUBTRACT* and *DIV* (perhaps making use of *COMPARE*).

9.3. Design a TM that accesses elements of a two-dimensional array, following the specification given in Table 9.5.

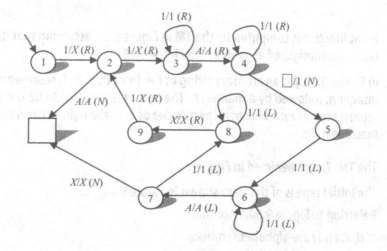

Figure 9.27. What function does this TM compute?

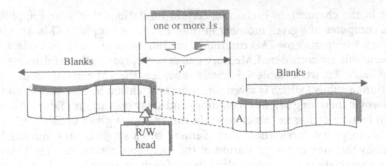

Figure 9.28. The input configuration for the TM in Figure 9.27

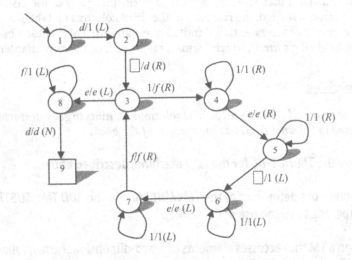

Figure 9.29. A TM that does language recognition and computation

9.4†. **What *function* is computed by the TM in Figure 9.27, assuming that its input tapes are configured as specified in Figure 9.28?**

In Figure 9.28, *y* is an arbitrary string of the form $1^n, n \geqslant 1$, representing the integer *n*, followed by a marker, *A*. The output is assumed to be the integer represented by *m*, where *m* is the number of 1s to the right of the *A* when the machine halts.

9.5. **The TM, *T*, is as depicted in Figure 9.29.**

The Initial tape is of the form shown in Figure 9.30.

Referring to Figure 9.30, note that:

   d, e, and *f* are alphabet symbols;

   x is a (possibly empty) sequence of 1s.

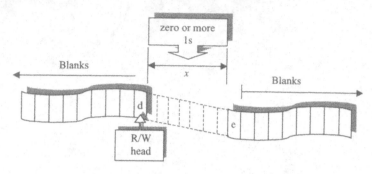

Figure 9.30. The input configuration for the TM in Figure 9.29

(a)[†] Briefly describe the relationship between the tape on entering the loop of states 3–7, and the tape when all iterations through the loop have been completed.

(b)[†] If the arc between state 3 and state 8 labelled "*e/e (L)*" was altered to "*e/1 (L)*", what function, expressed in terms of x, would the amended machine compute, assuming that $|x|$ (i.e. the number of 1s on the initial tape) represents an integer, and that the result is the number of 1s to the right of d when the machine halts?

(c) Delete state 1 from T, and by amending arcs as necessary, draw a machine, T_a, which accepts the language $\{d1^n e1^n : n \geq 0\}$. State 2 of T should become the start state of T_a and state 9 of T its halt state. Additional alphabet symbols may be required, but you should introduce no additional arcs.

Notes

1. Since *MULT* will use *ADD*, the sub-machines we define for *MULT* may need to take account of symbols used by *ADD*, such as C (the carry indicator), and so on.

Chapter 10
Turing's Thesis and the Universality of the Turing Machine

10.1 Overview

In this chapter, we investigate *Turing*'s thesis, which essentially says that the Turing machine (TM) is the most powerful computational device of all. We find evidence to support the thesis by considering a special TM called the *Universal* TM (UTM).

UTM is a TM that models the behaviour of any other TM, *M*, when presented with a binary code representing:

- *M*'s input
- *M* itself.

As defined in this chapter, *UTM* is a TM with 3 tapes (and three associated read/write heads).

This leads us on to consider multiple-tape (*k-tape*) TMs and the *non-deterministic* TMs, introduced as the recogniser for the type 0 languages in Chapter 7. We discover that:

- any non-deterministic TM can be represented by a deterministic, 4-tape TM, and then that
- any *k*-tape TM can be modelled by a standard, deterministic, single-tape TM.

We conclude with the observation that a significant outcome of the above, with respect to Part 1 of the book, is that the standard TM is the recogniser for the type 0 languages.

10.2 Turing's Thesis

We begin by continuing a discussion of the power of the TM that began in the preceding chapter. This leads to what is known as *Turing's thesis*. What you have to appreciate about a "thesis" is that it is not a "theorem". A *theorem* is a statement that can be proved to be true by a step-by-step argument that follows valid logical rules, just as the *repeat state theorem* for finite state recognisers (FSRs) was "proved" in Chapter 6. The activities we consider in Part 3 of this book are largely concerned with the proof of logical theorems. A *thesis*, on the other hand, is something that somebody asserts, for some reason, as being true. In other words, anybody can state any assertion is a thesis, but it serves little use as a thesis if some-one else immediately disproves it.

However, Turing's thesis, which was initially posited in the 1930s, has yet to be disproved. Not only has it not been *disproved*, but related work carried out by other mathematicians led to essentially the same conclusions as those reached by Turing himself.[1] For example, an American mathematician, Alonzo Church, working at around the same time as Turing, described the theory of *recursive* functions. Similarly, Emile Post, also a contemporary of Turing and Church, and, like Church, an American, described in 1936 a related class of abstract machines. These con-sisted of an infinite tape on which an abstract problem-solving mechanism could either place a mark or erase a mark. The result of a computation was the marked portion of the tape on completion of the process. These formalisms, and subse-quent developments such as Wang's machines (like TMs but with an alphabet of only blank and "*"), and the register machines of Sheperdson and Sturgis (closely resembling a simple assembler code-based system but with registers that can store arbitrarily large integers), have been shown to be:

(a) equivalent in computational power to *TMs*, and therefore,

(b) more powerful than any machine with limited storage resources.

Turing's thesis concerns the computational power of the primitive machines which he defined, which we call *TMs*. Essentially, in its broadest sense, Turing's thesis says this:

> *Any well defined information processing task can be carried out by some Turing machine.*

However, in a mathematical sense, the thesis says:

> *Any computation that can be realised as an effective procedure can also be realised by some Turing machine.*

This last statement of the thesis is probably not particularly helpful, as we have yet to specify what constitutes an *effective procedure*. Turing was working in an area of pure mathematics which had, for many years studied the processes of "doing" mathematics itself, i.e. the procedures that were used by mathematicians to actu-ally carry out the calculations and reasoning that mathematicians do. At the time Turing formulated the notion of TMs, the "computers" that his machines were intended to emulate were thus *humans engaged in mathematical activity*. An effective procedure, then, is a step-by-step series of unambiguous instructions that carry out some task, *but specified to such a level of detail that a machine could exe-cute them*. In this context, as far as we are concerned, Turing's thesis tells us this:

> *There is no machine (either real or abstract) that is more powerful, in terms of what it can compute, than the Turing machine.*

Therefore:

> *The Computational Limitations of the TM Are the Least Limitations of Any Real Computer.*

TMs are radically different from the computers with which you and I interact in our everyday lives. These real computers have *interfaces*; they may display detailed graphics, or even animations and video, or play us music, or enable us to speak to, or play games with, them. The computer may run an artificial intelligence program, such as an expert system that solves problems that even 20 years ago would have been impossible on anything but the largest mainframe. Yet, in a formal sense, everything that happens on the *surface*, so to speak, of the computer system, is merely a *re-presentation*, an alternative encoding, of the results of symbol manipulation procedures. Moreover, these procedures can be modelled by a simple machine that merely repeatedly reads a single symbol, writes a single symbol, and then moves to examine the next symbol to the right or left. Moreover, this simple abstract machine was formulated at a time when "real computers" were things like mechanical adders and Hollerith card readers.

The TM represents a formalism that allows us to study the nature of computation in a way that is independent of any single manufacturer's architecture, machine languages, high level programming languages or storage characteristics and capacity. TMs are thus a single formalism that can be applied to the study of topics such as the complexity of automated processes and languages (which is how we view them in the remainder of this part of the book).

In terms of languages, we have already seen the close relationship between TMs and the Chomsky hierarchy, and we refine this later in this chapter, and continue to do so in the next chapter. In Chapter 12, we consider in detail the notion of complexity as it applies both to TMs and to programs on real machines. However, an even more exciting prospect is suggested by our statement that the limitations of the TM are the least limitations of any real computer. For a start, only a real computer to which we could indefinitely add additional storage as required could approach the power of the TM. It is these limitations, and their ramifications in the real world of computing, that are the main concern of this part of the book.

In order to support the arguments in this chapter, we define a single TM that can model the behaviour of any other TM. We call this machine the *UTM*. The existence of *UTM* not only provides further evidence in support of Turing's thesis, that the TM is the most powerful computational machine, but also plays a part in finally establishing the relationship between computable languages and abstract machines. In the course of this, *UTM* also features in the next chapter, in a demonstration of an extremely significant result in computer science: the general unsolvability of what is called the *halting problem*.

Our *UTM* will accept as its input:

1. (a coded version of) a TM, *M*
2. (a coded version of) the marked portion of an input tape to *M*.

UTM then simulates the activities that *M* would have carried out on the given input tape.

Since *UTM* requires a coded version of the machine, *M*, and the input to *M*, our next task is to define a suitable scheme for deriving such codes.

10.3 Coding a TM and Its Tape as a Binary Number

We first consider a simple method of representing any TM as a binary code. We then see how the code can be used to code any input tape for the coded machine. The coded machine and the coded input can then be used as input to *UTM*.

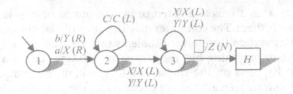

Figure 10.1. A simple TM

Table 10.1. The *quintuple* representation of the TM in Figure 10.1

$$(1, a, X, R, 2)$$
$$(1, b, Y, R, 2)$$
$$(2, C, C, L, 2)$$
$$(2, X, X, L, 3)$$
$$(2, Y, Y, L, 3)$$
$$(3, X, X, L, 3)$$
$$(3, Y, Y, L, 3)$$
$$(3, \square, Z, N, H)$$

10.3.1 Coding Any TM

To describe the scheme for coding any TM as a string of 1s and 0s we will use an example TM. As we proceed, you should appreciate that the coding scheme could be applied to any TM, consisting of any number of states and using any size of alphabet.

Our example machine, M, is shown in Figure 10.1.

An alternative representation of M (Figure 10.1) is shown in Table 10.1. You will note that each instruction (initial state, read symbol, write symbol, direction, and end state) is represented as a five-element list. In mathematical parlance, a five-element list is known as a *quintuple*. Table 10.1 thus provides the quintuples of the TM, M. This representation of TMs is fairly common. We use it here as it supports the subsequent discussion in this chapter, and the following two chapters.

We begin by associating the *states* of M (Figure 10.1) with strings of 0s as shown in Table 10.2. It should be clear that we could code the states of any TM using the scheme.

Next, as shown in Table 10.3, we code the *tape symbols* (*alphabet*) of the machine, which, for our machine M, is $\{a, b, C, X, Y, Z\}$ and the *blank* symbol. Again, we can see that this scheme could apply to a TM alphabet of any number of symbols.

We also code the direction symbols L, R, and N, as shown in Table 10.4.

You will note that some of the codes representing the states (Table 10.2) are identical to those used for the alphabet (Table 10.3) and direction symbols (Table 10.4). However, this does not matter; as we will see in a moment, our coding scheme ensures that they do not become mixed up.

Table 10.2. Coding the states of the machine of Figure 10.1 as strings of zeros

State	Code	Comment
1	0	*Start* state of any machine always coded as 0
H	00	*Halt* state of any machine always coded as 00
2	000	For the remaining states, we simply assign to each a unique string of 0s greater than 2 in length
3	0000	

Table 10.3. Coding the alphabet of the machine of Figure 10.1 as strings of 0s

Symbol	Code	Comment
□	0	*Blank* always coded as 0
a	00	For the rest of the alphabet, we simply assign to each any unique string of 0s greater than 1 in length
b	000	
C	0000	
X	00000	
Y	000000	
Z	0000000	

Table 10.4. Coding a TM's direction symbols as strings of 0s

Symbol	Code
L	0
R	00
N	000

We have coded the states, the alphabet, and the direction symbols. We now code the instructions (*quintuples*) themselves. Figure 10.2 shows how to code a particular quintuple of *M*. Once again, it is clear that any quintuple could be coded in this way.

Now, we simply code the whole machine as a sequence of quintuple codes, each followed by 1, and we place a 1 at the start of the whole code. The scheme is outlined in Figure 10.3.

The above should be sufficient to convince you that we could apply such a coding scheme to any TM, since any TM must have a finite number of states, tape symbols, and quintuples.

Such a binary code for a TM could be presented as the input to any TM that expects arbitrary binary numbers as its input. For example, we could present the code for the above machine, *M*, to the *MULT* TM (described in Chapter 9) as one of the two numbers to be multiplied. We could even present the coded version of *MULT* to the *MULT* machine as one of its input numbers, which would mean that *MULT* is being asked to multiply its own code by another number! All this may seem rather strange, but it does emphasise the point that codes can be viewed at many different levels. The code for *M* can be regarded as another representation of *M* itself (which is how *UTM* will see it), as a *string* in $\{0, 1\}^+$ (a non-empty string of 0s and 1s) or as a *binary number*. This is essentially no different from seeing Pascal source code as text (the text editor's view) or as a program (the compiler's view).

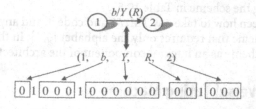

Figure 10.2. How one *quintuple* of the TM from Figure 10.1 is represented as a binary code (for coding scheme see Tables 10.2–10.4)

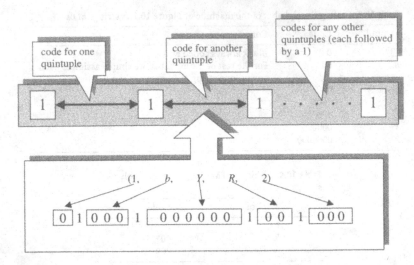

Figure 10.3. How a whole TM is coded as a binary number (the example *quintuple* is that of Figure 10.2)

Table 10.5. Coding a TM's input tape as a binary number

Code	Application to an input tape for the TM of Figure 10.1
We use the same codes for the tape symbols as we use for the machine's alphabet (Table 10.3)	
As for the quintuples (Figure 10.2), we place a 1 after the code for each symbol	
We place a 1 at the beginning of the coded input	

Since our *UTM* will require not only the coded version of a machine, but also a coded version of an input to that machine, we need to specify a scheme for coding any marked tape of the machine we are coding.

10.3.2 Coding the Tape

The coding scheme for the input is described in Table 10.5. Note that for a given machine, we use the same codes for the symbols on the input tape as those used to code the machine's alphabet (Table 10.3).

It should be clear that any input tape to any coded machine could be similarly represented using the scheme in Table 10.5.

We have now seen how to take any TM, *M*, and code it, and any of its input tapes, with a coding scheme that requires only the alphabet {0, 1}. In the next section, we use our coding scheme as an integral component of the architecture of our *UTM*.

10.4 The Universal Turing Machine

UTM is described here as a 3-*tape* deterministic TM. Though our TMs have hitherto possessed only one tape, it is quite reasonable for a TM to have any (fixed) number of tapes. We discuss such machines in more detail a little later in this

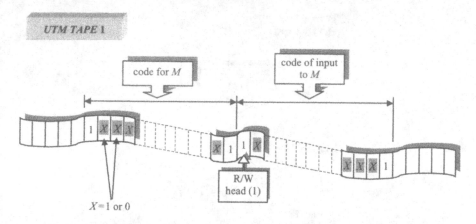

Figure 10.4. The set up of tape 1 of *UTM*. The machine, *M*, is coded as specified in Figure 10.3. The input to *M* is coded as specified in Table 10.5

chapter, when we also show that a "multi-tape" TM can be modelled by an equivalent single-tape machine. For the moment, it suffices to say that in a multi-tape TM, each tape is numbered, and each instruction is accompanied by an integer which specifies the number of the tape on which the read, write, and head movement operations in that instruction are to take place.

The 3-tape *UTM* receives as its input a coded version of a TM and essentially simulates the behaviour of the coded machine on a coded version of input to that machine. *UTM* therefore models the behaviour of the original machine on the original input, producing as its output a coded version of the output that would be produced by the original machine, which can then be decoded.

You may at this point be concerned that there is some "trickery" going on here. You might say that everything that *UTM* does is in terms of manipulating *codes*, not the "real" thing. If you *do* think that it is trickery, then it can be pointed out that every single program you have ever written is part of the same type of "trickery", as

- the source file
- the input
- the output

are all similarly coded and subsequently decoded when your program is run. So if you are not prepared to believe in the representational power of coding schemes, you ought not to believe that any of your programs have ever done anything meaningful!

Next, we specify the initial set-up of *UTM*'s three tapes. Following this, we briefly consider the operation of *UTM*. You will appreciate that the construction of *UTM* described would be capable of simulating the operation of any TM, *M*, given any suitable input tape to *M*.

10.4.1 *UTM*'s Tapes

Our version of *UTM* is configured so that tape 1 is as shown in Figure 10.4. As can be seen from Figure 10.4, tape 1 contains the code for the machine *M*, followed immediately by the code for *M*'s input.

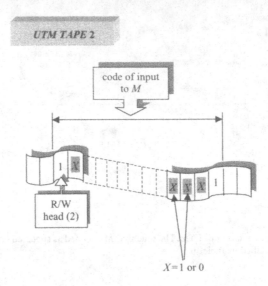

Figure 10.5. The set up of tape 2 of *UTM*. *UTM* initially copies the coded input (see Figure 10.4) from tape 1 onto tape 2. Tape 2 is then used as a "work tape" by *UTM*

UTM will use its second tape as a "work tape", on which to simulate *M*'s operation on its input by carrying out the operations specified by the coded quintuples on the coded data. Thus, the first thing *UTM* does is to copy the coded input from tape 1 to tape 2, at which point, tape 2 will be as specified in Figure 10.5.

Tape 3 of *UTM* will be used only to store a sequence of 0s denoting, at any stage, the current state of *M*. The organisation of tape 3 is depicted in Figure 10.6.

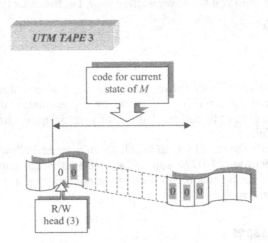

Figure 10.6. The set up of tape 3 of *UTM*. Tape 3 is used to hold the code (a sequence of 0s) for *M*'s current state. It is initialised to a single 0, as any machine's start state is coded as 0

To summarise:

Tape 1   stores the coded machine M, followed by the coded input to M

Tape 2   is used by *UTM* to work on a copy of M's coded input

Tape 3   is used to hold a sequence of 0s representing M's current state.

10.4.2 The Operation of *UTM*

UTM operates as specified by the procedure in Table 10.6. I leave it to you to appreciate how the procedure itself could be represented as a TM. For clarification of the procedure in Table 10.6, you are advised to refer to the coding scheme and the configurations of *UTM*'s three tapes described above.

A *UTM* based on the procedure in Table 10.6 is capable of carrying out the computation specified by any TM, given an initial set-up as described above. As input to such a machine, we code the TM according to our scheme, and then we can code an input tape to that TM using the same scheme. *UTM* will then simulate the computation performed by the original machine.

In particular:

- If M, the original machine, would produce a solution given a particular input tape, then *UTM* would produce a coded version of that solution (on its tape 2).

- If M would reach a state, s, in which it had no applicable quintuples for the current state and tape symbols, then *UTM* would halt with the code for the state s on tape 3, and tape 2 would contain a coded version of M's tape configuration at the time M stopped.[2]

- If M would reach a situation where it looped indefinitely, repeating the same actions over and over again, then *UTM* would also do the same. This is very significant, and highly relevant to a discussion in the next chapter.

We can now appreciate that we have described a three-tape deterministic TM, *UTM*, that can, if we accept Turing's thesis, carry out any *effective procedure*. We will see later that from our three-tape version we could construct a deterministic "standard" single-tape *UTM*.

10.4.3 Some Implications of *UTM*

The existence of *UTM* makes Turing's thesis that TMs are the pinnacle of functional power even more compelling. We have seen that a TM that can "run", as a kind of "program", any other TM needs to be no more sophisticated than a standard TM. In a sense, *UTM* actually embodies the *effective procedure* of *carrying out the effective procedure* specified by another TM.

Other designs of *UTM* may be suggested, based on different schemes for coding the machine M and its input. All are equivalent. *UTM* is a *TM*, and can itself be coded according to the scheme for one of the other *UTM*s. The code can then be presented, along with a coded version of an input tape, as input to that other *UTM*. In that sense also, *UTM* is truly *universal*.

In a more practical vein, *UTM* is a closer analogy to the stored program computer than the individual TMs we have considered so far. Our other TMs are essentially purpose built machines that compute a given function. *UTM*, with its ability

Table 10.6. The operation of the universal TM, *UTM*, in simulating the behaviour of a TM, *M*. For details of the coding scheme used, see Tables 10.2–10.5, and Figures 10.2 and 10.3. For the set-up of *UTM*'s 3 tapes, see Figures 10.4–10.6

Operation of UTM	Comments
1 Copy the code for *M*'s data from tape 1 to tape 2	Tape 2 is to be used as a work tape by *UTM*
2 Write a 0 on tape 3 to represent tape *M*'s start state	0 represents *M*'s start state 00 represents *M*'s halt state
3 while tape 3 does not contain exactly 2 zeros do	
Find a quintuple of *M* on tape 1 for *current state* on tape 3	*UTM* proceeds rightwards along tape 1, comparing the sequence of 0s in each quintuple of *M* that represent a current state, with the sequence of 0s on tape 3
	UTM needs to compare sequences of 0s to do this
if quintuple found then if *current symbol* code on tape 2 = quintuple *read symbol* code on tape 1 then on tape 2, overwrite the *current symbol* code with the quintuple *write symbol* code on tape 1	*UTM* may have to *shuffle* symbols, since each alphabet symbol is represented by a different length string of 0s (Chapter 7 contains examples of this type of activity)
move the read head on tape 2 according to the quintuple *direction* code on tape 1	*UTM* must move its tape 2 read/write head to the next 1 to the left or right (or not at all)
overwrite the *state* code on tape 3 with the quintuple *next state* code from tape 1	*UTM* replaces the string of 0s on tape 3 with the string of 0s representing the next state of *M*

endif
endif
endwhile	

to carry out computations described to it in some coded form, is a counterpart of real computers that carry out computations described to them in the form of codes called *programs*. However, once again it must be emphasised that a given computer could only be *as* powerful as *UTM* if it had unlimited storage capabilities. Even then, it could not be *more* powerful.

In the next section, we provide even more compelling evidence in support of Turing's thesis. In doing so, we close two issues that were left open earlier in this book. Firstly, in Chapter 7 we saw how a non-deterministic TM is the recogniser for the type 0 languages. Secondly, earlier in this chapter, we encountered a 3-tape *UTM*. We now see that the additional power that seems to be provided by such features is illusory: the basic, single-tape TM is the most powerful computational device.

10.5 Non-deterministic TMs

In Chapter 7, we saw that the *non-deterministic* TM is the abstract machine counterpart of the *type* 0, or *unrestricted* grammars, the most general classification of grammars of the Chomsky hierarchy. We saw a method for constructing non-deterministic TMs from the productions of type 0 grammars. Now we return to the notion of non-deterministic TMs to demonstrate that non-determinism in TMs provides us with no extra *functional* power (although we see in Chapter 12 that things change if we also consider the *time* that it might take to carry out certain computations). This result has implications for *parallel computation*, to which we return later in this chapter. Here, only a sketch is given of the method for converting any non-deterministic TM into an equivalent deterministic TM. However, some of the techniques introduced are fundamental to discussions in the remainder of this part of the book.

Now, if a TM is non-deterministic, and it correctly computes some function (such as accepting the strings of a language), that TM will be capable of reaching its halt state, leaving some kind of result on the tape. For a non-deterministic machine this is alternatively expressed as "there is some sequence of *transitions* that would result in the machine reaching its halt state from one of its start states". In terms of our discussion from earlier in this chapter, the phrase

   there is some sequence of *transitions*

can be alternatively expressed as:

   there is some sequence of *quintuples*.

Suppose, then, that we have a non-deterministic TM that sets off in attempting to solve some problem, or produce some result. If there *is* a result, and it can try all possible applicable sequences of quintuples *systematically* (we will return to this in a moment), then it will find the result. What we are going to discover is that there is a *deterministic* method for trying out every possible sequence of quintuples (shorter sequences first), and stopping when (or rather *if*) a sequence of quintuples is found that leads to a halt state of the original machine. Moreover, it will be argued that a *deterministic* TM could apply the method described. Analogously to *UTM*, the construction described represents a case where one TM executes the instructions of another, as if the other TM was a "program".

In order to simplify the description of our deterministic version of a non-deterministic TM, we shall assume that we have a special TM with four tapes.

We will then show how multi-tape machines can be modelled by single-tape machines, as promised earlier.

10.6 Converting a Non-deterministic TM into a 4-tape Deterministic TM

We now describe the conversion of a non-deterministic TM into a 4-tape deterministic TM. Let us call the non-deterministic TM N, and the corresponding deterministic 4-tape machine, D.

10.6.1 The Four Tapes of the Deterministic Machine, D

We begin by describing the organisation of each of D's four tapes, beginning with tape 1, which is shown in Figure 10.7.

As shown by Figure 10.7, tape 1 of D stores:

- a copy of the marked portion of the tape that N would have started with (we can assume for the sake of simplicity that N's computation would have begun with its read/write head on the leftmost symbol of its input)
- the label of N's current state, initially marked with N's start state label (let us assume that the start state is always labelled S)
- the label of N's halt state (we can assume that N has only one halt state: cf. Exercise 7.1, Chapter 7), which in Figure 10.7 is H.

Note that we assume that labels such as state names and alphabet symbols each occupy one *square* of a tape, i.e. are *single symbols*. In the formal world, there is no

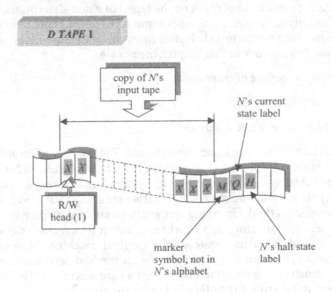

Figure 10.7. The set up of tape 1 of the deterministic, 4-tape TM, D (a deterministic version of a non-deterministic TM, N)

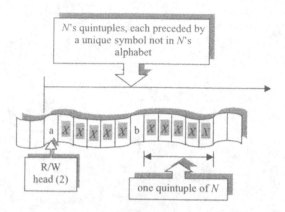

Figure 10.8. The set up of tape 2 of the deterministic, four-tape TM, *D* (a deterministic version of a nondeterministic TM, *N*). Tape 2 stores the quintuples of *N*

concept of a "character set" in which we "run out" of single symbol labels; there is a potentially infinite number of available single symbol labels. Thus, a TM of any size can be assumed to require state and alphabet labels each consisting of one symbol.

Now we can describe tape 2, schematically represented in Figure 10.8.

Figure 10.8 tells us that tape 2 of *D*, the deterministic four-tape TM, contains the quintuples of *N*, each occupying five consecutive tape squares, each preceded by a unique single symbol that is not a symbol of *N*'s alphabet. To illustrate, Figure 10.9 shows how a quintuple would be coded on tape 2 of *D*.

Next, tape 3 of *D*. Tape 3 is where *D* will repeatedly perform the systematic application of *N*'s quintuples, starting from a copy of *N*'s input tape (from tape 1, specified in Figure 10.7). *D*'s first activity, then, is to set up *tape* 3 as shown in Figure 10.10.

Note that *D* places special "markers" at either end of the copied input. These are necessary, as after the application of a sequence of *N*'s quintuples that leads to a dead end, *D* will have to erase the marked portion of tape 3, and copy the original input from tape 1 again. The potential problem is that *N*'s computation may result

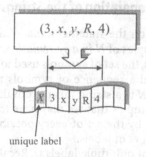

Figure 10.9. An example quintuple coded on tape 2 of *N* (see Figure 10.8)

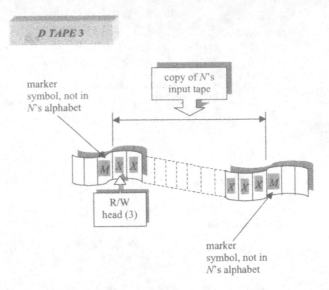

Figure 10.10. The set up of tape 3 of the deterministic, 4-tape TM, *D*. On tape 3, *D* will systematically apply the quintuples of the non-deterministic machine, *N*

in blank squares in between marked symbols, resulting in a tape that is no longer *packed*, so *D* needs a way of detecting the extreme ends of the marked portion of the tape.

The existence of the markers causes a slight complication in the operation of *D*. One of *N*'s quintuples may result in *D*'s tape 3 head moving onto the marker. When this happens, *D* must simply move the marker one square to the right or left, depending on whether the marker was encountered on a right or left move, respectively. *D* must replace the marker by a blank, and then overwrite the blank on the right or left, respectively, with the marker. *D* must also then move the head back one square to where the marker originally was (this corresponds to the position in the tape expected by the next applicable quintuple of *N*). Every time *D* makes a left or right move on tape 3, then, it must check if the next symbol is the marker; if so, it must take the appropriate action before attempting to apply the next quintuple of *N*.

10.6.2 The Systematic Generation of the Strings of Quintuple Labels

The whole process hinges on the statement, made more than once in the above, that *D* will apply the quintuples of *N* *in a systematic way*. As *N* can consist of only a *finite* number of quintuples, the set of symbols used to label the quintuples of *N* (on tape 2), is an *alphabet*, and a sequence of symbols representing labels for some sequence of quintuples of *N* is a string taken from that alphabet. So, the set of all possible sequences of quintuples that we may have to try starting from the initial tape given to *N* is denoted by the set of every possible non-empty string formed using symbols of the alphabet of labels.

Let us call the alphabet of quintuple labels *Q*. Recall from Chapter 2 that the set of every possible non-empty string formed using symbols of the alphabet *Q* is denoted by Q^+. Now, Q^+, for any alphabet Q, is not only *infinite* but also *enumerable*. Q^+ is a *countably infinite set*. Now, by *Turing's thesis*, if a *program* can enumerate Q^+, then a TM can do it.[3] That is exactly what we need as part of D:

- A sub-machine that will generate sequences of quintuple labels, one by one, shortest sequences first, so that the sequence of N's quintuples denoted by the sequence of symbols can be executed by D. Moreover, we must be sure that every possible sequence is going to be tried, if it proves necessary to do so.

To be prepared to try every possible sequence of quintuples is a rather brute force approach, of course. Many (in fact probably most) of the strings of labels generated will not denote "legal" sequences of N's quintuples for the given tape, or for *any* tape at all. Moreover, the method will try any sequence that has a prefix that has already led nowhere (there are likely to be an infinite number of these, also). However, because D tries the shorter sequences before the longer ones, at least we know that if there *is* a solution that N could have found, our machine D will eventually come across the sequence of quintuples that represents such a solution. We should also point out that if there is not a solution, then D will go on forever trying different sequences of quintuples (but then, if a solution does not exist, N would also be unable to find it!).

So, we finally come to tape 4 of D. Tape 4 is where D is going to repeatedly generate the next sequence of quintuple labels (strings in the set Q^+) that are going to be tried. Only an overview of one method will be given here. An exercise at the end of the chapter invites you to construct a TM to achieve this task.

To demonstrate the method, we will assume an alphabet $\{a, b, c\}$. The approach will be:

generate all strings of length 1:

a, b, c

generate all strings of length 2:

$aa, ab, ac, ba, bb, bc, ca, cb, cc$

generate all strings of length 3:

$aaa, aab, aac, aba, abb, abc, aca, acb, acc, baa, bab, bac, bba, bbb, bbc, bca, bcb, bcc,$
$caa, cab, cac, cba, cbb, cbc, cca, ccb, ccc$

⋮

You may notice from the order in which the strings are written that we obtain all of the strings of length n by copying each of the strings of length $n - 1$ three times (as there are three alphabet symbols). We then append a different symbol of the alphabet to each one. For example, from aa we generate $aaa, aab,$ and aac.

Let us assume that the machine initially has the three alphabet symbols somewhere on its tape. To generate the three strings of length 1 it simply copies each symbol to the right, each being followed by a *blank*, with the final blank followed by some marker (say "*").

The blanks following each of the strings in Figure 10.11 can be used both to separate the strings and to enable an auxiliary marker to be placed there, thus allowing the machine to keep track of the current string in any of its operations.

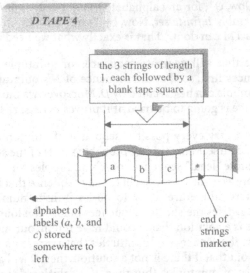

Figure 10.11. Tape 4 of D, when the set of quintuple labels is $\{a, b, c\}$, and all the strings of length equal to 1 have generated. This is part of D's systematic generation of all strings in $\{a, b, c\}^+$

The next stage, shown in Figure 10.12 involves copying the first string of length 1 (a) three times to the right of the *. The number of times the string is copied can be determined by "ticking off", one by one, the symbols of the alphabet, which, as shown in Figure 10.11, is stored somewhere to the left of the portion of tape being used to generate the strings.

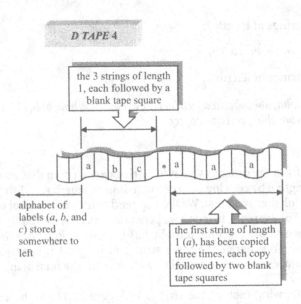

Figure 10.12. On tape 4, D is generating all strings in $\{a, b, c\}^+$ of length 2. The first stage is to copy the first string of length 1 (Figure 10.11) three times (since the alphabet $\{a, b, c\}$ contains 3 symbols)

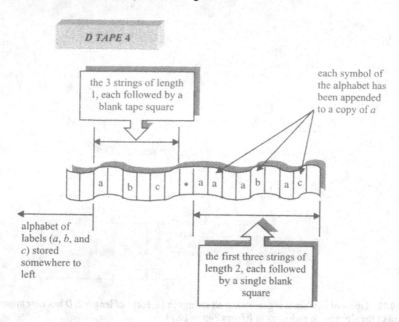

Figure 10.13. Tape 4 of *D*, when *D* is generating all strings in $\{a, b, c\}^+$ of length 2. *D* has now appended a symbol of the alphabet to each copy of *a* from Figure 10.12

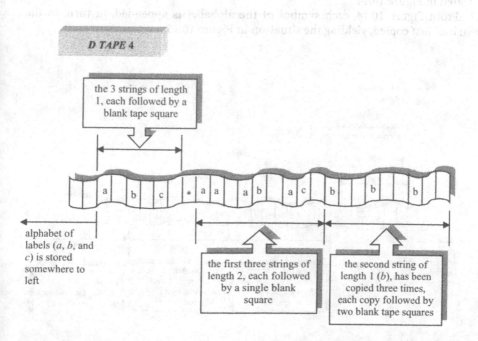

Figure 10.14. Tape 4 of *D*, when it is generating all strings in $\{a, b, c\}^+$ of length 2. From the situation in Figure 10.13, *D* has now added 3 copies of the second string of length 1 (*b*)

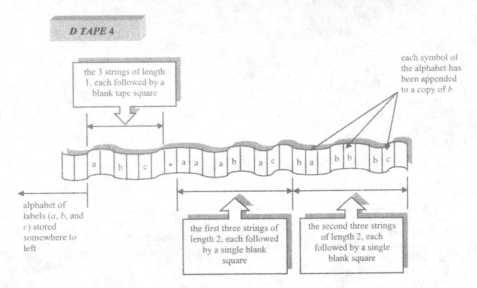

Figure 10.15. Tape 4 of D, when it is generating all strings in $\{a, b, c\}^+$ of length 2. D has now appended a symbol of the alphabet to each copy of b from Figure 10.14

From the situation in Figure 10.12, the next stage is to append each symbol of the alphabet, in turn, to the three copied as, giving the situation in Figure 10.13.

The next string of length 1, (b) is then copied three times to the right, as illustrated in Figure 10.14.

From Figure 10.14, each symbol of the alphabet is appended, in turn, to the strings just copied, yielding the situation in Figure 10.15.

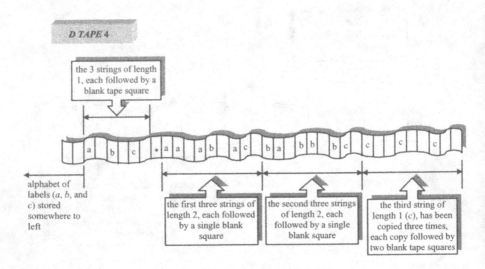

Figure 10.16. Tape 4 of D, when it is generating all strings in $\{a, b, c\}^+$ of length 2. From the situation in Figure 10.15, D has now added 3 copies of the second string of length 1 (c)

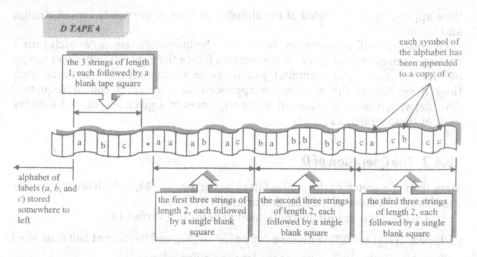

Figure 10.17. Tape 4 of D, when it is generating all strings in $\{a, b, c\}^+$ of length 2. D has now appended a symbol of the alphabet to each copy of c from Figure 10.16

The next stage from Figure 10.15, is to copy the next string of length 1 (c) three times to the right, leading to the set-up shown in Figure 10.16.

From Figure 10.16, each symbol of the alphabet is appended to each string that was copied in the previous stage, giving a situation shown in Figure 10.17.

Now, when the machine returns along the tape, then attempts to copy the next string of length 1, it encounters the "*". The machine thus detects that it has finished generating all strings of length 2. The machine can place a "*" at the end of the strings it created (Figure 10.17), as depicted in Figure 10.18.

From Figure 10.18, the machine is ready to begin the whole process again, i.e. copying each string between the two rightmost *s three times to the right, each

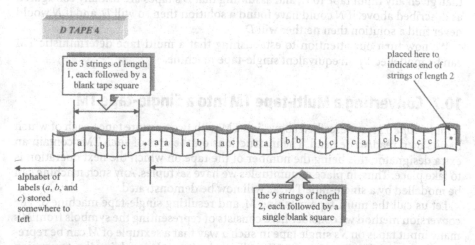

Figure 10.18. D has now generated all strings in $\{a, b, c\}^+$ of length 2, and has now appended a "*" to the end of the portion of tape containing the strings

time appending each symbol of the alphabet in turn to the newly copied strings, and so on.

Of course, it will be necessary for D to try the quintuple sequences each time a certain number of them have been generated (since the generation process can go on forever). It could, for example, try each new sequence as it was created, i.e. each time a symbol of the alphabet is appended to a recently copied sequence. Alternatively, it could generate all of the sequences of a given length, and then try each of those sequences in turn.

10.6.3 The Operation of D

Here then, is a sketch of the method used by the 4-tape TM, D, to deterministically model the behaviour of the non-deterministic TM, N.

D is assumed to be initially configured as described earlier, i.e.

Tape 1   (Figure 10.7) contains a copy of N's input, and its current halt state label

Tape 2   (Figure 10.8) contains the labelled quintuples of N

Tape 3   (Figure 10.10) is initially blank

Tape 4   contains the quintuple label symbols.

D operates as shown in Figure 10.19.

The most complicated process in the method specified in Figure 10.19 is the box that begins "Try sequence of quintuples …". This is expanded in Figure 10.20.

Sufficient detail has been presented for you to appreciate that the above construction could be applied to any similar situation.

10.6.4 The Equivalence of Non-deterministic and Four-tape Deterministic TMs

Thus far, you are hopefully convinced that any non-deterministic TM, N, can be represented by a 4-tape deterministic TM, D. D and N are equivalent, in the sense that, given any input tape to N, and assuming that D's tapes are initially configured as described above, if N could have found a solution then so will D, and if N would never find a solution then neither will D.

We now turn our attention to establishing that a multi-tape deterministic TM can be modelled by an equivalent single-tape machine.

10.7 Converting a Multi-tape TM into a Single-tape TM

A *multi-tape TM*, as described above, is a TM with two or more tapes, each of which has its own read/write head. The instructions on the arcs of such TMs contain an extra designator, this being the number of the tape on which the next operation is to take place. Thus, in place of quintuples we have *sextuples*. Any such machine can be modelled by a single-tape TM, as will now be demonstrated.

Let us call the multi-tape machine M, and resulting single-tape machine S. The conversion method we consider here consists of representing the symbols from M's many input tapes on S's single tape in such a way that a sextuple of M can be represented as several *quintuples* of S. We will use an example involving the conversion of a 3-tape machine into a single-tape machine to demonstrate the method.

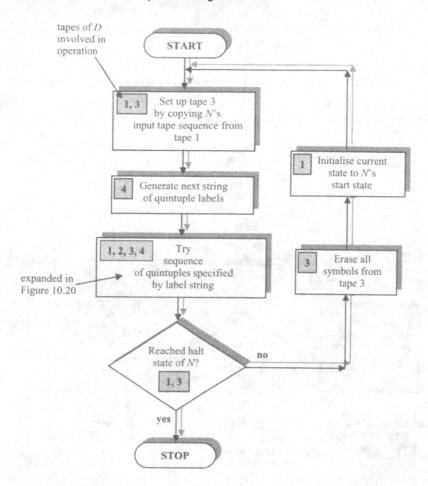

Figure 10.19. The operation of the 4-tape TM, *D*, that deterministically models the behaviour of a non-deterministic machine, *N*

10.7.1 Example: Representing Three Tapes as One

Let us suppose we have a three-tape machine, *M*, with its tapes initially set up as in Figure 10.21.

We use a scheme that systematically places the symbols from the three tapes of *M* on the single tape of *S*. From the three tapes above, we present the tape to *S* that is shown in Figure 10.22.

Thus, from left to right, and taking each tape of *M* in turn, we place the leftmost symbol from tape 1, followed by a blank (or a special marker if this symbol was pointed at by the read/write head of tape 1). This is followed by the leftmost symbol from tape 2, followed by a blank (or a special marker, if necessary, representing the read/write head of tape 2), and so on. It will be noted that tape 3 had the least marked squares, so pairs of blanks are left where tape 3 symbols would be expected on *S*'s tape. Similarly, for tape 1, two blanks were left so that the final symbol of tape 2, the tape of *M* with the most marked squares, could be placed on the single tape.

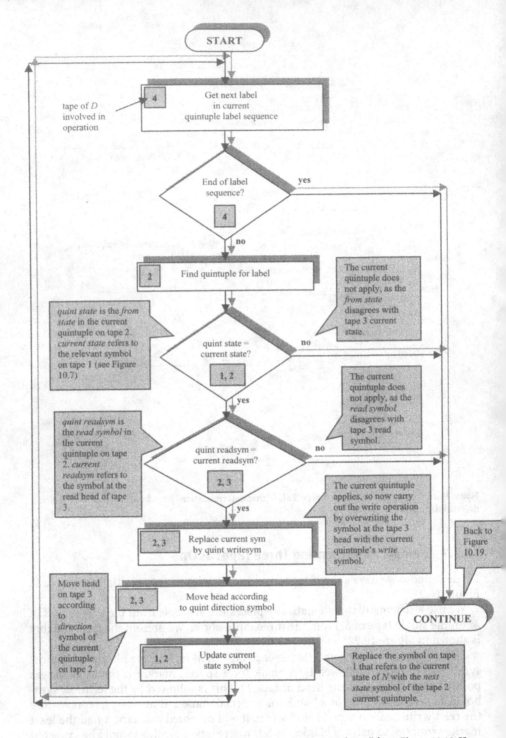

Figure 10.20. Expansion of the box that begins "Try sequence of quintuples ..." from Figure 10.19. How *D* models the behaviour of *N* specified by its quintuples

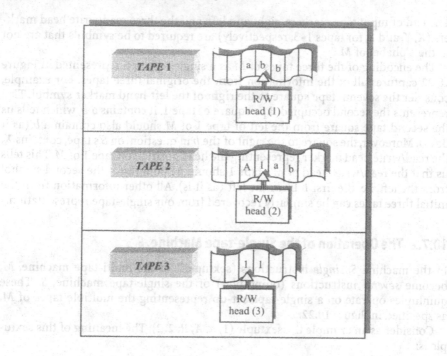

Figure 10.21. Example configurations of the three tapes of a hypothetical three-tape TM

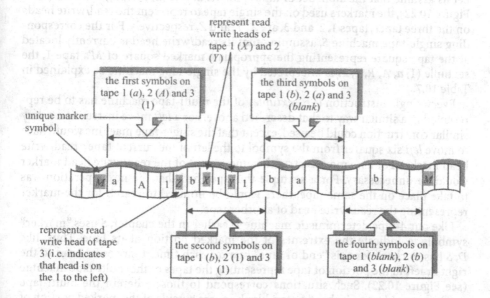

Figure 10.22. Representing the three tapes of Figure 10.21 on a single tape

The end of input tape markers, right and left, and the three read/write head markers (X, Y, and Z, for tapes 1–3, respectively) are required to be symbols that are not in the alphabet of M.

The encoding of the three tapes of M as a single tape of S represented in Figure 10.22 captures all of the information from the original three tapes. For example, consider the seventh tape square to the right of the left-hand marker symbol. This represents the second occupied tape square of tape 1. It contains a b, which tells us the second tape square from the left of tape 1 of M should also contain a b (as it does). Moreover, the square to the right of the b in question, on S's tape, contains X, the read/write head marker representing the head position on tape 1 of M. This tells us that the read/write head of M's tape 1 should be pointing at the second symbol from the left, i.e. the first b from the left (as it is). All other information from the initial three tapes can be similarly recovered from our single-tape representation.

10.7.2 The Operation of the Single-tape Machine, S

In the machine S, *single* instructions (sextuples) of the multi-tape machine, M, become *several* instructions (quintuples) of the single-tape machine, S. These quintuples operate on a single-tape set-up representing the multiple tapes of M, as specified in Figure 10.22.

Consider as an example the sextuple $(1, a, A, R, 2, 2)$. The meaning of this sextuple is:

> when in state 1, if the symbol at the read/write head of the current tape is a, replace it by A, move right one square on the current tape, switch processing to tape 2 and move into state 2.

Let us assume that the alphabet of tape symbols of S is $\{a, b, A, 1\}$, and that (as in Figure 10.22), the markers used on the single tape to represent the read/write heads on the three tapes, tapes 1, 2, and 3, are X, Y, and Z, respectively. For the corresponding single-tape machine S, assuming that its read/write head is currently located at the tape square representing the appropriate marked square of M's tape 1, the sextuple $(1, a, A, R, 2, 2)$ is interpreted by the single-tape machine as explained in Table 10.7.

Every single instruction (i.e. *sextuple*) of the multi-tape machine has to be represented in a similar way to that described above. For a *left* move instruction a very similar construction could be used, except that the single-tape machine would have to move *left* six squares from the symbol to the left of the current tape's read/write head marker. For a *no* move instruction, movement of the read/write head marker would be unnecessary. For a sextuple that specified that the next operation was to take place on the same tape, there would be no need to search for the marker representing the read/write head of another tape.

Like our 4-tape deterministic machine D earlier in the chapter, S uses "marker" symbols to indicate the extremes of the marked portion of its tape. Thus, like D, S has to check for this "end of tape" marker each time it moves its head to the right or left in the portion of tape representing the tapes of the multi-tape machine (see Figure 10.22). Such situations correspond to those wherein the multi-tape machine would move its head onto a blank square outside of the marked portion of one of its tapes. In such circumstances, S must *shuffle* the "end of tape marker" to the left or right by $2k$ squares, where k is the number of tapes of the simulated

Table 10.7. An overview of how a single-tape TM, S, simulates the execution of the sextuple $(1, a, A, R, 2, 2)$ of a 3-tape TM, M. The 3-tapes of M are assumed to have been coded onto a single tape, as described in Figures 10.21 and 10.22. The *current state* of M is assumed to be 1

M instruction $(1, a, A, R, 2, 2)$	Corresponding behaviour of S
"If the symbol at the read/write head of tape 1 is a, replace it by A ...	Since (see text), the read/write head of S is pointing at the symbol on its tape that represents the current symbol on M's tape 1, if that symbol is a, S replaces it with A
... move right one square on tape 1, ...	We have three tapes of M coded onto one tape of S. Each tape square of M is represented by two tape squares of S (one for the symbol, one to accommodate the marker that indicates the position of one of M's read/write heads). Thus, moving "one square" on a tape of M involves moving *six* squares on S's tape. S also has to move the marker (X) representing M's tape 1 read/write head so that it is in the appropriate place (i.e. next to the symbol representing one move right on M's tape 1)
... switch processing to tape 2 and move into state 2."	S must search for the marker that represents the read/write head of M's tape 2 (in our example, Y)

machine, M. For example, a "move right" on a single tape representing a ten-tape TM would involve moving 20 squares to the right (shuffling the "end of tape marker" right appropriately if necessary), and we would of course need 10 distinct read/write head markers, and so on.

You should be able to see how the above could be adapted for multi-tape machines with any number of tapes. You are asked to add more detail to the construction in an exercise at the end of this chapter.

10.7.3 The Equivalence of Deterministic Multi-tape and Deterministic Single-tape TMs

The preceding discussion justifies the following statement:

> *Any deterministic multi-tape TM, M, can be replaced by a deterministic single-tape TM, S, that is equivalent to M.*

In the preceding discussion, we have seen that it can be a convenience to define multi-tape, rather than single-tape machines. Recall also the previous chapter, when we described TMs such as *MULT* and *DIV*. These machines featured as "sub-machines" other TMs we had defined, such as *ADD*, and *SUBTRACT*. We were forced to use different areas of the tape to deal with the computations of different sub-machines, which led to some tricky copying and shuffling. A multi-tape approach would have enabled us to use certain tapes in a similar way to the "registers" or "accumulators" of the CPU, to store temporarily the results of sub-machines that would be subsequently needed.

A second, and related, feature of multi-tape machines is that they permit a *modularity* of "memory organisation", a memory "architecture", so to speak, which has more in common with the architecture of the real computer system that the single-tape TM. This can be appreciated by considering TMs such as the deterministic multi-tape machines we specified in the previous section, where the *program* (i.e. the quintuples of another machine) was stored on one tape, and the *data* were stored on another. Moreover, the machine used a further tape as a "working memory" area, and another tape as a kind of *program counter* (the tape where the sequences of quintuple labels were generated).

The modularity facilitated by a multi-tape TM architecture can be useful for specifying complex processes, enabling what is sometimes called a "separation of concerns". For example, there should be no need for *MULT* to be concerned with the internal workings of *ADD*. It needs only to gain access to the result of each addition. It would therefore be appropriate if *MULT* simply copied the two numbers to be added onto a particular tape, then passed control to *ADD*. The latter machine would then leave the result in an agreed format (perhaps on another tape), where it would be collected by *MULT* ready to set up the next call to *ADD*, and so on.

That a multi-tape TM can be replaced by an equivalent single-tape construction means that we can design a TM that possesses any number of tapes, in the knowledge that our machine could be converted into a single-tape machine. We took advantage of this, of course, in the specification of *UTM*, earlier in this chapter.

10.8 The Linguistic Implications of the Equivalence of Non-deterministic and Deterministic TMs

We began by showing that from any non-deterministic TM we could construct an equivalent four-tape deterministic TM. We then showed that from any multi-tape deterministic machine we could construct an *equivalent* deterministic TM. We have thus shown the following:

For any non-deterministic TM there is an equivalent deterministic TM.

Moreover, in Chapter 7 we showed that the non-deterministic TM is the corresponding abstract machine for the *type* 0 languages (in terms of the ability to *accept*, but not necessarily to *decide* those languages, as we shall see in Chapter 11).

We can now say:

The deterministic TM is the recogniser for the phrase structure languages (PSLs), *in general.*

This is so because:

Any phrase structure grammar (PSG) can be converted into a non-deterministic TM, which can then be converted into a deterministic TM

(via the conversion of the non-deterministic TM into an equivalent 4-tape deterministic TM, and *its* conversion into an equivalent deterministic single-tape TM).

10.9 Exercises

For exercises marked †, solutions, partial solutions, or hints to get you started appear in "Solutions to Selected Exercises" at the rear of the book.

10.1. Write a program representing a general TM simulator. Your machine should accept as its input the quintuples of a TM, and then simulate the behaviour of the TM represented by the quintuples on various input tapes. In particular, you should run your simulator for various TMs found in this book, and the TMs specified in Exercise 7.2 of Chapter 7.

Hint: the advice given for Exercise 5.6 of Chapter 5, in "Solutions to Selected Exercises" is also very relevant to this exercise.

10.2. Define a TM to generate $\{a, b, c\}^+$, as outlined earlier in the chapter. Then convince yourself that for any alphabet, A, A^+ is similarly *enumerable*, i.e. we can define a TM to print out its elements in a systematic way.

10.3[†]. Sketch out the part of the single-tape machine, S, that deals with the sextuple $(1, a, A, R, 2, 2)$ of the three-tape machine, M, described earlier in the chapter.

Notes

1. Reading related to the work referenced in this section can be found in the "Further Reading" section.
2. This assume that the "while" loop in Table 10.6 terminates if no suitable quintuple can be found.
3. A pseudo-code program was given in Chapter 2 to perform the same task.

Chapter 11
Computability, Solvability, and the Halting Problem

11.1 Overview

This chapter investigates the limitations of the Turing machines (TM). The limitations of the TM are, by Turing's thesis (introduced in the last chapter), also the limitations of the digital computer.

We begin by formalising the definition of various terminology used hitherto in this book, in particular, we consider the following concepts of computability:

- functions
- problems
- the relationship between solvability and decidability.

A key part of the chapter investigates one of the key results of computer science, the unsolvability of the *halting* problem, and its theoretical and practical implications. These implications include:

- the inability of any program to determine in general if any program will terminate given a particular assignment of input values
- the existence of decidable, semidecidable and undecidable languages.

The chapter concludes with a specification of the (Turing-) computable properties of the languages of the Chomsky hierarchy (from Part 1 of the book).

11.2 The Relationship Between Functions, Problems, Solvability, and Decidability

We have said much in the preceding chapters about the power of the TM. We spend much of the remainder of this chapter investigating its *limitations*. In Part 2 of this book there have seen many references to *functions* and *solving problems*. However, no clear definition of these terms has been given. We are now familiar enough with TMs and Turing's thesis to be able to provide these definitions.

11.2.1 Functions and Computability

A *function* associates input values with output values in such a way that for each particular arrangement of input values there is but one particular arrangement of output values. Consider the arithmetic *add* (+) operator. This is a *function* because for any pair of numbers, x and y, the result of $x + y$ is always associated with a number z, that represents the value obtained by adding x and y. In the light of observations made in the preceding two chapters, it should be appreciated that, since any TM can be represented in binary form, it suffices to define the values (input and output) of functions in terms of *numbers*. This applies equally to the language processing machines considered in this book as to those that carry out numerically oriented operations.

We know that we can represent any of our abstract machines for language *recognition* (*finite state recognisers* (FSRs), *pushdown recognisers* (PDRs)) as TMs (see Chapter 7), then represent the resulting TM in such a way that its operation is represented entirely as the manipulation of binary codes. Thus, the notion of *function* is wide-ranging, covering associations between *strings* and *truth values* ("true" if the string is a sentence of a given language, "false" if it is not), numbers and other numbers (*MULT*, *DIV*, etc.), numbers and truth values (*COMPARE*), and so on.

Calling a function an *association* between input and output values represents a very high level view of a function. We are usually also interested in *how* the association between the input and output values is made, i.e. the *rules* by which we derive a particular arrangement of output values from a particular arrangement of input values. More specifically, we are interested in characteristics of the *program* that embodies the "rules" and produces the output values when presented with the input values.

The last chapter discussed the relationship between TMs and *effective procedures*. We can now say that the latter are those procedures that are:

- deterministic
- finite in time and resources (space)
- mechanical
- represented in numeric terms.

It has been argued above that such effective procedures are precisely those that can be carried out by TMs, and, in the light of the previous chapter, we can say by one TM, which we call the *universal TM* (UTM).

Having a clearer idea of the notions "function" and "effective procedure" enables us to define what we call the *computable functions*, in Table 11.1.

So, Turing's thesis asserts that the *effectively computable functions* and the *TM-computable functions* are the same.

11.2.2 Problems and Solvability

Associated with any function is a *problem*. In abstract terms, the problem associated with a function is a *question*, which asks "what is the result of the function for

Table 11.1. "Computable functions" and "TM-computable functions"

The *computable functions* are those functions for which there is an *effective procedure* for deriving the correct output values for each and every assignment of appropriate input values. They can therefore be called the *effectively computable functions*

A *TM-computable function* is a function that can be computed by some TM

these particular input values". For the "+" function, then, an associated problem is "what is $3 + 4$?". It was suggested above that the sentence recognition task for a given language could be regarded as a function associating strings with truth values. The associated problem then asks, for a given string, "what is the truth value associated with this string in terms of the given language?". This is more usually expressed: "is this string a sentence of the language or not?".

A problem is *solvable* if the associated function is computable. We know now that "computable" is the same as saying "TM-computable". So, a problem is solvable if it can be coded in such a way that a TM could compute a solution to it in a finite number of transitions, or a program run on an unlimited storage machine could do the same.

A problem is *totally solvable* if the associated function can be computed by some TM that produces a correct result for *every* possible assignment of valid input values. An example of this is the "+" operator. We could define a TM, such as the ADD machine, that produces the correct result for any pair of input values, by adding the numbers and leaving the result on its tape. Another totally solvable problem is the *membership problem* for the context sensitive language $\{a^i b^i c^i: i \geqslant 1\}$. In Chapter 7, we defined a TM that printed T for any string in $\{a, b, c\}*$ that was a member of the language, and F for any such string which was not. The membership problem for $\{a^i b^i c^i: i \geqslant 1\}$ is thus *totally solvable*.

A problem is *partially solvable* if the associated function can be computed by some TM that produces a correct result for each possible assignment of input values for which it halts. However, for some input values, the TM will not be able to produce any result. The critical thing is that on any input values for which the TM halts, the result it produces is correct. This concept is best considered in the context of language recognition, and you may already have appreciated the similarity between the concept of *solvability* and the concepts of *deciding* and *accepting* languages discussed in Chapter 7. This similarity is not accidental, and we return to it later in the chapter. First, in the next section we consider one of the most important partially solvable problems of computer science.

A problem is *totally unsolvable* if the associated function cannot be computed by any TM. It is worth mentioning at this point that in discussing the concepts of solvability we never exceed the characteristics of effective procedures as defined above: we are still considering deterministic, finite, mechanical, numeric processes. For an abstract, simply stated, yet totally unsolvable problem consider this:

Define a TM that has as its input a non-packed tape (i.e. an arbitrary number of blank squares can appear between any two marked squares) on which there are *a*s, *b*s, and *c*s, in any order. The machine should halt and print T if the leftmost marked symbol on its tape is *a*, and should halt and print F otherwise. The read/write head is initially located on the rightmost marked tape square.

The above machine cannot be defined. Suppose we design a TM to address the above problem that moves left on its tape until it reaches an *a*, then prints T and halts. Does this mean the problem is *partially* solvable? The answer to this is "no", since partial solvability means that the TM must produce a correct result *whenever* it halts, and our "cheating" TM would not (though it would "accidentally" be correct some of the time). A TM to even partially solve the above problem cannot be designed. I leave it to you to convince yourself of this.

11.2.3 Decision Problems and Decidability

Problems that ask a question expecting a yes/no answer are sometimes known as *decision problems*. In seeking to solve a decision problem we look for the existence of a program (TM) to resolve each of a set of questions, where the formulation of each question depends on the assignment of specific values to certain parameters. Examples are:

- *given two regular grammars, R_1 and R_2, are they equivalent?*
- *given a number, N, is N prime?*
- *given a TM, M, and a tape T, will M halt eventually when started on the tape T?*

A decision problem is *totally solvable* if we can construct a TM that will provide us with the correct "yes/no" response for each and every possible instantiation of the parameters. Thus, for example, the second decision problem above is totally solvable only if we can construct a TM that decides, for any number, if that number is prime or not. We could do this if we wished, so the second problem is totally solvable.

Of the other two problems, only one is totally solvable, and this is the first. In Chapter 4, we saw what we now know can be called *effective procedures* for converting any regular grammars into non-deterministic FSRs (non-DFSRs), for converting non-DFSRs into deterministic FSRs (DFSRs), and for converting DFSRs into their minimal form. If two regular grammars are equivalent, their associated minimal machines will be exactly the same, if the states of one minimal machine are renamed to coincide with those of the other. If the states of one of the minimal machines cannot be renamed to make the two minimal machines exactly the same, then the two grammars are not equivalent. We could design a (complicated!) TM to do all of this, and so the equivalence problem for regular grammars is *totally solvable*.

11.3 The Halting Problem

Let us look again at the third of the three *decision problems* presented in the immediately preceding section:

- given a TM, *M*, and a tape *T*, will *M* halt eventually when started on the tape *T*?

This problem is known as the *halting problem*, and was briefly alluded to in the discussion of the power of the FST, at the end of Chapter 8. All of us who have written programs have experienced the practical manifestation of this problem. We run our program, and it seems to be taking an overly long time over its computation. We decide it is stuck in an "infinite loop", and interrupt the execution of the program. We have usually made some error in the design or coding of a loop termination condition. Even commercially supplied programs can sometimes suffer from this phenomenon. Some unexpected combination of values leads to a non-terminating repetition of a sequence of instructions. It would be useful if someone could design a program that examined other programs to see if they would terminate for all of their permissible input values. Such a program would form a useful part of the compiler, which could then give out an error message to the effect that a given program would not terminate for certain inputs.

As we have established, a computer can never be more powerful than a TM, so if the halting problem, as represented in TM terms, is unsolvable, then by extension it is unsolvable even by a computer with infinite memory. We see next that the halting problem is indeed unsolvable in the general sense, though it is *partially solvable*, which is where we begin the investigation.

11.3.1 *UTM$_H$* Partially Solves the Halting Problem

Consider again our *UTM*, as described in Chapter 10. *UTM* simulates the computation of any machine, *M*, on any tape *T*. From the construction of *UTM* we can see (as also pointed out in Chapter 10), that this simulation is so faithful to *M* as to enter an indefinite loop if that is what *M* would do when processing the tape in question. However, one thing is certain: if *M* would have reached its halt state on tape *T*, then *UTM* will also reach its halt state given the coded version of *M* and the coded version of *T* as *its* input. We would therefore simply have to modify the *UTM* procedure (Table 10.6) so that it outputs a "1" and halts as soon as it finds the code "00" on tape 3 (indicating that the simulated machine *M* has reached its halt state).

Let us call the new version of *UTM*, described immediately above, *UTM$_H$*. As we have defined it, *UTM$_H$*, *partially* solves the halting problem. Given any machine *M* and tape *T*, appropriately coded, *UTM$_H$* will write a 1 (on tape 2) and halt if *M* would have reached its halt state when presented with the tape *T* as its input. If *M* would have stopped in an auxiliary state, because no transition was possible, then *UTM$_H$* will not reach its halt state. We could further amend *UTM$_H$* to output a 0, then move to its halt state, if no applicable quintuple of *M*'s could be found, and therefore *M* would have halted in a non-halt state. However, if *M* would loop infinitely when given tape *T*, then, as defined, *UTM$_H$* has no alternative but to do the same.

It is clear that the critical situations for the halting problem are those in which *M* loops infinitely. If, in addition to our amendments specified above, we could show how to amend *UTM$_H$* so that it halts and writes a 0 when *M would* have looped infinitely on tape *T*, we would be showing that the halting problem is *totally* solvable.

This last amendment is difficult to imagine, since various conditions can result in infinite loops, and a "loop" in a TM could actually comprise numerous states, as a loop in a program may encompass numerous statements. At this point, then, it seems pertinent to try a different approach.

11.3.2 *Reductio ad Absurdum* Applied to the Halting Problem

Let us assume that we have already amended *UTM$_H$* so that it deals with both the halting *and* non-halting situations. This new machine, which we will simply call *H*, is specified schematically in Figure 11.1.

Figure 11.1 defines *H* as a machine that prints *Y* and *N* rather than 1 or 0. Note that the two arcs entering *H*'s halt state are shown as having the blank as the current symbol. We assume that at that point *H* has erased everything from its tape, and would thus write *Y* or *N* on a blank square before terminating, leaving the read/write head pointing at the result.

So, *H* is assumed to:

- print *Y* and halt if *M* would halt on the input tape *T* or
- print *N* and halt if *M* would not halt on the tape *T*.

In other words, *H* is assumed to *totally solve* the halting problem.

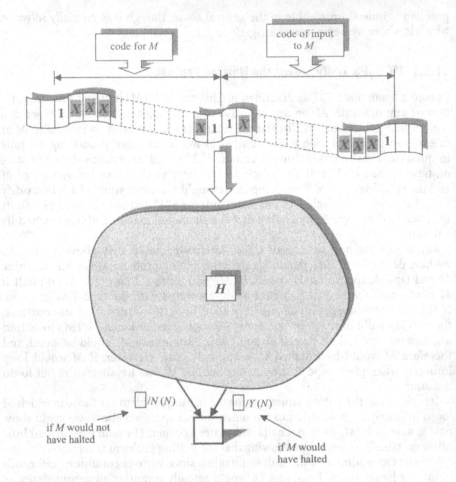

Figure 11.1. The TM, H, is assumed to solve the *halting problem*. Its input is a TM M, and an input tape to M, coded as binary numbers, as described in Chapter 10

Next, we make what appears to be a strange amendment to H, to create a new machine called H_1, depicted in Figure 11.2.

In order to produce H_1, we would need to make straightforward amendments to H's behaviour in its halting situations. H_1 would:

- print Y and halt if M would not halt on the input tape T or
- loop infinitely if M would halt on the tape T.

You should be able to see that, if H existed, it would be straightforward to make the necessary amendments to it to produce H_1.

The next thing we do probably seems even stranger: we modify H_1 to produce H_2, depicted in Figure 11.3.

H_2 copies the code for the machine M, then presents the two copies to H_1. Our scheme codes both machine and tape as sequences beginning and ending with a 1, between these 1s being simply sequences of 0s separated by 1s, so a coded machine could clearly be interpreted as a coded tape.

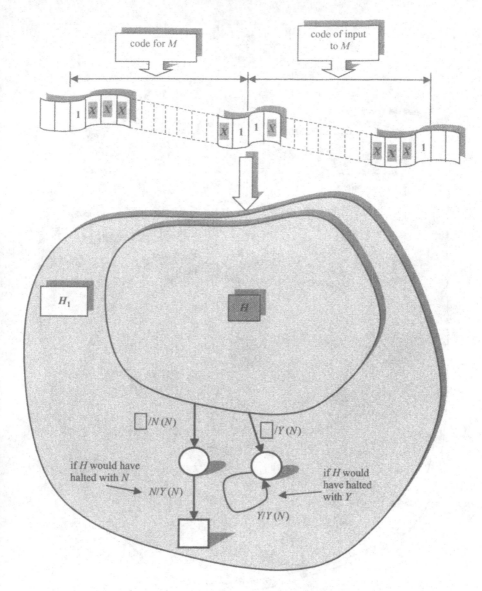

Figure 11.2. The TM, H of Figure 11.1, is amended to produce the TM H_1 that appears to solve the *"not halting"* problem

Thus, instead of simulating the behaviour of M on its input tape, H_2 now simulates the behaviour of the machine M on a description of M itself (!). H_2's result in this situation will then be to:

- print Y and halt if M would not halt on a description of itself or
- loop infinitely if M would halt on a description of itself.

The next step is the strangest of all. We present H_2 with the input specified in Figure 11.4.

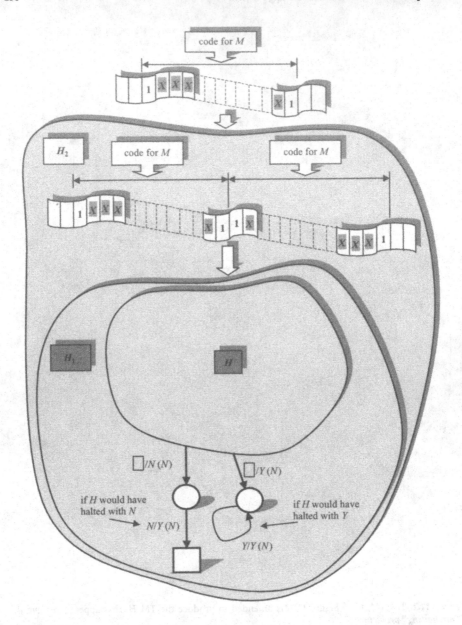

Figure 11.3. H_2 makes H_1 (of Figure 11.2) apply the machine M to the code for M

Figure 11.4 shows us that H_2 has been presented with the binary code for itself to represent the machine, M. H_2 will first copy the code out again, then present the two copies to H_1. To H_1, the left-hand copy of the code for H_2 represents a coded TM, the right-hand copy the coded input to that TM. Thus, H_2's overall effect is to simulate its own behaviour on an input tape that is a coded version of itself. H_2 uses only the alphabet $\{0, 1, Y, N\}$, as it was derived from H_1, which also uses the same alphabet. Therefore, H_2 could be applied to a description of itself in this way.

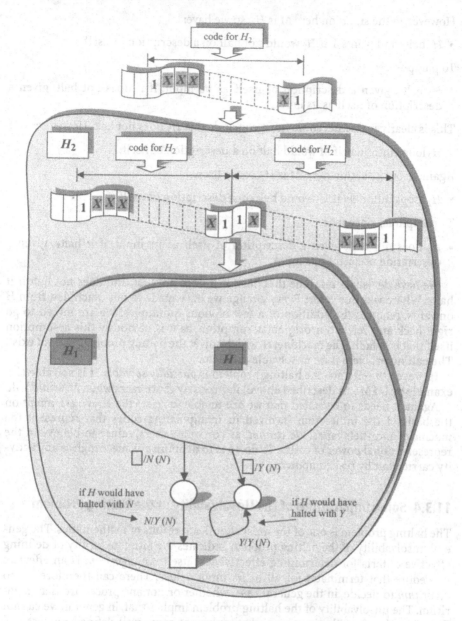

Figure 11.4. H_2, of Figure 11.3, is given its own code as input

11.3.3 The Halting Problem Shown to Be Unsolvable

Now we ask:

- Does H_2, configured according to Figure 11.4, halt?

Well, suppose H_2 *does* halt:

- H_2 halts and prints Y if M would not halt on a description of itself.

However, in the situation here M is H_2, so we have:

- H_2 halts and prints Y if H_2 would not halt on a description of itself.

To paraphrase:

- H_2 halts, given a description of itself as its input, if it does not halt, given a description of itself as its input.

This is clearly nonsense, so we must conclude that H_2 does not halt. However:

- H_2 loops infinitely if M would halt on a description of itself.

Again, in our situation here, M is H_2, so we have:

- H_2 loops infinitely if H_2 would halt on a description of itself.

To paraphrase once more:

- H_2 does not halt, given a description of itself as its input, if it halts, given a description of itself as its input.

We have devised a machine that halts if it does not halt, and does not halt if it halts. Where was our error? Every change we have made to our machines, from H onwards, requires the addition of a few obvious quintuples. We are forced to go right back and reject our original assumption, as it is obviously this assumption itself that is at fault. The machine, H, which solves the halting problem, cannot exist. The halting problem is an unsolvable problem.

However, to reiterate, the halting problem is *partially solvable*. It is solvable (for example by UTM_H, as described above) in precisely those cases when M *would* halt.

Again, it needs to be stated that we are unable to reject the above argument on the basis of the indirection involved in manipulating codes that represent the machines and their tapes. We *cannot*, as computer users, refuse to believe in the representational power of codes. To do so is to denounce as meaningless any activity carried out by real computers.

11.3.4 Some Implications of the Unsolvability of the Halting Problem

The halting problem is one of the most significant results of mathematics. The general unsolvability of the halting problem indicates the ultimate futility of defining effective criteria for determining effectiveness itself. An *algorithm* is an effective procedure that terminates for all of its input values. There can therefore be no *algorithm* to decide, in the general case, whether or not any procedure *is* an algorithm. The unsolvability of the halting problem implies that, in general, we cannot determine by algorithmic means whether or not some well-defined process will terminate for all of its values.

The halting problem also has practical implications for the enterprise of proving the correctness of programs. A program is said to be *partially* correct if whenever it terminates it produces a correct result. A program is said to be *totally* correct if it is partially correct *and* it terminates for each and every possible input value. The goal of proving programs totally correct is thus one that cannot be attained by algorithmic means, as to be able to do so would require a solution to the halting problem.

Many unsolvable problems of computer science reduce to the halting problem. That there is no computational device more powerful than the standard TM implies that effective procedures for analysing the behaviour of other effective procedures are doomed ultimately to simulating those procedures, at the expense of being caught in the never-ending loops of the simulated process, if such loops arise.

As an example of a problem that reduces to the halting problem, consider the *equivalence* problem for TMs, or for programs. Two TMs, M_1 and M_2, are equivalent if for each and every input value, they both yield the same result. Now assume that E is a TM that solves the equivalence problem for any two TMs. Suppose, for argument's sake, that E simulates the behaviour of each machine. It performs the task specified by the next applicable quintuple from M_1, followed by the next applicable quintuple from M_2, on coded representations of an equivalent input tape for each machine. Suppose that the M_1 simulation reaches its halt state first, having produced a result. E must now continue with the M_2 computation until that ends in a result. But what happens if M_2 would not terminate on the tape represented? In order to output F (i.e. the two machines are not equivalent), E would need to determine that M_2 was not going to halt. *E would need to solve the halting problem.*

The halting problem also manifests itself in many decision problems associated with programming. Some of these are:

- will a given program terminate or not? (the halting problem)
- do two programs yield the same answers for all inputs?
  (this is essentially the TM equivalence problem that is shown to be unsolvable immediately above)
- will a given program write a given value in a given memory location?
- will a given machine word be used as data or as an instruction?
- what is the shortest program equivalent to a given program?
- will a particular instruction ever be executed?

The exercises ask you to represent the above *decision problems* of programming as TM decision problems that can be proved to lack a general solution. In all cases, the method of proof is the same: we show that if the given problem was solvable, i.e. there was a TM to solve it, then that TM would, as part of its activity, have to solve the halting problem.

The proof of the halting problem includes what seems like a peculiar form of *self-reference*. A formal system of sufficient representational power is capable of being used to represent itself. If it is capable of representing itself then it can be turned in on itself, so to speak, to lead to peculiar contradictions. Turing's results are analogous to similar results in the domain of logic established by the great mathematician Kurt Gödel. Here we simplify things greatly, but essentially Gödel showed that a system of logic that is capable of generating all true statements (what is called *completeness*) can be used to generate a statement P, which, in essence refers to itself, saying "P is false".[1] Such a statement can be neither true nor false, and is therefore inconsistent (if it is true, then it is false, and vice versa). In a way, such a statement says "I am false". In a similar way, our machine H_2, when given its own code as input, says: "I halt if I do not halt, and I do not halt if I halt"!

The general unsolvability of the halting problem and Gödel's findings with respect to completeness leave computer science in a curious position. A machine powerful enough to compute everything we may ask of it is powerful enough to deal with representations of itself. This leads to *inconsistencies*. If we reduce the power of the machine so that inconsistencies cannot occur, then our machines are *incomplete*, and will not be able to compute everything that we may reasonably expect of them. We return to the relationship between computability and formal logical systems at the end of the book.

It is worth briefly discussing some wider implications of the halting problem. The inability of algorithmic processes to solve the halting problem does *not* mean that machines are *necessarily* less intelligent than humans. You may think that *we* can look at the texts of our programs and predict that this or that particular loop will not terminate, as we often do when writing programs. This does not mean we can solve the halting problem *in general*, but only in certain restricted situations. There must come a point at which the size and complexity of a program becomes such that we are forced to "execute" (for example, on paper) the program. There is no reason to disbelieve that, for the restricted cases in which we *can* determine that a program would not halt, we could also write a program (and thus construct a TM) to do the same. The general halting problem embodies a level of complexity that is beyond the capabilities of machines *and* human beings.

11.4 Computable Languages

In Chapter 7 we saw how from any phrase structure grammar (PSG) we could construct a non-deterministic TM to *accept* the language generated by that grammar. In Chapter 10 we saw how any *non-deterministic TM* could be converted into an equivalent *deterministic TM*. On this basis, we concluded that the TM is the recogniser for the phrase structure languages (PSLs) in general. The PSLs include all of types 0, 1, 2, and 3 of the *Chomsky hierarchy*. However, as the earlier chapters of this book have shown, the TM is a more powerful computational device than we actually *need* for the more restricted types (context free and regular). Nevertheless, the PSLs are *computable* languages, using the term "computable", as we do now, to mean *TM-computable*. A computable language, then, is a formal language that can be *accepted* by some TM.

A question that arises, then, is "are there *formal languages* that are not computable?" In the next section, we shall encounter a language that is not computable, in that there can be no TM to accept the language. This, in turn, means that there can be no PSG to generate that language (if there was we could simply follow the rules in Chapter 7 and construct from the grammar a TM recogniser for the language). This result will enable us to refine our hierarchy of languages and abstract machines, and fully define the relationship between languages and machines.

With respect to formal languages, the analogous concept to *solvability* is called *decidability*.

A language is *decidable* if some TM can take any string of the appropriate symbols, and determine whether *or not* that string is a sentence of the language. An alternative way of putting this is to say that the *membership problem* for that language is totally solvable.

A language is *acceptable*, which we might also call *computable*, if some TM can determine if any string that *is* a sentence of the language is indeed a sentence, but for strings that are not sentences cannot necessarily make any statement about that string at all. In this case, the membership problem for the language is *partially* solvable.

At the end of this chapter, we summarise the decidability properties of the various types of formal language we have encountered so far in this book. For now we return to the question asked above: *are there formal languages that are not computable?* If a language is non-computable, then it is not acceptable, and by Turing's thesis there is no algorithmic process that can determine, for *any* appropriate string, if that string is in the language.

11.4.1 An Unacceptable (Non-computable) Language

We now establish that there is indeed a non-computable formal language. As you will see, the "proof" of this is inextricably bound up with the halting problem, and also takes advantage of our TM coding scheme from the previous chapter.

Consider the following set X:

$$X = \{1x1: x \in \{0,1\}^+\},$$

X is the set of all strings starting and ending in 1, with an arbitrary non-empty string of 0s and 1s in between. Now each string in this set either can, or cannot, represent the code of a TM. By considering a slight adaptation of the TM (sketched out in Chapter 10) that generates all strings in $\{a, b, c\}^+$ as part of its simulation of a non-deterministic machine, we know that the strings of this set could be generated systematically.

Now consider a subset Y of X defined as follows:

$$Y = \{x: x \in X \text{ and } x \text{ is the code of a TM with alphabet } \{0, 1\}\}.$$

It would be straightforward to define a TM, or write a program that, given any string from X, told us whether or not that string was a valid TM code, i.e. of the format shown in Figure 11.5.

Note from Figure 11.5 that we are only interested in codes that represent *deterministic* machines, i.e. machines that have no more than one *quintuple* with the

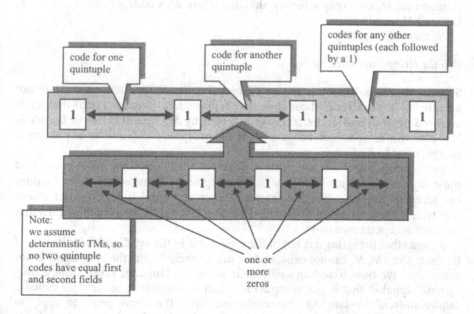

Figure 11.5. The form of a valid binary code to represent a TM, according to the coding scheme of Chapter 10

same "current state" and "current symbol". We are also interested only in binary TMs, but we have already argued that all TMs can be reduced to binary form (cf. *UTM*). This means that our "current symbol" and "write symbol" quintuple fields (i.e. the second and third fields of our quintuple representations) will not be allowed to contain more than two 0s.

The set X can be generated systematically, and we can check for each string whether or not it represents a code for a TM. Therefore, we can see that there is an *effective procedure* for generating the set Y. Note that we are only interested in whether the codes are in the correct form to represent some TM, not in what the TM does. Most of the TM codes in Y would represent machines that do nothing of real interest, or, indeed, nothing at all.

Given any code from the set Y we could recover the original machine. We can assume, for argument's sake, that for the alphabet, the code 0 represents 0, 00 represents 1, and all other codes are as we defined them in Chapter 10. If y is a string in the set Y, we will call the machine whose code is y, M_y.

Now consider the following set, Z:

$$Z = \{x : x \in Y \text{ and } M_x \text{ does not accept } x\}.$$

Every string in Y represents a coded TM with alphabet $\{0, 1\}$, and thus every string in Y can be presented as input to the machine whose code is that string. Thus, Z is the set of all binary alphabet machines that do not accept the string represented by their own code.

Now, we ask:

- is Z an *acceptable* language?

Suppose that it is. In this case, some TM, say M_z, is the acceptor for Z. We can assume that M_z uses only a binary alphabet. Thus, M_z's code will be in the set Y. Let's call M_z's code z.

We now ask,

- *is the string z accepted by M_z?*

Suppose that it is. Then z is not in Z, since Z contains only strings that are not accepted by the machine whose code they represent. It therefore seems that if z is accepted by M_z, i.e. is in the language accepted by M_z, then it is not in the set Z. However, this is nonsense, because M_z is the acceptor for Z, and so any string it accepts must be in Z.

We therefore reject our assumption in favour of its negation. We suppose that *the string z is not accepted by M_z*. M_z is the acceptor for Z, and so any string M_z does not accept is not in Z. But this cannot be correct, either! If M_z does not accept the string z, then z is one of the strings in Z, as it is the code for a machine (M_z) that does not accept its own code.

It seems that the string z is not in the set Z if it is in the set Z, and is in the set Z if it is not! The TM, M_z cannot exist, and we must conclude that the language Z is not acceptable. We have found an easily defined formal language taken from a two-symbol alphabet that is not computable. Such a language is sometimes called a *totally undecidable* language. We could generalise the above proof to apply to alphabets of any number of symbols, in effect showing that there are an infinite number of such totally undecidable languages.

11.4.2 An Acceptable, but Undecidable, Language

One problem remains to be addressed, before the next section presents an overall summary of the relationship between formal languages and automata. We would like to know if there are computable languages that are necessarily acceptable, but not decidable. For reasons that will be presented in the next section, any such languages must be *properly* type 0 in the Chomsky hierarchy, since the more restricted types of language in the hierarchy are *decidable*.

Here is a computable language that is acceptable, but not decidable. Given the sets X and Y, as specified above, we define the set W, as follows:

$$W = \{x: x \in Y \text{ and } UTM_H \text{ halts when it applies } M_x \text{ to } x\}^2.$$

The set W is the set of codes of *TM*s that halt when given their own code as input. W is an *acceptable* language. W is not a *decidable* language because a TM that could tell us which valid TM codes were not in W (i.e. represent the code for machines that do not halt when given their own code as input) would have to solve the halting problem. Languages, like W, that are acceptable but not decidable are sometimes called *semidecidable* languages.

11.5 Languages and Machines

As implied above, associated with any language is what is called the membership problem for that language. Given a grammar G and a string s, the membership problem says: "is s a sentence of the language $L(G)$?". From our discussion of problems

Table 11.2. Decidability properties for the languages of the Chomsky hierarchy

Grammar type	General decidability	Justification
Regular (type 3)	Decidable	As for context free grammars (CFGs)
Context free (type 2)	Decidable	If necessary, remove ε from grammar using the method described in Chapter 5, then continue as for context sensitive grammars
Context sensitive (type 1)	Decidable	All productions are of the form $x \to y$ where $|x| \le |y|$. Thus, each step in a derivation either increases or leaves unchanged the length of the sentential form
		Construct a non-deterministic TM to generate in parallel all sentential forms that are not greater in length than the target string s. If s is one of the generated strings, print T and halt, if it is not, print F and halt
		Convert the non-deterministic TM into an equivalent deterministic TM (see Chapter 10)
Unrestricted (type 0)	Semidecidable	Some productions may be of form $x \to y$ where $|x| > |y|$. The sentential form may thus lengthen and shorten during a derivation. There may therefore be no obvious point at which to give up if we do not manage to generate s, the target string
		However, we know from Chapter 7 that all type 0 languages are *acceptable*

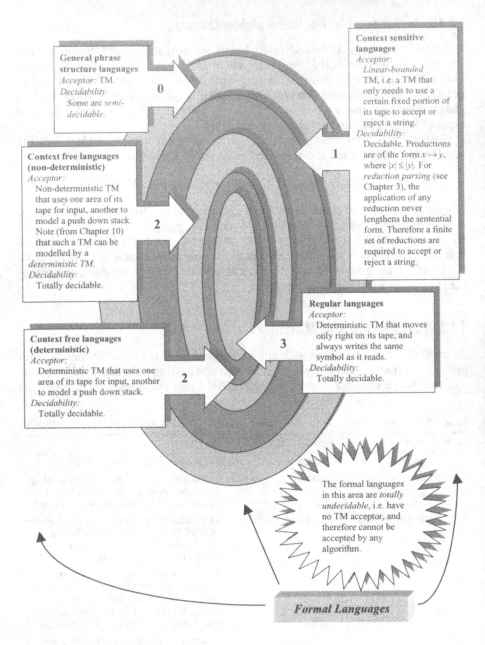

Figure 11.6. Languages and machines (the numbers denote the type of language in the *Chomsky Hierarchy* – see Chapter 2)

and solvability earlier in this chapter, it should be clear that the membership problem for a language is *solvable* if the associated membership *function* for that language is (TM-)computable.

For all the PSLs, the associated membership function is computable. All PSLs are computable (acceptable) languages. However, we have seen that some type 0 languages are *acceptable*, but not *decidable*. However, we shall see that the other types (1, 2, and 3) are all decidable.

Table 11.2 summarises the decidability properties of the *regular, context free, context sensitive*, and *unrestricted* languages. In fact, since we would like to show only that types 1, 2, and 3 are *decidable*, we can specify the same (very inefficient!) method for all three types, assuming some initial minor amendments to the type 2 and 3 grammars.

In terms of formal languages, as presented in part one of this book, we described each class of language in the Chomsky hierarchy, working from the regular languages up through the hierarchy. We simultaneously worked upwards from the FSR to the TM, by adding extra functionality to one type of abstract machine to yield a more powerful type of machine. For example, we showed how a machine that is essentially an FSR with a stack (a PDR) was more powerful than an FSR. We finally developed the TM, and showed that we need go no further. There is no more powerful device for computation or language recognition than the TM.

We might, as an alternative, have approached the subject by first presenting the TM, and then discussing how the TM actually gives us more power than we need for the more restricted languages and computations (such as *adding* and *subtracting*, which we saw being performed by FSTs in Chapter 8). Then, the abstract machines for the more restricted languages could have been presented as TMs that were restricted in some appropriate way. Figure 11.6 presents this alternative but equivalent view of computable languages and abstract machines.

11.6 Exercises

For exercises marked †, solutions, partial solutions, or hints to get you started appear in "Solutions to Selected Exercises" at the rear of the book.

11.1†. The printing problem asks if a TM, *M*, will print out a given symbol, *s*, in *M*'s alphabet, when started on a specified input tape. Prove that the printing problem is, in general, unsolvable.

   Hint: this is a problem that reduces to the halting problem, as discussed earlier.

11.2. Show that the following (all taken from this chapter) are all unsolvable problems. In each case first transform the statement of the problem so that it is couched in terms of TMs rather than programs. Then show that in order to solve the given problem, a TM would have to solve the halting problem.

   (a) will a given program write a given value in a given memory location?

   (b) will a given machine word be used as data or as an instruction?

   (c) what is the shortest program equivalent to a given program?

   (d) will a particular instruction ever be executed?

11.3. Express the totally unsolvable problem in Section 11.2.2 in terms of programs, rather than TMs.

Hint: consider a non-terminating program such as an operating system, for which the input can be regarded as a stream of characters.

Notes

1. Rather like the person who says: "Everything I say is a lie".
2. UTM_H was introduced earlier in the chapter. It halts and prints 1 when the simulated machine would have halted on the given tape, and thus *partially* solves the halting problem.

Chapter 12
Dimensions of Computation

12.1 Overview

In previous chapters, we have referred to computational power in terms of the task modelled by our computations. When we said that two processes are *equivalent*, this meant that they both modelled the same task. In this chapter, we refine our notions of computational power, by discussing its three dimensions:

function: *what* is computed

space: *how much storage* the computation requires

time: *how long* the computation takes

In particular, we introduce a simple model of time into our analysis, in which each transition of a machine is assumed to take one *moment*. This leads us to discuss key concepts relating to both abstract machines and algorithms:

- the implications of parallel processes
- predicting the running time of algorithms in terms of some measure of the "size" of their input, using big O notation.

We discover that algorithms can be classified as having

- linear
- logarithmic
- polynomial or
- exponential

running times.

A theme that runs through most of the chapter is an as yet unsolved problem of computer science. In essence, this problem asks if, in general, parallel processes can be modelled by similarly efficient serial processes.

We conclude by setting the scene for Part 3 of the book: *Computation and Logic*.

12.2 Aspects of Computation: Space, Time, and Complexity

Earlier in this book, we have seen that the limitations, in terms of computational power, of the *Turing machine* (TM) are also the limitations of the modern digital computer. We argued that, simple as it is, the TM is actually *more* powerful than *any* real computer unless we allow for an arbitrary amount of storage to be added to the real computer, as required, during the course of its activity. Even then, the computer would not be *more* powerful than the TM. However, claims about the power of the TM must be appreciated in terms of the TM being an *abstract machine*. In Chapter 8, when discussing the FST, we presented so-called *memory machines*, that "remember" an input symbol for a fixed number of state transitions before that symbol is output. In a sense, by doing this we were using an implicit model of *time*, and the basic assumption of that model was that the transition between two states of a machine took exactly one unit of this time, which we called a *moment*.

Turing's thesis, as discussed in Chapter 10, reflects a *functional* perspective on computation: it focuses on *what* the TM can perform. However, as users and programmers of computers, we are not only interested in the functions our computer can carry out. At least two other fundamental questions concern us, these being:

- how much memory and storage will our task require (*space* requirements)?
- how long will it take for our task to be completed (*time* requirements)?

In other words, the dimensions of computation that occupy us are *function, space* and *time*. It is clear then that reference to computational *power*, a term used freely in the preceding chapters, should be qualified in terms of those dimensions of computational power to which reference is being made.

Now, in terms of *space* requirements, our binary TMs of Chapter 9 seem to require an amount of memory (i.e. occupied tape squares) that does not differ significantly from that required by a computer carrying out a similar task on a similar set of data. For example, consider the descriptions of the memory accessing TMs towards the end of Chapter 9. Apart from the binary numbers being processed, the only extra symbols that our machines required were special markers to separate one number from the next. However, these markers were needed because we had implicitly allowed for variable memory "word" (or array element) length. If we had introduced a restriction on word length, as is the case in most memory schemes, we could have dispensed with the need for such markers. However, if we *had* dispensed with the markers between memory items, we would have had to "program in" the knowledge of where the boundaries between items occurred, which in terms of TMs means we would have had to introduce more states to deal with this.

The discussion above seems to imply that we can reduce the number of states in a (well-designed) TM *at the expense of introducing extra alphabet symbols*, and we can reduce the number of alphabet symbols *at the expense of introducing extra states*. In fact, this relationship has been quantified, and it has been shown that

- any TM that has more than two states can be replaced by an equivalent TM that has only two states.

More importantly, from our point of view, it has been shown by the information theorist Claude Shannon, that

- any TM that uses an alphabet of more than two symbols can be replaced by an equivalent TM that uses only two.

The fact that any TM can be replaced by an equivalent TM using only a two-symbol (i.e. *binary*) alphabet is of fundamental importance to the practical use of computers. It indicates that any shortfalls in functional power that real computers have are certainly not a consequence of the fact that they only use a two-symbol alphabet. Due to the reciprocal relationship outlined above between the number of states in a TM and the number of alphabet symbols it needs, the product of the number of states and the number of alphabet symbols is often used as a measure of the *complexity* of a TM. There are, however, other aspects of TMs that could be said to relate to their complexity. One of these aspects is non-determinism, which, as we discovered in the previous chapter, we can dispense with (in *functional* terms). However, as we see next, there are implications for the *space* and *time* dimensions of computation if we remove non-determinism from our machines.

12.3 Non-deterministic TMs Viewed as Parallel Processors

We saw earlier in this chapter that computation is concerned not only with *function*, but also with *space* and *time*. When we make statements about the *equivalence* of TMs, as we have in this chapter, we are usually talking about equivalence of *function*. For example, we say that the result of the "non-deterministic to deterministic TM conversion" of Chapter 10 is "equivalent" to the original machine. However, there are different ways of viewing the operation of a non-deterministic machine. One is to imagine that the machine attempts to follow a sequence of transitions until it finds a solution or reaches a "dead end", as none of its available instructions can be applied. In the latter case, we have assumed that it then tries an alternative sequence. When combined with a notion of *time* that is essentially a measure of the number of *transitions* made by the machine, this perspective reflects a *serial* model of computation. Our earlier discussions regarding *non-deterministic finite state recognisers* (non-DFSRs) (Chapter 4) and *pushdown recognisers* (PDRs) (Chapter 5) also reflect this view, when considering concepts such as *backtracking*, for example.

As an alternative to a serial model of execution, we might assume that all possible sequences of transitions to be considered by the non-deterministic TM are investigated in *parallel*. When the machine is ready to start, imagine that a signal arises, analogously to the way in which the internal "clock" governs the state changes of the CPU. On this signal, each and every applicable transition is carried out simultaneously (each transition being allowed to take with it its own copy of the input tape, since each sequence of transitions may result in different configurations of the tape). Figure 12.1 shows an example starting situation with three applicable transitions.

Figure 12.1 is intended to represent the situation that arises on the first "*clock* tick", which we assume results in the execution of the three applicable transitions in parallel (the result of each being shown by the dashed arrows). Note that the new tape configuration for the three cases includes an indication of which state the machine has reached after making the relevant transition. A description of the marked portion of a tape, the location of the read/write head and the current state of a machine at any stage in a TM's computation is sometimes called an *instantaneous description*.

The next clock tick would result in the *simultaneous* execution of:

- each applicable transition from state 2 given the tape configuration on the left of Figure 12.1
- each applicable transition from state 3 given the tape configuration in the centre of Figure 12.1

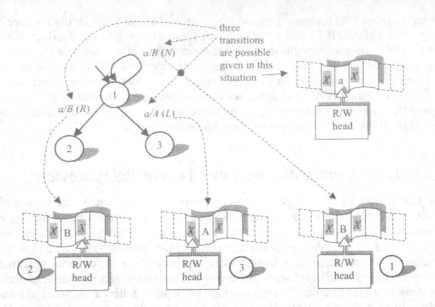

Figure 12.1. The transitions of a non-deterministic TM viewed as parallel processes

- each applicable transition from state 1 given the tape configuration on the right of Figure 12.1.

Then the next clock tick would result in the simultaneous execution of each applicable transition from each of the situations created at the previous clock tick.

Clearly, if a solution exists the above method will find it. In fact, the first sequence of transitions we find that leads us to a solution will be the shortest such sequence that can be found.[1] Now consider a deterministic four-tape version of the non-deterministic machine (as described in Chapter 10). This machine executes all possible sequences of quintuples of the non-deterministic machine, but it *always explores shorter sequences first*. Thus, the deterministic 4-tape machine is also bound to discover the shortest sequence of quintuples that will lead to a solution.

12.3.1 Parallel Computations and Time

Now our model of *time* enters the analysis. If there is a solution that requires the execution of n instructions (i.e. n *quintuples*), the parallel machine will find it in n moments. In our example above, if any applicable sequence of transitions leads directly to the halt state from any of the resulting configurations shown, our parallel machine would have taken two moments to find the solution. The first transitions would be simultaneously executed in one moment, then all of the applicable transitions from the three resulting configurations would similarly be executed in the next moment.

However, consider how long it takes our *deterministic* version to find a solution that requires the execution of n instructions. The deterministic machine will have tried:

- all single instructions
- all sequences of two instructions

⋮

- some or all of the sequences of n instructions.

Suppose the alphabet of quintuple labels (as in the example in Chapter 10), is $\{a, b, c\}$, i.e. we are talking of a three-quintuple machine. The deterministic machine generates strings of these by generating all strings of length 1, all strings of length 2, and so on. There would be 3 strings of length 1, 9 strings of length 2, and 27 strings of length 3, which are:

> $aaa, aab, aac, aba, abb, abc, aca, acb, acc, baa, bab, bac, bba, bbb, bbc, bca, bcb, bcc,$
> $caa, cab, cac, cba, cbb, cbc, cca, ccb, ccc.$

At each stage, the strings for the next stage were derived by appending each of the three symbols of the alphabet to each of the strings obtained at the previous stage. Thus, for our example here, there would be 81 (i.e. 27×3) strings of length 4, then 243 (81×3) strings of length 5, and so on.

In general, the number of strings of length n from an alphabet of k symbols is k^n (in the example above, where $n = 4$ and $k = 3$ we had 3^4, or 81, strings).

To return to our main discussion, consider a 3-quintuple non-deterministic parallel TM, N. Suppose the shortest solution that this machine could find, given a certain *input configuration*, involves four instructions. N will take four moments to find it. Now consider the deterministic version, D, constructed using our method described earlier. We will simplify things considerably by considering the execution time of D in terms of the number of quintuples of N it tries. We will assume that it takes one moment to try each quintuple. We will also ignore the fact that a large number of quintuple sequences will be illegal and thus aborted by D before the entire sequence has been applied. It is reasonable to ignore these things, as D has to *generate* each sequence of quintuple labels, even if the corresponding sequence of quintuples subsequently turns out to be inapplicable. This machine first tries:

> 3 sequences of instructions of length 1 (3 moments)
> 9 sequences of instructions of length 2 (18 moments)
> 27 of length 3 (81 moments)
> between 1 and 81 sequences of length 4 (between four and 324 moments).

Adding all these up, we get $3 + 18 + 81 + 4 = 106$ moments in the best case (i.e. when the first four-quintuple sequence tried results in a solution). In the worst case (when all of the four-quintuple sequences have to be tried, the last one resulting in the solution) we get $3 + 18 + 81 + 324 = 426$ moments. In general, given k quintuples in the non-deterministic machine, and a shortest solution requiring n instructions, the parallel TM, N, will take n moments to find the solution, while the corresponding deterministic machine, D, will take somewhere between:

> $(k \times 1) + (k^2 \times 2) + \cdots + (k^{n-1} \times n{-}1) + (k^n \times n)$
> $= k + 2k^2 + \cdots + n - 1k^{n-1} + nk^n$ moments in the worst case,

and

$$(k \times 1) + (k^2 \times 2) + \cdots + (k^{n-1} \times n-1) + n$$
$$= k + 2k^2 + \cdots + n - 1k^{n-1} + n \text{ moments in the best case.}$$

Now, as we have already seen, even simple TMs, such as those we have seen in this book, consist of many more than three quintuples. Moreover, solving even the simplest problems usually requires the application of sequences considerably more than four-quintuples long. For example, if a 10-quintuple non-deterministic parallel TM (a very simple machine) found the solution in 11 moments (a very short solution), the best the corresponding deterministic TM constructed by our method could hope to do (as predicted by the above formula) is find a solution in 109 876 543 221 moments. To consider the extent of the difference in magnitude here, let us suppose a moment is the same as one-millionth of a second (a reasonably slow speed for a modern machine to carry out an operation as simple as a TM transition). Then:

- the non-deterministic machine executes the 11 transitions in slightly over 1/100 000 of a second,

while

- its deterministic counterpart takes around 30.5 h to achieve the same solution!

The above example may seem rather contrived, but consider that a parallel algorithm is, say, updating a complex graphical image, and millions of individual operations are being carried out simultaneously. Such computations would be impossible to carry out in a reasonable time by a serial processor. We would not wish to wait 30.5 h for an image in an animation to change, for example.

12.4 A Brief Look at an Unsolved Problem of Complexity

The preceding discussion leads us to an as yet unsolved problem of computer science. You may, on reading the discussion, think that the disparity between the non-deterministic TMs and their four-tape deterministic counterparts is entirely due to the rather simple-minded conversion scheme we have applied. You may think that there is a more "efficient" way to model the behaviour of the non-deterministic machine than by attempting to execute every possible sequence of its quintuples. We have shown that for any non-deterministic TM we can construct a deterministic TM that does the same job. The question that now arises is: can we construct one that does the same job as efficiently? In order to discuss whether this question can be answered, we take a slightly more detailed look at complexity.

12.5 A Beginner's Guide to the "Big O"

As far as we are concerned here, the theory of complexity examines the relationship between the length (number of symbols) of the input to a TM and the number of instructions it executes to complete its computation. The same applies to programs, where the relationship concerned is between the number of input elements and the number of program instructions executed. In considering this relationship, we usually take account only of the components of the program that dominate the computation in terms of time expenditure.

12.5.1 Predicting the Running Time of Algorithms

Suppose we are to sum all of the elements in a two-dimensional array of numbers, then print the result, and we can only access one element of the array at a time. We *must* visit each and every element of the array to obtain each value to be added in to the running total. The time taken by the program is *dominated* by this process. If the array is "square", say, $n \times n$, then we are going to need n^2 accessing operations. The times taken by the other operations (for example, adding the element to the running total in the loop, and printing out the result) we can ignore, as they do not change significantly whatever the size of n.

In overall terms, our array summing program can be said to take an amount of time approximately proportional to $n^2 + c$. Even if we had to compute the array subscripts by adding in an offset value each time, this would not change this situation (the computation of the subscripts would be subsumed into c). Obviously, since c is a constant measure of time, the amount of time our algorithm takes is dominated by n^2 (for large values of n). n^2 is called the *dominant expression* in the equation $n^2 + c$. In this case, then, we say our array processing program takes time *in the order of* n^2. This is usually written, using what is called *big O* – "Oh", not nought or zero – notation, as $O(n^2)$. Similarly, summing the elements of a one-dimensional array of length n would be expected to take time $O(n)$. There would be other operations, but they would be assumed to take constant time, and so n would be the dominant expression. In a similar way, summing up the elements of a three-dimensional $n \times n \times n$ array could be assumed to take time $O(n^3)$.

In estimating the running time of algorithms, as we have said above, we ignore everything except for the dominant expression. Table 12.1 shows several running time expressions and their big O equivalents.

Finally, if it can be demonstrated that an algorithm will execute in the least possible running time for a particular size of input, we describe it as an *optimal* algorithm.

Table 12.1. Some expressions and their "big O" equivalents. In the table, $c_1 \ldots c_j$ are constants, n is some measure of the size of the input

Expression	Big O equivalent	Type of running time
c	$O(1)$	Constant
$c_1 + c_2 + \cdots + c_j$	$O(1)$	Constant
n	$O(n)$	Linear
cn	$O(n)$	Linear
n^2	$O(n^2)$	Polynomial
cn^2	$O(n^2)$	Polynomial
$c_1 n^{c_i}$	$O(n^{c_i})$	Polynomial
c^n	$O(c^n)$	Exponential
$n + c_1 + c_2 + \cdots + c_j$	$O(n)$	Linear
$n^c + n^{c-1} + n^{c-2} + \cdots + n$	$O(n^c)$	Polynomial (if $c > 1$)
$(1/c_1)n^{c_i}$	$O(n^{c_i})$	Polynomial (if $c_2 > 1$)
$(n^2 + n)/2$	$O(n^2)$	Polynomial
$(n^2 - n)/2$	$O(n^2)$	Polynomial
$\log_2 n$	$O(\log_2 n)$	Logarithmic
$n \times \log_2 n$	$O(n \times \log_2 n)$	Logarithmic

A large part of computer science has been concerned with the development of optimal algorithms for various tasks. The following discussion briefly addresses some of the key issues in algorithm analysis. We discuss the main classes of algorithmic running times, these being *linear, logarithmic, polynomial,* and *exponential.*

12.5.2 Linear Time

If the running time of an algorithm is described by an expression of the form $n + c$, where n is a measure of the size of the input, we say the running time is $O(n)$ – "of the order of n" – and we describe it as a *linear time* algorithm. The "size" of the input depends on the type of input with which we are dealing. The "size" of a string, would be its *length*. The "size" of an array, list or sequence would be the number of elements it contains.

As an example of a linear time algorithm, consider the sequential search of an array of n integers, sorted in ascending order. Table 12.2 shows a Pascal-like program to achieve this task. As stated earlier, the components assumed to require constant running time are ignored. The key part of the algorithm for the purposes of running time estimation is indicated in the table.

If you are alert, you may have noticed that the algorithm in Table 12.2 does not always search the whole of the array (it gives up if it finds the number or it becomes obvious that the number is not there). For many algorithms, a single estimate of the running time is insufficient, since properties of the input other than its size can determine their performance. It is therefore sometimes useful to talk of the *best, average,* and *worst* case running times. The best, worst, and average cases for the algorithm in Table 12.2, and their associated running times, are shown in Figure 12.2. You will note that though the average case involves $n/2$ searches, as far as big O is concerned, this is $O(n)$. We ignore coefficients, and $n/2 = 1/2n$, so the "1/2" is ignored.

Since an abstract machine can embody an algorithm, it is reasonable, given that model of time is based on the number of transitions that are made, to talk of the efficiency of the *machine*. A deterministic FSR (DFSR), for example, embodies an $O(n)$ algorithm for accepting or rejecting a string, if n is the length of the string.

Table 12.2. *Linear search* of a one-dimensional array, *a*, of *n* integers, indexed from 1 to *n*, sorted in ascending order

{target *is an integer variable that holds the element to be found*}	
done := false	
foundindex := 0	{*If* target *is found,* foundindex *will hold its index number. If not found,* foundindex *will remain at* 0}
i := 1	
while i ⩽ n and not(done) do	{*the number of times this loop iterates, as a function of n, the size of the input, is the key element in the algorithm's running time*}
if target = a[i] then	{*found*}
done := true	
foundindex := i	
else	
if target < a[i] then	{*we are not going to find it*}
done:= true	
endif	
endif	
i := i + 1	
endwhile	

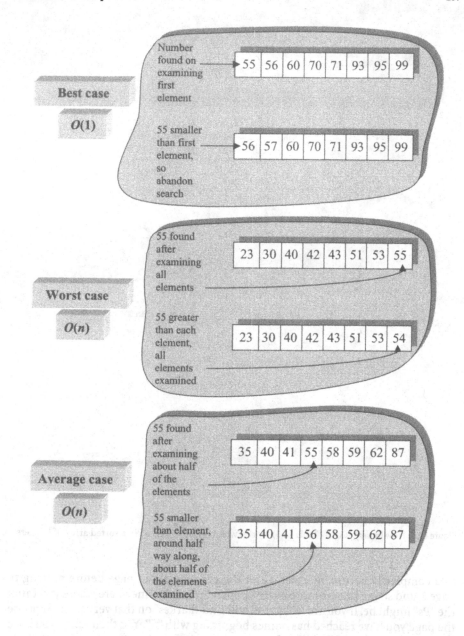

Figure 12.2. The best, worst and average cases for linear search of a sorted array of integers, when the number being searched for is 55

12.5.3 Logarithmic Time

Consider again the algorithm of Table 12.1. As stated earlier, it performs a linear search of a sorted array. To appreciate why this algorithm is not optimal, consider a telephone directory as being analogous to a sorted array. If I asked you to look up "Parkes" in the directory, you are hardly likely to begin at the first entry on page 1

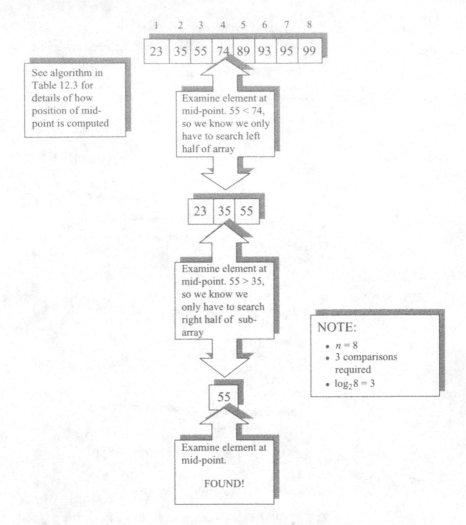

Figure 12.3. How binary search (Table 12.3) works, when trying to find 55 in a sorted array of integers

and completely peruse, in order, all of the entries on that page before moving to page 2, and so on. You are more likely to open the book somewhere where you think the "Ps" might be. If you are lucky, you will find "Parkes" on that very page. Suppose the page you have reached has names beginning with "S". You then know you have gone too far, so you open the book at some point before the current page, and so on. The process you are following is a very informal version of what is known as *binary search*. We consider this in more detail now.

Binary search searches for a given item within a sorted list (such as an array), as does linear search, discussed above. However, it is much more efficient. Figure 12.3 schematically represents the operation of binary search on a particular example. Binary search is represented algorithmically in Table 12.3.

So what is the running time of binary search? As hinted in Figure 12.3, it turns out to be $O(\log_2 n)$, i.e. the running time is proportional to the binary logarithm of

Table 12.3. *Binary search* of a one-dimensional array, *a*, of *n* integers, indexed from 1 to *n*, sorted in ascending order

{target *is an integer variable that holds the element to be found*}	
low := 1	
high := n	{low *and* high *are used to hold the lower and upper indexes of the portion of the array currently being searched*}
foundindex : = 0	{*If* target *is found,* foundindex *will hold its index number. If not found,* foundindex *will remain at* 0}
done := false	
while low ≤ high and not(done) do	
mid := (low + high) div 2	{*Calculate mid-point of current portion of array* (div *is integer division*)}
if target = a[mid] then	{*found*}
foundindex : = mid	
done := true	
else	
if target < a[mid] then	{*need to search left half of current portion*}
high := mid−1	
else	
low := mid + 1	{*need to search right half of current portion*}
endif	
endif	
endwhile	

the number of input items. Many students have a problem in grasping logarithmic running times, so we will spend a little time considering why this is so.

Firstly, for those who have forgotten (or who never knew) what logarithms are, let us have a brief revision (or first time) session on "logs". Some of you may be used to base 10 logarithms, so we will start there. If x is a number, then $\log_{10} x$ is the number y, such that $10^y = x$. For example, $\log_{10} 100 = 2$, since $10^2 = 100$. Similarly, $\log_{10} 1000 = 3$, since $10^3 = 1000$. We can get a reasonable estimate of the $\log_{10}$ of a number by repeatedly dividing the number by 10 until we reach 1 or less. The number of divisions we carry out will be the estimate of the $\log_{10}$ value. Take 1000, for example:

$$1000/10 = 100 - \text{step } 1$$
$$100/10 = 10 - \text{step } 2$$
$$10/10 = 1 - \text{step } 3.$$

We require three steps, so $\log_{10} 1000$ is approximately (in fact it is *exactly*) 3. Let us consider a less convenient example. This time we will try to estimate $\log_{10} 523$:

$$523/10 = 52.3 - \text{step } 1$$
$$52.3/10 = 5.23 - \text{step } 2$$
$$5.23/10 = 0.523 - \text{step } 3.$$

Here, the result of the final step is *less than* 1. This tells us that $\log_{10} 523$ is more than 2 but less than 3 (it is in fact 2.718501689, or thereabouts!). That this approach is a gross simplification does not concern us here; for the purposes of algorithm analysis it is perfectly adequate (and as we shall shortly see, very useful in analysing the running time of binary search).

The same observations apply to logarithms in any number base. If x is a number, $\log_2 x$ is the number y, such that $2^y = x$. For example, $\log_2 64 = 6$, as $2^6 = 64$. We can use the same technique as we used for $\log_{10}$ to estimate the $\log_2$ of a number (but this time we divide by 2). Let us consider $\log_2 10$:

$10/2 = 5$ – step 1

$5/2 = 2.5$ – step 2

$2.5/2 = 1.25$ – step 3

$1.25/2 = 0.625$ – step 4.

So we know that $\log_2 10$ is more than 3 but less than 4 (it is actually around 3.322).

It turns out that our method for estimating the $\log_2$ of a number is very relevant in analysing the running time of binary search. As you can see from Figure 12.3, binary search actually reduces by half the number of items to be searched at each stage. Assume the worst case, which is that the item is found as the last examination of an element in the array is made, or the searched for item is not in the array at all. Then the number of items examined by binary search, beginning with an array of n items will closely follow our method for estimating the $\log_2$ of n:

examine an item, split array in half [$n/2$ – step 1]

examine an item, split half array in half [$(n/2)/2$ – step 2]

$\vdots$

… until only an array of one element remains to be searched. Hence, the number of items in an n-element array that need to be examined by binary search is around $\log_2 n$ in the worst case. All of the other operations in the algorithm are assumed to be constant and, as usual, can be ignored. The running time of binary search is thus $O(\log_2 n)$.

In general, any algorithm that effectively halves the size of the input each time it loops will feature a logarithm in the running time. The importance of logarithmic running time is that it represents a great saving, even over a linear time algorithm. Consider the difference in performance between linear search, discussed above, and binary search if the number of items to be searched is 1 000 000. In cases tending towards the worst, linear search may have to examine nearly all of the 1 000 000 items, while binary search would need to examine only around 20.

Binary search is justifiably a famous algorithm. Also among the better known of the logarithmic algorithms is a sorting algorithm known as *quicksort*. Quicksort sorts an array of integers, using the approach schematically represented for an example array in Figure 12.4. As can be seen, the algorithm works by placing one item (the *pivot*) into its correct position, then sorting the two smaller arrays on either side of the pivot using the same (quicksort) procedure. However, quicksort is not just a $\log_2 n$ algorithm, but actually an $O(n \times \log_2 n)$ algorithm, as hinted in Figure 12.4. As we shall see shortly, many established sorting algorithms are $O(n^2)$, and thus $n \times \log_2 n$ is a considerable improvement. An n^2 sorting algorithm would require about 225 000 000 units of time to sort 15 000 items, while quicksort could achieve the same job in 210 000 time units.

12.5.4 Polynomial Time

Programs that take times such as $O(n^2)$ and $O(n^3)$, where n is some measure of the size of the input, are said to take *polynomial time* in terms of the input data.

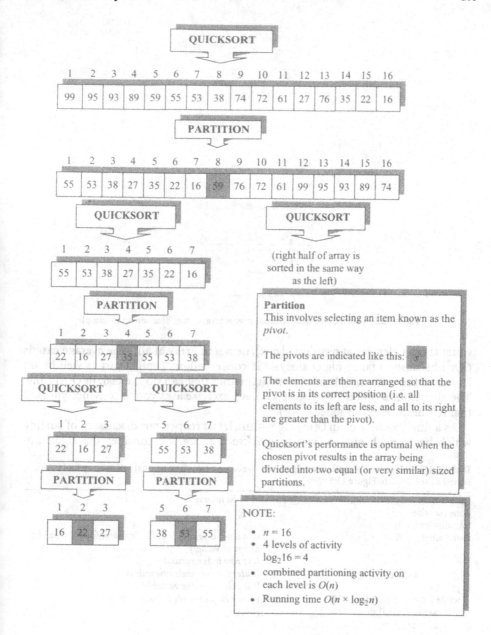

Figure 12.4. How *Quicksort* sorts an array of 16 integers into ascending order

Consider an algorithm related to our linear search, above. This time, the algorithm searches for an element in a *two*-dimensional ($n \times n$) array, sorted as specified in Figure 12.5. This can be searched in an obvious way by the program in Table 12.4. This algorithm has a running time $O(n^2)$ for both the average (target element somewhere around position $a[n/2, n/2]$ – or search can be abandoned at this point) and

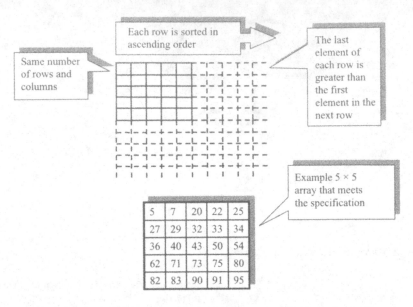

Figure 12.5. A sorted two-dimensional $n \times n$ array – specification and example

worst (target element at position $a[n, n]$, or not in array at all). The average case is $O(n^2)$ because as far as big O analysis is concerned, $n^2/2$ might just as well be n^2. However, there is a simple but clever algorithm that can do the task in $O(n)$ time. The algorithm is outlined in Figure 12.6. An exercise asks you to convince yourself that its running time is indeed $O(n)$.

As a final example of an $O(n^2)$ algorithm, let us reopen our discussion of sorting which began with our analysis of *quicksort*, above. We consider the algorithm

Table 12.4. The obvious, but inefficient, algorithm to search a two-dimensional $n \times n$ array, a of integers, sorted as specified in Figure 12.5

```
{target is an integer variable that holds the element to be found}
done := false
foundindexi := 0
foundindexj := 0                           {If target is found, foundindexi and foundindexj hold
                                            its index values}
i := 1                                      {i is row index variable}
while i ≤ n and not(done) do                {outer loop controls row index}
  j := 1                                    {j is column index variable}
  while j ≤ n and not(done) do              {inner loop controls column index}
    if target = a[i, j] then                {found}
      done := true
      foundindexi := i
      foundindexj := j
    else
      if target < a[i, j] then              {we are not going to find it}
        done := true
      endif
    endif
    j := j + 1
  endwhile
  i := i + 1
endwhile
```

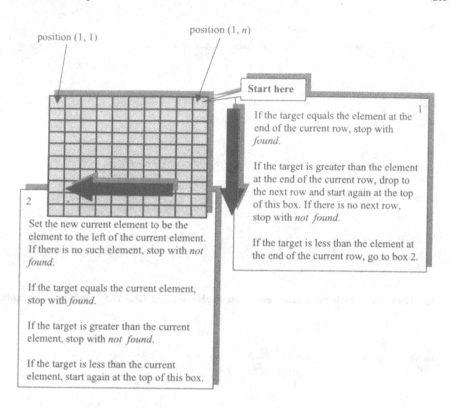

Figure 12.6. A sketch of an $O(n)$ algorithm to search for an element in an $n \times n$ array that meets the specification in Figure 12.5

known as *bubble* (or sometimes *exchange*) sort. This algorithm is interesting partly because it has a best case running time of $O(n)$, though its worst case is $O(n^2)$. The bubble sort algorithm is shown in Table 12.5. Figure 12.7 shows a worst case for the algorithm based on a six-element array. The algorithm gets its name from the way the smaller elements "bubble" their way to the "top".

Considering the bubble sort algorithm (Table 12.5) and the trace of the worst case (Figure 12.7), it is perhaps not obvious that the algorithm is $O(n^2)$. The outer loop iterates $n - 1$ times, as pointer i moves up the array from the second to the last element. However, for each value of i, the second pointer, j, moves from position n down to position i. The number of values taken on by j then, is not simply n^2, but rather $(n - 1) + (n - 2) + \cdots + 1$. Let us write this the other way round, i.e.

$$1 + \cdots + (n - 2) + (n - 1).$$

Now, there is an equation that tells us that

$$1 + 2 + \cdots + n = (n^2 + n)/2.$$

You may like to verify this general truth for yourself. However in our case, we have a sum up to $n - 1$, rather than n. For $n - 1$, the equation should be expressed:

$$((n - 1)^2 + (n - 1))/2 = (n^2 + 1 - 2n + n - 1)/2$$
$$= (n^2 - n)/2.$$

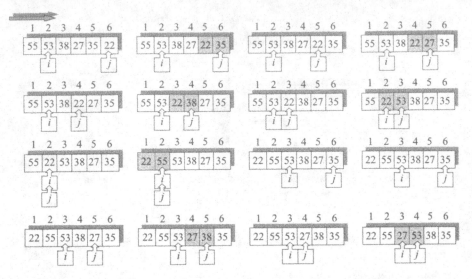

Figure 12.7a. An example worst case scenario for *Bubble sort* (see Table 12.5), resulting in a running time $O(n^2)$.

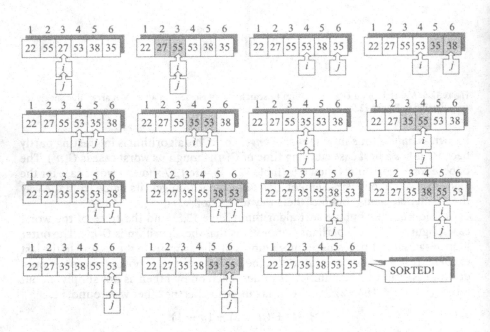

Figure 12.7b. Bubble sort continues from Figure 12.7a

This tells us that, in the worst case for bubble sort, the inner loop goes through $(n^2 - n)/2$ iterations. $(n^2 - n)/2 = 1/2n^2 - 1/2n$. Ignoring the coefficients and the smaller term, this reduces to a big O of $O(n^2)$.

An exercise asks you to define the best case scenario for bubble sort, and to argue that in the best case it has a running time $O(n)$.

Table 12.5. The *bubble* (or *exchange*) *sort* procedure that sorts an array, *a* of *n* integers in ascending order. The operation of this algorithm is shown in Figure 12.7

```
i := 2
sorted := false
while i ≤ n and not(sorted) do
        sorted := true                    {if sorted is still true at the end of the outer loop, the array is
                                           sorted and the loop terminates}
        for j := n downto i do            {downto makes the counter j go down from n to i in steps of 1}
          if a[j] < a[j−1] then
            sorted := false
            {the following three statements swap the elements at positions j and j−1}
            temp := a[j]
            a[j] := a[j−1]
            a[j−1] := temp
          endif
        endfor
     i := i + 1
endwhile
```

We complete the discussion of polynomial time algorithms with a look at an $O(n^3)$ algorithm. The algorithm we consider is known as Warshall's algorithm. The algorithm applies to *directed graphs* (an abstract machine is one form of directed graph). The algorithm computes what is known as the *transitive* closure of a directed graph. It takes advantage of the fact that if there is an arc from one node, *A*, to another, *B*, and an arc from node *B* to node *C*, there is a path from node *A* to *B*, and so on. We shall apply Warshall's algorithm to the problem of seeing if all non-terminals in a regular grammar feature in the derivation of a terminal string. To do this we will use the following grammar, *G*:

$$S \rightarrow aB \mid aC$$
$$B \rightarrow cB \mid cD$$
$$C \rightarrow cC \mid fD \mid d$$
$$D \rightarrow dB \mid dD.$$

We represent *G* using techniques from Chapter 4, as the FSR in Figure 12.8. We then represent this as a structure known as an *adjacency matrix*, shown in Table 12.6. The matrix is indexed by node names. The existence of an arc from any node, *X*, to

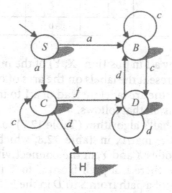

Figure 12.8. A *finite state recogniser* (see Chapter 4)

Table 12.6. An *adjacency matrix*, M, for the finite state recogniser in Figure 12.8, indexed by state names. If M(X, Y) = 1, there is an arc from X to Y in the represented machine

	S	B	C	D	H
S	0	1	1	0	0
B	0	1	0	1	0
C	0	0	1	1	1
D	0	1	0	1	0
H	0	0	0	0	0

Table 12.7. Warshall's algorithm for producing a *connectivity matrix* from an n (n adjacency matrix, M, representing a directed graph. The connectivity matrix tells us which nodes in the graph can be reached from each of the nodes

```
for k := 1 to n do
  for i := 1 to n do
    for j := 1 to n do
      if M[i, j] = 0 then
        if M[i, k] = 1 and M[k, j] = 1 then
          M[i, j] = 1
        endif
      endif
    endfor
  endfor
endfor
```

Table 12.8. The *connectivity matrix*, C, derived by applying Warshall's algorithm (Table 12.7) to the adjacency matrix of Table 12.6. If C(X, Y) = 1, there is a path (i.e. series of one or more arcs) from X to Y in the represented machine (Figure 12.8). Note that the matrix tells us that the machine cannot reach its halt state from states B or D

	S	B	C	D	H
S	0	1	1	1	1
B	0	1	0	1	0
C	0	1	1	1	1
D	0	1	0	1	0
H	0	0	0	0	0

a node Y is indicated by a 1 in position (X, Y) of the matrix. Note that the matrix we are using does not represent the labels on the arcs of our machine. It could do, if we wished (we simply use the arc label instead of a 1 to indicate an arc), but we do not need this in the discussion that follows.

We now apply Warshall's algorithm (Table 12.7) to the matrix in Table 12.6. The algorithm results in the matrix in Table 12.8, which is known as a *connectivity matrix*. For any two nodes X and Y, in the connectivity matrix, if position (X, Y) is marked with a 1, then there is a path from X to Y in the represented graph. For example, the existence of a path from S to D in the FSR (Figure 12.8) is indicated by the 1 in position (S, D) of the matrix. Note that there is no corresponding path from D to S, hence 0 is the entry in position (D, S).

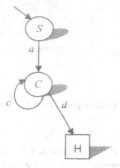

Figure 12.9. The finite state recogniser of Figure 12.8 with its useless states and associated arcs omitted

Warshall's algorithm is clearly $O(n^3)$, where n is the number of states in the FSR, since it contains an outer n-loop, encompassing another n-loop which, in turn, encompasses a further n-loop.

Let us return to our problem. We are interested in any state (apart from H) from which there is not a path to the halt state (such states are useless, since they cannot feature in the acceptance of any strings). In Table 12.8, the row for any such state will have a 0 in the column corresponding to the halt state. In Table 12.8, these states are B and D. They serve no purpose and can be removed from the FSR along with all of their ingoing arcs. This results in the FSR shown in Figure 12.9. The corresponding productions can, of course, be removed from the original grammar, leaving:

$$S \rightarrow aC$$
$$C \rightarrow cC \mid d.$$

The purpose of the above has been to illustrate how an algorithm that runs in time $O(n^3)$ can help us solve a problem that is relevant to Part 1 of this book. There are other algorithms that can achieve the task of removing what are known as *useless* productions from regular and context free grammars (CFGs). We will not consider them here. There are books in the Further Reading Section that discuss such algorithms. There is an exercise based on the approach above at the end of this chapter.

12.5.5 Exponential Time

Now, if the dominant expression in an equation denoting the running time of a program is $O(x^n)$, where $x > 1$ and, again, n is some measure of the size of the input, the amount of execution time rises dramatically as the input length increases. Programs like this are said to take *exponential time* in their execution.

Let us briefly consider the *subset algorithm* (Table 4.6 of Chapter 4) that removes non-determinism from a finite state recogniser, M, leaving a deterministic version of the same machine, M^d. The deterministic machine is such that each of its states represents a distinct subset of the states of M. In fact, in the worst case, we could have to create a new state for every single subset of set of states of M. Consider the FSR in Figure 12.10, which represents a worst case scenario for the subset algorithm. If you apply the subset algorithm to this machine (as you are asked to do in an exercise at the end of this chapter), you should produce eight states. There are eight subsets of the set $\{1, 2, 3\}$ (including the empty set).

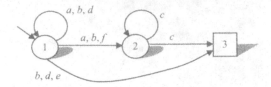

Figure 12.10. A finite state recogniser that represents a worst case situation for the *subset algorithm* of Chapter 4

In general, for a set of n elements, there are 2^n subsets. Thus, in the worst case, the running time of the subset algorithm is, *at the very least*, $O(2^n)$, where n is the number of states in the original non-deterministic machine. This is an algorithm that can have an *exponential running time*. For example, a worst case FSR with 30 states might require the algorithm to create 1 073 741 824 new states for the deterministic machine. A machine that can count up to 1 000 000 in a second would take around 18 min to count up to 1 073 741 824. However, consider a worst case machine with 75 states. Our counting machine would take over 1 197 000 000 *years* to count up to the number (2^{75}) of states that the new FSR might require!

We now return, as promised, to the example of the four-tape deterministic version of the non-deterministic TM from earlier in this chapter. Simplifying somewhat, we can regard this as taking time $O(nk^n)$, where k is the number of quintuples in the non-deterministic machine, in order to find a solution of length n. We saw the striking difference between the time taken by the non-deterministic parallel machine, which was $O(n)$, i.e. of *linear* relationship to the length of the solution, and that taken by the deterministic version, for even small values of n (3 and 11), and small values of k (3 and 10).

We have considered the difference in the time taken by two machines (a non-deterministic TM and its four-tape deterministic counterpart) in terms of the length of a given solution. However, the disparity between the time taken by the two machines becomes even greater if we also consider the relationship between the length of the *input* and the length of the solution. For example, suppose the non-deterministic machine solves a problem that has minimum length solution n^3, where n is the length of the input. Then the non-deterministic machine will find that solution in time $O(n^3)$. Suppose $n = 11$, then, simplifying things grossly, our deterministic machine will find a solution in the best case in a time $O(k^{n^{-1}})$, which represents a time of 10^{1330} "moments". Remember, we indicated above that 2^{75} is a pretty big number. The point being made here is that when the non-deterministic machine requires *polynomial* time to find the shortest solution, the deterministic machine constructed from it by our method requires *exponential* time.

12.6 The Implications of Exponential Time Processes

If you find *big O* a little daunting, think about it in this way: for our hypothetical problem, our non-deterministic machine, N, has 11 squares marked on its input tape. Since $n = 11$, and we know that the shortest solution is n^3 instructions in length, N is thus searching for a solution of 11^3 instructions. If N is a parallel machine, as described above, it will find the solution in 11^3 moments.

Now consider the behaviour of D, the deterministic version of N. To get to the first solution of length 11^3, D will have to first generate all sequences of quintuple labels of length $11^3 - 1$. If there are 10 quintuples, there are $10^{11^3 - 1}$ distinct sequences of labels of length $11^3 - 1$, i.e. 10^{1330}.

Let us now consider a simpler example, so that we can appreciate the size of the numbers that we are dealing with here. Suppose the shortest sequence of instructions that a particular 10-quintuple non-deterministic machine can find is 101 instructions, for an input of 10 symbols. Before finding the 101 quintuple sequence that yields the solution, the deterministic version will generate all sequences of quintuples of length 100. There are 10^{100} of these sequences, since there are 10 quintuples. The time taken by our deterministic machine is around 10^{100} moments, by the most gross simplification imaginable (for example, even if we assume that the generation of each sequence takes only a moment, and ignore that our machine will have generated all the sequences of length 99, of length 98, and so on).

At this point we need to buy a faster deterministic machine. We go to the TM store, and purchase a machine that does 10^{20} operations in a *year*. This machine does in excess of *three million million operations each second*. If we used it to find the solution to the problem above that took our original deterministic "one-million-instructions-a-second" model 30.5 h to find, it would take our new machine about 3/100 of a second! However, our new machine would still take in the order of 10^{80} *years* to solve our new problem, while our non-deterministic machine, at a similar instruction execution rate, would take $1/30\,000\,000\,000$ of a *second*. Given that, by some estimations, the Earth is at this time around 400 million years old, if we had started off our deterministic machine at the time the Earth was born, it would now be about

$$\frac{1}{250\,000\,000\,000\,000\,000\,000\,000\,000\,000\,000\,000\,000\,000\,000\,000\,000\,000\,000\,000\,000\,000\,000\,000}$$

of the way through the time it needs to find the solution.

12.7 Is P Equal to NP?

From a purely formal point of view, the amount of time taken to solve a problem is not important. The requirements are that a solution be found in a *finite* number of steps, or, put another way in a *finite* period of time. Now, a time period like 10^{80} years may be to us an incredibly long period of time, but nevertheless it is finite. The purpose of the construction we used to create our deterministic machine was to show functional *equivalence*, i.e. that whenever there is a solution that the non-deterministic machine *could* find, the deterministic machine *will* find it (eventually!).

In a more practical vein, in creating a deterministic machine that models one that is non-deterministic, we may also wish to maintain invariance between the two machines in the other dimensions of computational power (*space* and *time*) as well as invariance in *functionality*. Given a non-deterministic TM that finds solutions in polynomial time, we would ideally like to replace it by a deterministic machine that also takes polynomial time (and not *exponential* time, as is the case for the construction method we have used in this chapter). Clearly, this can be done in many cases. For example, it would be possible to model a search program (of the types discussed earlier) deterministically by a polynomial time TM.

The above discussion leads to a question of formal computer science that is as yet unanswered. The question is:

> *For each polynomial time non-deterministic TM can we construct an equivalent polynomial time deterministic TM?*

This is known as the *"Is P Equal to NP?"* problem, where *P* is the set of all polynomial time TMs, and *NP* is the set of all non-deterministic polynomial time TMs. As yet, no solution to it has been found. Given our analogy between non-deterministic TMs and parallel processing, the lack of a solution to this problem indicates the possibility that there are parallel processes that run in *polynomial* time, that can only be modelled by serial processes that run in *exponential* time. More generally, it suggests that there may be some problems that can be solved in a reasonable amount of time only if solved non-deterministically. True non-determinism involves *random* execution of applicable instructions, rather than the systematic execution processes considered above. This means that a truly non-deterministic process is not guaranteed to find a solution even if there is one.

There may therefore be problems that are such that either,

a solution *may* be found in *polynomial* time by a non-deterministic program (but it *may not* actually find a solution even if one exists) or

a solution *will* be found in *exponential* time by a deterministic program, if a solution exists, but it may take a prohibitive amount of time to discover it.

12.8 Observations on the Efficiency of Algorithms

We discovered earlier in this part of the book that there are well-defined processes that are deceptively simple in formulation and yet cannot be solved by algorithmic means. In this chapter, we have also discovered that certain processes, despite being *algorithmic* in formulation, may, in certain circumstances be *practically* useless. The caveat "in certain circumstances" is a very important part of the preceding sentence. Algorithms such as the subset algorithm can be very effective, for relatively small sizes of input. In any case, the real value of the subset algorithm is in theoretical terms; its existence demonstrates a property of finite state recognisers. In that sense, discussion of its running time is meaningless. We *know* that it will (eventually) produce a solution, *for every possible assignment of valid input values.* Hence, it represents a general truth, and not a practical solution. The same applies to many of the other algorithms that have supported the arguments in Parts 1 and 2 of this book.

However, as programmers, we know that efficiency is an important consideration in the design of our programs. An optimal algorithm is desirable. Analysis of the running time of algorithms using the big O approach is a useful tool in predicting the efficiency of the algorithm that our program embodies. This chapter has endeavoured to show certain ways in which our knowledge of formal computer science can support and enrich the enterprise of programming. This has hopefully demonstrated, at least in part, the fruitful relationship between the worlds of abstraction and practicality.

12.9 End of Part 2

This brings us to the end of Part 2 of the book. In Part 3, we consider formal logical systems, and discover how to carry out reasoning in such systems. In terms of this book, much of what we will see is a reformulation of various aspects of what has come before. The foundation of computer science, like any other science, is logic.

Throughout this book we have implicitly used the systems of logic to convince ourselves that things are true. We have exploited the principle of *reduction ad absurdum* on several occasions, for example. Without logic, all argument relating to formal computer science would be mere conjecture. As Parts 1 and 2 of this book have demonstrated, languages and computation are inextricably linked. However, formal logical systems *are* languages, and reasoning in formal logical systems *is* computation. It is the aim of Part 3 of this book to investigate this.

12.10 Exercises

For exercises marked †, solutions, partial solutions, or hints to get you started appear in "Solutions to Selected Exercises" at the rear of the book.

12.1[†]. Justify that the average and worst case running times for the algorithm specified in Figure 12.6 are both $O(n)$.

12.2. Identify the best case scenario for the bubble sort algorithm (Table 12.5). Convince yourself that in such a case, the algorithm has a running time $O(n)$.

12.3. In connection with the above application of Warshall's algorithm to the regular grammar problem:

  (a) The solution assumes an adjacency matrix representation of the equivalent finite state recogniser. Sketch out an algorithm to produce an adjacency matrix from a regular grammar, and analyse its running time.

  (b) Sketch out the part of the solution that involves searching the final matrix to detect which states do not lie on any path to the halt state, and analyse its running time.

  Note: see the comments for part (c) in the Sample solutions section.

  (c)[†] The approach being considered here is not guaranteed to identify all useless states. Identify why this is so, and suggest a solution.

  Hint: our method suggests detecting all states which do not lie on a path to a halt state. However, there may be states from which the halt state can be reached that are nevertheless useless.

12.4[†]. Apply the subset algorithm of Chapter 4 (Table 4.6) to the FSR in Figure 12.10.

12.5. Design an algorithm to carry out the *partitioning* of an array as required by the *quicksort* sort routine (see earlier in chapter and Figure 12.4). A simple way to choose the *pivot* is to use the first element in the array. The running time of your algorithm should be $O(n)$. Convince yourself that this is the case.

Notes

1. The shortest solution sequence may not be unique. In this case, our parallel machine will find all of the shortest solutions simultaneously. It will simplify subsequent discussions if we talk as if there is only one shortest solution.

Computation and Logic

Part 3

Computation and Logic

Chapter 13
Boolean Logic and Propositional Logic

13.1 Overview

This chapter presents two basic types of logic: *Boolean* and *propositional*.
For Boolean logic, we consider:

- the Boolean operators and Boolean expressions
- how *truth tables* can be used to prove statements in Boolean logic
- the role of Boolean logic in computing.

We then turn to propositional logic, and consider:

- propositions and their meaning
- the meanings of *logical implication* and *equivalence*
- the rules of reasoning and equivalence in propositional logic
- problem solving in propositional logic, using rules of inference and truth tables.

We end by discussing the shortcomings of propositional logic if viewed as a tool for modelling general reasoning.

13.2 Boolean Logic

Boolean logic is so named because it was formulated in the 1800s by George Boole, an English-born mathematician and Professor of mathematics at Cork University in Ireland. As we shall see, Boolean logic is a fundamental part of computer science; it is represented in the very foundations of digital circuitry and is a component of the more advanced logics we consider later.

In this section, we first consider the basic operators of Boolean logic, then we see how to form Boolean *expressions*, featuring Boolean *variables*. We see how we can use *truth tables* to compute the truth value of a Boolean expression. Finally, we look briefly at the role of Boolean logic in computer science.

13.2.1 Boolean Logic Operators

Boolean expressions evaluate to either 1 (which represents *true*), or 0 (representing *false*). The same applies to any variables that feature in Boolean expressions.

Table 13.1. The basic Boolean operators (*A* and *B* are Boolean *variables* or an arbitrary Boolean *expression*)

Expression	Operator name	Meaning	Alternative
$A . B$	and	1 if $A = B = 1$, 0 otherwise	$AND, \&, \wedge$
$A + B$	or	1 if $A = 1$ and/or $B = 1$, 0 otherwise	$OR, |, \vee$
$\sim A$	not	0 if $A = 1$, 1 if $A = 0$	$NOT, NEG,$

Table 13.2. The truth table for the *and* operator (.)

A	B	$A.B$
0	0	0
0	1	0
1	0	0
1	1	1

Table 13.3. The truth table for the *or* operator (+)

A	B	$A + B$
0	0	0
0	1	1
1	0	1
1	1	1

Table 13.4. The truth table for the *not* operator ($\sim$)

A	$\sim A$
0	1
1	0

We now consider the basic operators of Boolean logic. Suppose that *A* and *B* are Boolean variables. Table 13.1 describes the basic operators. There are many different ways of writing the basic operators. For example, the "*and*" operator is sometimes written as an ampersand (&). Table 13.1 shows some of the variations you may encounter.

Any Boolean expression contains a finite number of variables, each of which can be either 0 or 1. Thus, the entire range of truth values for an expression can be written out in tabular form. The truth tables for the basic operators from Table 13.1 are given in Tables 13.2 (*and*), 13.3 (*or*) and 13.4 (*not*).

We now consider a simple example of the use of Boolean logic as a tool for problem solving. The example is instructive, since it suggests one of the most common applications of Boolean logic, i.e. the modelling of digital circuitry. We return to this point later.

13.2.2 Boolean Logic for Problem Solving

In this section, we consider how to use Boolean logic and truth tables to represent and solve a simple problem. The problem is as follows:

A device functions according to four signals, *A*, *B*, *C*, and *D*. For the device to function correctly, one or both of signals *A* and *B* must be off, and either signal *C* must be on or signal *D* must be off or both signal *C* must be on and signal *D* must be off. An engineer's task is to repeatedly measure the signals entering the device and report immediately if the measurements taken indicate that the device is not working.

Table 13.5. How the engineer's problem (see text) is expressed as a Boolean expression

Problem statement	Boolean version
... one or both of signals A and B must be off ...	$\sim(A \cdot B)$ ...
... and either signal C must be on or signal D must be off or both signal C must be on and signal D must be off	.... $(C + \sim D)$

Full statement:
$\sim(A \cdot B).(C + \sim D)$

We now represent the above as a Boolean expression. Table 13.5 shows the parts of the above statement and the resulting Boolean expression for each part. The final expression describing the constraints on the inputs entering the device is:

$$\sim(A \cdot B) \cdot (C + \sim D).$$

According to the problem statement, the engineer must repeatedly measure the ingoing signals, and be able to detect immediately if the combination of signals taken indicates that the device is not working. Thus, if we provide the engineer with the truth table for the above expression, he or she can simply look up the truth value associated with the current observed combination of signals. The truth table for the expression is shown in Table 13.6. Note that as there are four signals $(A - D)$, and each signal can be either on or off, there are 2^4, i.e. 16, possible combinations of signals.

Note also how in Table 13.6 we make it easier to derive the truth table by breaking the statement down into component parts (called X and Y in the table), writing down the truth values for those, then using those values in deriving the values for the whole expression.

Equipped with the truth value in Table 13.6, our engineer can readily state if a particular combination of signals indicates that the device is functioning or not.

Table 13.6. The truth table for the expression $\sim(A \cdot B) \cdot (C + \sim D)$

A	B	C	D	(X) $A \cdot B$	$\sim X$	(Y) $C + \sim D$	(Result) $\sim X \cdot Y$
0	0	0	0	0	1	1	1
0	0	0	1	0	1	0	0
0	0	1	0	0	1	1	1
0	0	1	1	0	1	1	1
0	1	0	0	0	1	1	1
0	1	0	1	0	1	0	0
0	1	1	0	0	1	1	1
0	1	1	1	0	1	1	1
1	0	0	0	0	1	1	1
1	0	0	1	0	1	0	0
1	0	1	0	0	1	1	1
1	0	1	1	0	1	1	1
1	1	0	0	1	0	1	0
1	1	0	1	1	0	0	0
1	1	1	0	1	0	1	0
1	1	1	1	1	0	1	0

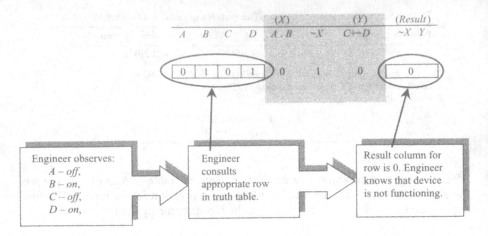

Figure 13.1. The engineer's observation (the row is from the truth table in Table 13.6)

For example, if the engineer observes the following combination:

$$A - off,$$
$$B - on,$$
$$C - off,$$
$$D - on,$$

he or she can immediately tell that the device will not be functioning (there is a 0 in the final column of the truth table), as indicated in Figure 13.1.

13.3 Boolean Logic and Computing

One of the main applications of Boolean logic in computing, and one that it hinted by our engineer example, above, is in the actual hardware of the digital computer. Boolean operators are represented as tiny components that, on receiving the appropriate number of electronic signals as input, produce a signal as output that represents the result of the Boolean operator applied to the input. Connecting together these so-called logic *gates* creates the logic circuitry of the digital computer.

Figure 13.2 shows how the basic Boolean operators for *or, and,* and *not* are represented as logic gates. Figure 13.3 shows how the expression for the engineer's circuit, i.e. $\sim(A.B).(C+\sim D)$, is represented using logic gates.

In reality, digital computer circuitry is often fashioned using what are known as *universal* logic gates. The most common of these are *NAND* gates (short for "Not AND") and *NOR* gates "Not OR". Figure 13.4 shows the graphical representation of *NAND* and *NOR* gates. The useful thing about such universal gates is that entire logic circuits can be built using only one type of gate. Here, we show only how the three basic Boolean operators *or, and,* and *not* can be modelled by *NAND* gates (Figures 13.5–13.7). Tables 13.7–13.9 show the equivalence of the circuits from Figures 13.5–13.7, and their respective basic operators. That we can model the basic operators in this way is sufficient to demonstrate that any circuit using these operators can be represented using the *NAND* gate. An exercise asks you to establish analogous results with respect to the *NOR* gate.

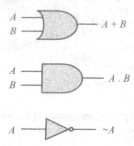

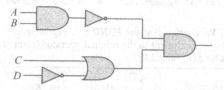

Figure 13.2. The logic gates representing the basic Boolean functions, *or* (+), *and* (.), and *not* (~)

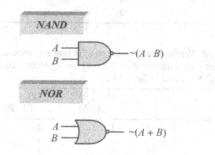

Figure 13.3. A logic gate representation of the engineer's circuit $\sim(A.B).(C+\sim D)$

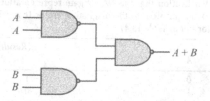

Figure 13.4. The *NAND* and *NOR* logic gates

Figure 13.5. The *NAND* gate representation of the Boolean *or* operator (+)

There are other well-established applications of Boolean logic that we will not discuss in detail in this book. One of these applications is the use of Boolean expressions in programming languages. Many of the algorithms earlier in this book feature such expressions in conditional statements such as "if" and "while". Yet another application can be found in information retrieval systems, such as

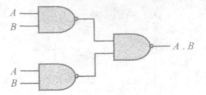

Figure 13.6. The *NAND* gate representation of the Boolean *and* operator (.)

Figure 13.7. The *NAND* gate representation of the Boolean *not* operator ($\sim$)

Table 13.7. Verification that the *NAND* gate representation of the *or* operator (+) is equivalent to the original operator (compare with the truth table for *or* in Table 13.3)

A	B	(X) $\sim(A.A)$	(Y) $\sim(B.B)$	(Result) $\sim(X.Y)$
0	0	1	1	0
0	1	1	0	1
1	0	0	1	1
1	1	0	0	1

Table 13.8. Verification that the *NAND* gate representation of the *and* operator (.) is equivalent to the original operator (compare with the truth table for *and* in Table 13.2)

A	B	(X) $\sim(A.B)$	(Y) $\sim(A.B)$	(Result) $\sim(X.Y)$
0	0	1	1	0
0	1	1	1	0
1	0	1	1	0
1	1	0	0	1

Table 13.9. Verification that the *NAND* gate representation of the *not* operator ($\sim$) is equivalent to the original operator (compare with the truth table for *not* in Table 13.4)

A	(Result) $\sim(A.A)$
0	1
1	0

world wide web search engines, into which one can enter a query such as:

```
("Boolean" AND "logic") AND NOT("information
retrieval" OR "IR")
```

to mean "retrieve all documents in which the terms 'Boolean' and 'logic', but not 'information retrieval' or 'IR'", appear. Boolean search has been a feature of information retrieval systems for many years, long before the emergence of the web.

Our encounter with Boolean logic in this book has been brief. We have not considered such topics as the manipulation of Boolean expressions, i.e. the application of the rules of Boolean *algebra*. Our purpose here has been to introduce the basic Boolean operators, as they can be found in the two logical systems that are considered in the remainder of this book. The more powerful of these, the *first order predicate logic* (FOPL), is the subject of Chapters 14 and 15. The other logical system, the *propositional logic*, is considered in the rest of this chapter.

13.4 Propositional Logic

Propositional logic, as its name suggests, deals with *propositions*. A proposition is a statement that is either true or false. For example, the preceding sentence is a proposition. So is the sentence preceding this one. And the sentence that is four sentences before this one. And so on!

In this section, we consider propositional logic as a system for representation, reasoning and problem solving. We consider rules of inference for propositional logic. Finally, we consider the shortcomings of propositional logic as a general tool for reasoning. The following two chapters are then devoted to the study of a logical system that overcomes the limitations of propositional logic.

13.4.1 Propositions

As stated earlier, a proposition is a statement that is either true or false. The following are examples of propositions:

> 6 *is a prime number*,
>
> 19*th June* 2000 *was a Saturday*,
>
> *My wife is called Anna.*

The first two of the above three propositions are false, and the last is true (if uttered by me, or anyone else who happens to have a wife called Anna). Propositions such as those above are called *atomic* propositions, since they are treated as a single entity and not subdivided in any way.

We can form complex propositions (sometimes called propositional *sentences*) by connecting together atomic propositions using *logical connectives*, including those we encountered in our investigation of Boolean logic above. However, in propositional logic we usually use the following alternative symbols (see Table 13.1):

> *and* is represented as ∧,
>
> *or* is represented as ∨,
>
> *negation* ("*not*") is the same (∼).

Thus, the following is a true proposition:

> (6 *is a prime number* ∧ 19*th June* 2000 *was a Saturday*) ∨ *My wife is called Anna.*

It is true because the "*and*" part is bracketed, meaning that it is evaluated before the "*or*". Note that to call such a statement true we have placed an *interpretation* on what the atomic propositions mean. However, we have only done this in order to assign a truth value to each atomic proposition. Thus, deciding that the whole complex proposition is true could simply be done by looking in the result column for the row in the Boolean truth table for the proposition in which

6 *is a prime number* = 0,
19*th June* 2000 *was a Saturday* = 0,
My wife is called Anna = 1.

However, we do not usually use propositional logic in this way. Later, we consider the rules of inference for propositional logic. These rules enable us to derive additional true statements from statements we already know to be true.

Note that, in addition to the above three operators, classical propositional logic also permits the use of the "implies" operator, which is written "→". This is discussed in the next section.

13.4.2 Implication and Equivalence

Consider the relationship between signals C and D of our Boolean logic problem earlier in this chapter:

$$C \vee \sim D \quad (\textit{note: the } \vee \textit{ connective has been used instead of "+").}$$

Since C and D are independent, we could write:

$$\sim D \vee C.$$

We see later that the above transformation is valid for any propositions "ored" together.

For the statement $\sim D \vee C$ to be true, either D must be false, or C must be true, or both. Another way of expressing this (you may have to think about this a little to convince yourself) is to say "if D is true then C is true". This captures the fact that D is true means that $\sim D$ is false, which further means that C must be true for the statement $\sim D \vee C$ to be true.

The statement "if p is true then q is true", for any pair of propositions, p, q, can also be expressed as p *implies* q. In propositional logic, p *implies* q is written:

$$p \rightarrow q.$$

An intuitive example will now be given to clarify the meaning of the implication operator.

Suppose we have the two propositions:

it is Sunday,
it is my day off work.

Let us call the former p, and the latter, q. Then if we write $p \rightarrow q$, we are asserting that *if* it is Sunday *then* it is my day off work. When $p \rightarrow q$ we say that p is a *sufficient condition* for q. Consider our example. The day being Sunday is a sufficient condition for it being my day off work. Also, when $p \rightarrow q$ we say that q is a *necessary condition* for p. By this we mean that if q is false then p *must also be false*. For our example, this means that if it is not my day off then it is not Sunday, which is clearly true if the original implication is true. Note the difference between *sufficient* and *necessary* conditions. I could have days off on other days apart from Sunday.

Table 13.10. The truth table for the
implication operator ($\rightarrow$). Note that $p \rightarrow q$
(*p implies q*) is the same as $\sim p \vee q$

p	q	$p \rightarrow q$
0	0	1
0	1	1
1	0	0
1	1	1

$p \rightarrow q$ does not rule this out (if it did, then the statement would mean that I only ever have Sundays as days off).

On the basis of the example discussed immediately above we can derive the *truth table* for propositions of the form $p \rightarrow q$. First, suppose that it is not Sunday. Then whether or not it is my day off is immaterial. In other words, $p \rightarrow q$ is true if p is false. Now suppose that it is Sunday. If I am at work, i.e. it is not my day off, then $p \rightarrow q$ is false. But if I am not at work, then it is true. These observations lead to the truth table for statements of the form $p \rightarrow q$ shown in Table 13.10. It can be seen that this is exactly the same as the truth table that would be constructed for $\sim p \vee q$, which accords with the discussion at the start of this section.

Above, we found that, for a statement $p \rightarrow q$, p is a *sufficient* condition for q, while q is a *necessary* condition for p. Now consider the two propositions,

it is noon,

it is 1200 h.

Letting $p =$ *it is noon*, and $q =$ *it is 1200 h*, it is certainly true to say that $p \rightarrow q$, since if it is noon it is 1200 h. However, it is equally true to say $q \rightarrow p$, since if it is 1200 h it is noon. Each proposition is a necessary condition of the other. A moment's thought will satisfy you that the same cannot be said of the propositions in our "Sunday" example, above. In other words we could write:

$$p \rightarrow q \wedge q \rightarrow p.$$

For any two propositions p, q, if the above is the case we say that the two propositions are *equivalent*. The equivalence operator is sometimes written "$\leftrightarrow$", to represent "implies and is implied by", while the symbol "$\equiv$" is also frequently used. Yet another form used is the string "iff", which means "if and only if". The truth table for $p \rightarrow q \wedge q \rightarrow p$ (and thus for $p \leftrightarrow q$) can be seen in Table 13.11. You will note that the expression is true only when p and q are both true, or both false.

We have considered logical implication in some detail here. It has been my experience that students often find it difficult to fully grasp its meaning. However, if any

Table 13.11. The truth table for the *equivalence* operator. Note that $p \leftrightarrow q$ (*p is equivalent to q*) is the same as $p \rightarrow q \wedge q \rightarrow p$

p	q	(X) $p \rightarrow q$	(Y) $q \rightarrow p$	(X ∧ Y) $p \leftrightarrow q$
0	0	1	1	1
0	1	1	0	0
1	0	0	1	0
1	1	1	1	1

doubt remains, it is always worth remembering that $p \rightarrow q$ *is equivalent to* $\sim p \vee q$ (i.e. $p \rightarrow q \leftrightarrow \sim p \vee q$). Many find the $\sim p \vee q$ form easier to comprehend.

Equivalence plays a fundamental role in logical reasoning, for if two statements are equivalent, we can simply replace one by the other. For example, we now know that if we have a statement of the form $p \rightarrow q$, we can replace it by $\sim p \vee q$, or vice versa. Equivalences are an important feature of many of the rules for reasoning in propositional logic. We turn to these rules next.

13.4.3 Rules of Inference

Propositional logic is a tool for representation and reasoning. In order to carry out reasoning, we require rules that enable us to transform logical statements while preserving their truth value, and rules to derive further true statements on the basis of statements we know to be, or have established as being, true. We now consider such rules, before turning to their application in problem solving.

The rules that we will use are presented in Table 13.12. The rules of equivalence (rules 1–8) can be verified by constructing truth tables for each A and its corresponding B. Rules 1–3 are trivial and will not be discussed further.

Rules 5 and 6 are called de Morgan's laws, after Augustus de Morgan, an Indian-born English Mathematician and contemporary of George Boole. We have already encountered one of de Morgan's laws, in the context of languages (sets) in Chapter 6 (Figure 6.7). You may also have observed that we implicitly used de Morgan's laws when we used *NAND* and *NOR* gates in Boolean logic, earlier in this chapter.

Intuitively, one can appreciate the truth of rules 5 and 6. If it is false that p and q are both true (A side of rule 5), then it must be the case that either or both of them are false. Similarly, for rule 6, if it is true that it is false that one or both of p and q is true, then it is certainly true that p and q are both false. The justification of rules 7 and 8 is left to the exercises.

We now turn to the rules of inference of Table 13.12. These differ from the rules of equivalence in that they enable us to detach part of a propositional expression and assert that part as an independent fact. Rule 11 tells us that if we have the

Table 13.12. Equivalence rules and rules of inference for propositional logic. In the following, p, q, and r are *any* propositional statements. A and B are *equivalent*, and thus interchangeable

Rule no.	A	B	Comment
1	$\sim \sim p$	p	
2	$p \wedge q$	$q \wedge p$	
3	$p \vee q$	$q \vee p$	
4	$p \rightarrow q$	$\sim p \vee q$	
5	$\sim(p \wedge q)$	$\sim p \vee \sim q$	de Morgan's law
6	$\sim(p \vee q)$	$\sim p \wedge \sim q$	Same as above
7	$p \wedge (q \vee r)$	$(p \wedge q) \vee (p \wedge r)$	Distribution of $\wedge$
8	$p \vee (q \wedge r)$	$(p \vee q) \wedge (p \vee r)$	Distribution of $\vee$

Rules 11–13 are *rules of inference*. p and q are any propositional statements

Rule no.	Given	Assert	Comment
11	$p \rightarrow q, p$	q	*modus ponens*
12	$p \rightarrow q, \sim q$	$\sim p$	*modus tollens*
13	$p \wedge q$	p (or q)	

statement $p \rightarrow q$, and we know p to be true, we can assert q. This rule models much of the reasoning we carry out in day to day life. To consider our Sunday example from earlier in this chapter, we have essentially the following implication:

if it is Sunday then I have a day off work.

It is rule 11 that we are applying when we wake up on a Sunday, and think "it is Sunday, therefore I have a day off work". In this case, we have, $p \rightarrow q$ (the statement above), and we observe that p is true, so we independently assert q ("I have a day off work") as a true statement. Rule 11 is called *modus ponens*, which is Latin and literally means *method of affirming*. This rule features in most systems of logical reasoning.

Rule 11 uses the fact that in $p \rightarrow q$, p is a *sufficient* condition for q (as was discussed above). Rule 12, on the other hand, uses that fact, also discussed above, that in $p \rightarrow q$, q is a *necessary* condition for p. If q turns out to be false, we can assert that p is false. An example of an intuitive application of the rule in everyday life (related to our Sunday example), might be if we discovered that we were scheduled to work on the 13th April, and then said, "well, the 13th of April is not a Sunday, then". The Latin phrase *modus tollens*, i.e. *method of denying*, is used to refer to this rule.

Rule 13 is easy to understand. If $p \wedge q$ is a true statement, then obviously both p and q are true so we can assert either of them as true statements. The converse is also true. If two separate propositions, p and q are known to be true, we can assert $p \wedge q$.

It should be appreciated that, as stated in Table 13.12, in the rules the letters p, q, etc. stand for propositional *statements*, not necessarily single atomic propositions. For example, rule 5 could be applied to the propositional statement

$$\sim((((a \rightarrow b) \vee c) \wedge (d \vee e)),$$

to yield

$$\sim((a \rightarrow b) \vee c) \vee \sim(d \vee e).$$

Then rule 5 could be applied again to the $\sim(d \vee e)$ part, yielding

$$\sim((a \rightarrow b) \vee c) \vee (\sim d \wedge \sim e),$$

to which rule 4 could be applied, giving:

$$((a \rightarrow b) \vee c) \rightarrow (\sim d \wedge \sim e),$$

and so on.

Having convinced ourselves of the validity of the rules, we now turn to their application in reasoning.

13.4.4 Problem Solving and Reasoning in Propositional Logic

In this section, we briefly consider the application of the propositional logic rules of inference in Table 13.12 in problem solving. Consider the following:

if it rains or I oversleep, I do not walk to work, and if I do not oversleep I have breakfast. Given that today I did not have breakfast, show that I did not walk to work.

The problem representation for the above is shown in Table 13.13. Table 13.14 shows the rule applied, and the corresponding result, at each stage of the solution.

Table 13.13. How to represent and solve a problem in propositional logic

Step	Description	Results
1	Represent each atomic statement by a unique propositional name: *it rains* *I oversleep* *I walk to work* *I have breakfast*	r s w b
2	Use the propositional connectives and the basic propositions from step 1 to represent the given facts about the domain: *if it rains or I oversleep, I do not walk to work, and if I do not oversleep I have breakfast. Given that today I did not have breakfast…*	$((r \vee s) \to {\sim}w) \wedge ({\sim}s \to b) \wedge {\sim}b$
3	Represent the statement to be proved as a propositional statement: *… I did not walk to work*	${\sim}w$
4	Use the rules and equivalences from Table 13.12 to derive the proposition of step 3 from the proposition at step 2	See Table 13.14

Table 13.14. Solving the problem of Table 13.13, using the rules and equivalences from Table 13.12. The statement to be proved is ${\sim}w$. Unless otherwise indicated, the rule indicated is being applied to the statement at the current step

Step	Statements	Rule applied
1	$((r \vee s) \to {\sim}w) \wedge ({\sim}s \to b) \wedge {\sim}b$	13
2	$((r \vee s) \to {\sim}w) \wedge ({\sim}s \to b)$	13
3	${\sim}s \to b$	13 (to 1)
4	${\sim}b$	13 (to 1)
5	$(r \vee s) \to {\sim}w$	4
6	${\sim}(r \vee s) \vee {\sim}w$	6
7	$({\sim}r \wedge {\sim}s) \vee {\sim}w$	3
8	${\sim}w \vee ({\sim}r \wedge {\sim}s)$	8
9	$({\sim}w \vee {\sim}r) \wedge ({\sim}w \vee {\sim}s)$	13
10	${\sim}w \vee {\sim}s$	3
11	${\sim}s \vee {\sim}w$	4
12	$s \to {\sim}w$	12 (to 3 and 4)
13	s	11 (to 12 and 13)
14	${\sim}w$	

Two points should be noted at this stage. Firstly, you will observe that at most stages of the proof there are many applicable rules. This implies that proving things in propositional logic can be computationally expensive. Secondly, in carrying out a proof we never remove anything from our collection of facts. Everything we prove by the application of the rules becomes an additional fact in our representation. We return to these points in the following two chapters.

13.4.5 Using Truth Tables to Prove Things in Propositional Logic

An alternative to using the rules of inference to solve a problem in propositional logic is to use truth tables. Referring back to our oversleeping problem, and in particular the representations in Tables 13.13 and 13.14, we are trying to prove that *given*

$$((r \vee s) \to \sim w) \wedge (\sim s \to b) \wedge \sim b,$$

then

$$\sim w$$

is true. Another way of saying it is that we wish to prove that *if* $((r \vee s) \to \sim w) \wedge (\sim s \to b) \wedge \sim b$ *is true, then so is* $\sim w$. As we know, we express this as:

$$(((r \vee s) \to \sim w) \wedge (\sim s \to b) \wedge \sim b) \to \sim w.$$

To prove our statement, we thus draw a truth table for it.

Now, if indeed our statement is true, the *result* column of the corresponding truth table will consist entirely of 1s. In other words, the overall statement is true whatever the assignments of truth values to its atomic propositions. We call such a statement a *theorem*.

The truth table for our oversleeping problem is shown in Table 13.15. You will note that it is indeed the case that the result column is entirely made up of 1s.

To take a slightly simpler example, let us consider rule 11 of Table 13.12 (*modus ponens*). We wish to prove that if $p \to q$ is true, and p is true, then so is q, i.e. we want to prove:

$$((p \to q) \wedge p) \to q.$$

The corresponding truth table is shown in Table 13.16. You can see that *modus ponens* is indeed a theorem of propositional logic. The useful thing about a theorem is that it is universally applicable, i.e. can be used in any propositional representation. For example, the rule of *modus ponens*, if correctly applied to any true statements in a propositional representation, will always yield an additional true statement.

Table 13.15. A truth table to solve the problem of Tables 13.13 and 13.14. Note how the "Result" column consists entirely of 1s

r	s	w	b	$r \vee s$	(X) $r \vee s \to \sim w$	(Y) $\sim s \to b$	(Z) $\sim b$	(Q) $\sim w$	(P) $X \wedge Y \wedge Z$	(Result) $P \to Q$
0	0	0	0	0	1	0	1	1	0	1
0	0	0	1	0	1	1	0	1	0	1
0	0	1	0	0	1	0	1	0	0	1
0	0	1	1	0	1	1	0	0	0	1
0	1	0	0	1	1	1	1	1	1	1
0	1	0	1	1	1	1	0	1	0	1
0	1	1	0	1	0	1	1	0	0	1
0	1	1	1	1	0	1	0	0	0	1
1	0	0	0	1	1	0	1	1	0	1
1	0	0	1	1	1	1	0	1	0	1
1	0	1	0	1	0	0	1	0	0	1
1	0	1	1	1	0	1	0	0	0	1
1	1	0	0	1	1	1	1	1	1	1
1	1	0	1	1	1	1	0	1	0	1
1	1	1	0	1	0	1	1	0	0	1
1	1	1	1	1	0	1	0	0	0	1

Table 13.16. Using a truth table to prove that the rule of *modus ponens* is a theorem of propositional logic

p	q	(X) $p \to q$	(Y) $X \wedge p$	$(Result)$ $Y \to q$
0	0	1	0	1
0	1	1	0	1
1	0	0	0	1
1	1	1	1	1

The exercises ask you to carry out a similar process for rule 12 of Table 13.12 (the rule of *modus tollens*) and also some theorems that we have not included in Table 13.12.

13.5 Observations on Propositional Logic

Propositional logic is a powerful formal system, that enables us to represent mathematical and more general properties, and then reason about them. If our initial representation is logically consistent, every statement we derive from it, as a result of applying rules such as those in Table 13.12, will be a true statement.

We have seen how to solve problems in propositional logic by using both rules of inference (Tables 13.13 and 13.14) and truth tables (Tables 13.15 and 13.16). In some senses, the former represents a *linguistic* approach, the latter a *computational* approach.

The application of the rules of inference is similar to the application of *production rules* that we considered earlier in the book. As such, it suffers from the same problems as that of *parsing*. At each stage, there are numerous applicable rules, and while the solution itself (expressed as a sequence of rules which yields the statement to be proved), once discovered, may be relatively short, the *process* of discovering that solution may be very expensive, in computational terms.

Consider our oversleeping problem from earlier. The solution (Table 13.14) can effectively be regarded as involving 13 steps (if we assume that a *step* is the application of one rule of inference). Let us assume, for argument's sake, that at each stage, four rules were applicable. Revisiting discussions from Part 2 of the book, our solution was taken from a collection of 4^{13} possible sequences of rules, only some of which would yield our target statement. Once again, we could imagine the application of a rule as requiring one *moment* in time. As discussed in Chapter 12, a *parallel* machine (at each stage applying all applicable rules simultaneously) would take only 13 moments to reach the solution, assuming our 13-step solution is the shortest. A serial machine might require 67 108 864 (4^{13}) moments to find the solution. Even so, a very fast serial machine could do it in a few seconds.

However, the propositional representation of Tables 13.13 and 13.14 is very small, and a solution of 13 steps is very short. As you may realise, we are dealing, once again, with a computational process that tends towards an *exponential* running time (see Chapter 12). For our example, when we start to consider problems that require longer (but still reasonably long) solutions, we run into problems. Consider, for example, a problem for which the solution is of length 40 (not at all unreasonable). A serial machine that can apply 1 000 000 000 000 rules in a second might take over 38 000 years before it reaches the solution. Hence, much effort in computational logic is expended on obtaining more efficient procedures for problem solving. Intuitively, we feel that such procedures ought to be available. After all, when I solved the oversleeping problem in order to include the solution in this book, I did not, as a precursor to arriving at the "length 13 solution", attempt anywhere near the 16 777 216 sequences of length 12. It seems that *human* problem solvers gain much from being able to simultaneously view a system on many levels of abstraction, and in particular to be able to take a *meta-level* view, as it were. In any case, there *are* more efficient procedures for reasoning in propositional logic than the mere *brute force* application of applicable rules at each stage; but we are not concerned with these here.

Let us now consider the *truth table* approach to propositional logic problem solving. In the parlance of Part 2 of this book, such an approach is an *effective procedure*, and, in the parlance of Part 1, represents a *decision program* for the given problem. We represent the problem in a way similar to that in Tables 13.15 and 13.16, we then write down all possible assignments of truth values for the atomic propositions involved. We then complete the truth tables (in mechanical terms, this can be regarded as simply a process of using the truth tables for the basic operators as "look up" tables). If our statement is indeed true, the "result" column of our truth table will consist entirely of 1s. If this is not the case, then our statement is false. It is clear that, in the terminology of Chapter 11, this represents a *totally solvable problem*. It should also be clear that the procedure is such that a Turing machine (TM) could be constructed to carry it out. It is, also in the language of Chapter 11, a *Turing-computable function*.

The truth table approach seems very promising until, once again, we explore its computational dimensions in more detail. If there are n atomic propositions in our statement, there are 2^n rows in our truth table (cf. Table 13.15, where there are 4 atomic propositions, and there are $2^4 = 16$ rows). For a statement containing 75 atomic propositions, our "1 000 000 000 000 instructions a second" machine would take nearly 1200 years to complete any column of the table (ignoring the cost of actually filling in the assignments of truth values for the atomic propositions in the first place. Moreover, as a data structure, the truth table would be rather large. Assuming each character occupies one byte of storage, and ignoring the need to store additional information to delimit parts of the table, and so on, we would need in excess of 35 000 000 000 000 GB of storage to hold it. Once again, we are faced with a simple, well-defined algorithmic procedure that is practical only for small input values, but *theoretically* is highly important.

Unfortunately there are more problems with propositional logic than its computational complexity. Propositional logic is not even powerful enough to represent facts about numbers that we take for granted, such as

for every number, x, there is a number, y, such that $y = x + 1$.

Propositions, as stated earlier, are *atomic*. We cannot make a proposition range over a whole set of objects ("for every number, x"), as is required by our statement. Nor can we say that a proposition holds for some objects but not necessarily all ("there is a number, y"). The following two chapters are devoted to a logical system that is capable of such things.

13.6 Exercises

For exercises marked †, solutions, partial solutions, or hints to get you started appear in "Solutions to Selected Exercises" at the rear of the book.

13.1†. Draw the *NOR* gate representation of the Boolean *and, or* and *not* operators. In each case, verify your representation by constructing a truth table. For the *and* operator, consider the relationship between your representation and de Morgan's law for the intersection of two sets from Chapter 6 (Figure 6.7).

13.2. With reference to the engineer's problem, above, draw the circuit using:

(a)† *NAND* gates only

(b) *NOR* gates only.

In both cases, use *truth* tables to prove that your circuits are correct.

13.3. Verify rules 7 and 8 from Table 13.12 by constructing a truth table. Justify by intuitive argument that the rules are correct.

13.4. Use a truth table to show that the rule of *modus tollens* (rule 12 of Table 13.12) is a theorem of propositional logic.

Hint: see Table 13.16.

13.5. Use truth tables to prove that the following are theorems:

(a) $(p \to q \wedge q \to r) \to (p \to r)$

(b) $(p \wedge 1) \leftrightarrow p$

(c)† $(p \wedge p) \leftrightarrow p$

(d) $(p \vee p) \leftrightarrow p$

(e) $(p \vee 0) \leftrightarrow p$

(f) $(p \wedge 0) \leftrightarrow 0$

(g) $(p \vee q \wedge r) \leftrightarrow ((p \vee q) \vee r) \leftrightarrow (p \vee (q \vee r))$

(h) $(p \wedge q \wedge r) \leftrightarrow ((p \wedge q) \wedge r) \leftrightarrow (p \wedge (q \wedge r))$.

What is the significance of the above proofs?

Chapter 14
First Order Predicate Logic

14.1 Overview

The previous chapter discussed Boolean and propositional logic. We ended by stressing the shortcomings of propositional logic as a general representation and reasoning system.

In this chapter, we consider *first order predicate logic* (FOPL[1]). In particular, we discuss:

- the building blocks of FOPL: *predicates, functions, constants,* and *variables*
- the *quantifiers*; the existential quantifier, $\exists$, and the universal quantifier, $\forall$
- the rules for forming well-formed FOPL formulae (FOPL *sentences*)
- how to use FOPL as a tool for problem solving and reasoning.

14.2 Predicate Logic

As we saw towards the end of the previous chapter, there are many things we cannot really express in propositional logic. The basic problem is that a proposition is *atomic* (a term we have encountered many times before in this book), and cannot be "picked" apart in any way. *FOPL* overcomes many of the problems of propositional logic, though as we shall see, it has certain problems of its own.

In this section, we consider the building blocks of predicate logic. We first consider *predicates*, which enable us to specify *relationships* or *properties*. We then consider the way in which predicates can be combined to form what are called *well-formed formulae* (*abbreviated to WFFs*, which some people pronounce as the sound of a dog barking) or sometimes *sentences*. Just as when we considered languages and propositional logic, a predicate logic *sentence* is simply a sequence of symbols assembled according to the rules of the grammar (in this case the grammar of predicate logic, which we will also consider later). Later, we consider two additional features of FOPL that are used to express logical statements that apply to some things, or to a whole set of things. These additional features are the source of the expressive power of FOPL.

14.2.1 Predicates

Suppose some person, say *Algernon*, owns a dog, say *Dorian*. We could express this is propositional logic as "*Algernon owns Dorian*". In predicate logic, we might express it as follows:

owns(Algernon, Dorian).

We might equally express it as:

owned_by(Dorian, Algernon).

As you can probably tell, we intend the former statement to be interpreted as "*Algernon owns Dorian*" and the latter to be interpreted as "*Dorian is owned by Algernon*". We may also wish to represent facts such as:

dog(Dorian) and
person(Algernon).

In the above,

owns
owned_by
dog and
person,

are all *predicates*, while

Dorian and
Algernon

appear as *arguments* to the predicates. What is more, we are clearly using the terms *Dorian* and *Algernon* to denote specific objects, and thus they are *constants*. We can also use *variables* as arguments to predicates. For example, we might write:

owns(Algernon, x)

to state that Algernon owns something. In the remainder of this book, we will use strings with a capitalised letter to denote constants. Numbers are also constants. Strings that are not capitalised, and are not predicate or function names (see later) denote variables.

A very important point to note is that predicates have no meaning in themselves. In the above examples, I chose names for the predicates and arguments that were meant to suggest their interpretation. However, another interpretation could be made:

Let *Algernon* stand for me (Alan)
Let *Dorian* stand for my car
Let *owns* stand for *drives.*

Then

owns(Algernon, Dorian)

states that Alan drives his car. Later, we discuss the interpretation of FOPL in more detail.

Just as for Boolean expressions and propositions, a predicate applied to its arguments represents a statement that is either *true* or *false*. Moreover, we can combine predicates into larger statements by using the logical connectives we encountered in the previous chapter. So, we could write:

dog(Dorian) ∧ *person(Algernon)* ∧ *owns(Algernon, Dorian)*,

which, given our initial interpretation, would mean:

Dorian is a dog, Algernon is a Person, and Algernon owns Dorian.

In addition to predicates, we define ourselves, we use the standard arithmetic comparison operators (=, <, ≤, and so on). Since they are simply predicates (that apply to numbers), we could write such operators in our expressions as if they were predicates (such as *owns* from above). For example, we could express the fact that 6 is greater than 3 by writing:

$$> (6, 3).$$

However, it is more familiar to see such operators used in *infix* form (i.e. *between* their two arguments), and this is the way we will use them. So the statement immediately above would be written in the usual way, i.e.

$$6 > 3.$$

14.2.2 Functions

In addition to predicates, predicate logic also features *functions*. For example, consider the statement *Algernon was born in the same place as Fred's Father*. This could be expressed:

equal(birthplace_of(Algernon), birthplace_of(father_of(Fred))).

This example features two functions, *birthplace_of* and *father_of*, which can be intuitively interpreted as:

birthplace_of(x) – denotes the town in which *x* was born
father_of(x) – denotes *x*'s father.

Note the difference between predicates and functions. A predicate is a statement that is either true or false, whereas a function denotes a value of some kind. Thus, in predicate logic, a *function call* (i.e. a function applied to the appropriate number of arguments) can appear only as an argument to a predicate. It cannot represent a statement in its own right, as that would not make sense as a logical statement. For example, suppose that Algernon was born in Birmingham. The statement:

birthplace_of(Algernon)

represents the constant *Birmingham*. *Birmingham* is not a statement with a true or false value. However,

equal(birthplace_of(Algernon), Birmingham)

is a valid statement, since its "value" can be either *true* or *false*.

Note also that *nested* function calls are allowed, to any depth of nesting. For example, *birthplace_of(father_of(Fred))*, denotes the application of the function *birthplace_of* to the result of the application of *father_of* to the argument *Fred*, i.e. denotes *the birthplace of the father of Fred*.

Finally, you may have noticed that the *equal* predicate, as used above, is meant to represent the equality relationship, which we usually write as the infix comparison operator "=". Thus, our initial statement would perhaps more usually be written:

$$birthplace\_of(Algernon) = birthplace\_of(father\_of(Fred)).$$

As for the arithmetic comparison operators, we also use arithmetic functions in FOPL expressions. For example, we might say 10 divided by 2 is greater than 3 plus 1:

$$greater(half\_of(10), one\_more\_of(3)).$$

Since it is clear that we intend *half_of* to mean *divided by* 2 and *one_more_of* to mean *plus* 1. We could write the statement more conventionally as

$$greater(10/2, 3+1), \text{ or even more usually as } 10/2 > 3+1.$$

You will note from the above that each function name has an "*_of*" suffix (for example, *father_of*). This is simply a convention we will use to distinguish between the names of functions and predicates. Remember that the names have no meaning. Meaning has to be assigned to the terms in our representations, which, as indicated above, will feature in our later discussion.

Next, we turn to the rules that define syntactically correct predicate logic expressions.

14.2.3 "Sentences" Revisited: Well-formed Formulae

The rules for creating WFFs, or what we sometimes call *sentences*, are simple. First of all, a predicate applied to its arguments (which as we have seen can be *variables*, *constants*, and *function calls*) is called an *atom*. Given this, we can now specify the rules that define a predicate logic sentence.

A WFF, or sentence, of predicate logic is defined as follows:

R1   an *atom* is a *sentence*

R2   if *s* is a sentence, then ~*s* is a sentence

R3   if *s* and *t* are sentences, then so is *s* → *t*

R4   if *s* and *t* are sentences, then so is *s* ∧ *t*

R5   if *s* and *t* are sentences, then so is *s* ∨ *t*.

The above rules constitute a *recursive definition*. As indicated by Exercise 7.3 of Chapter 7, such rules also serve as a *grammar* with which to generate sentences, as did the grammars in Part 1 of this book.

Let us consider an example of the application of the above rules. In the example, the names of the objects have deliberately been chosen to be free of meaning, so the purely syntactic nature of the rules can be appreciated. Assume, we have the predicates *pred_1* (one argument) and *pred_2* (two arguments). We have the following variables: *x*, *y*, and *z*. The constants C1 and C2 are also available to us. Finally, we

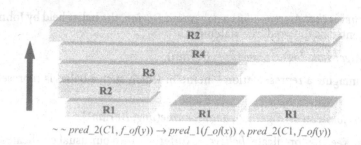

$$\sim\sim pred\_2(C1, f\_of(y)) \rightarrow pred\_1(f\_of(x)) \wedge pred\_2(C1, f\_of(y))$$

Figure. 14.1. How the rules for WFFs were applied to obtain statement (S4) (*see text*)

have the function f_of(one argument). The following are suitable atoms:

$pred\_1(f\_of(x))$ (A1)

$pred\_2(C1, f\_of(y))$.   (A2)

R1 tells us that each of the above is a sentence. Thus, by R2,

$\sim pred\_2(C1, f\_of(y))$ (S1)

is a sentence. Then, since (A1) is a sentence, by R3,

$\sim pred\_2(C1, f\_of(y)) \rightarrow pred\_1(f\_of(x))$ (S2)

is also a sentence. (S2) and (A2) are sentences, so, by rule R4,

$\sim pred\_2(C1, f\_of(y)) \rightarrow pred\_1(f\_of(x)) \wedge pred\_2(C1, f\_of(y))$ (S3)

is a sentence. Finally, since (S3) is a sentence, so, by R2, is:

$\sim\sim pred\_2(C1, f\_of(y)) \rightarrow pred\_1(f\_of(x)) \wedge pred\_2(C1, f\_of(y))$.   (S4)

You may have noticed that, if regarded as a grammar, the rules are *ambiguous* (see Chapter 2). This is considered further in Exercise 14.1 at the end of this chapter. For now, if we consider the order in which the rules were applied to generate (S3), the structure of the statement is as indicated by the square parentheses inserted in the following version of the statement, and also depicted in Figure 14.1:

$$\sim[[\sim pred\_2(C1, f\_of(y)) \rightarrow pred\_1(f\_of(x))] \wedge pred\_2(C1, f\_of(y))].$$

14.2.4 The "First Orderness" of First Order Logic

One very important point should be stressed before we proceed. You may have observed from above that the definition of an *atom* refers to predicate arguments being "variables, constants, and function calls". We have also noted earlier that function calls can be nested, i.e. one or more arguments to a function can themselves be function calls. What we *cannot* have as an argument to a predicate is a predicate. This is *not* allowed. That is why it is called *first order* logic. A logical system in which predicates are allowed to have predicates as their arguments is known as higher order logic. Higher order logics are used to represent such things as temporal reasoning (reasoning about changing states of affairs over time) and belief systems. One can intuitively appreciate the need for higher order logic in modelling beliefs, since the objects of beliefs are themselves logical statements. For example,

if *John believes that Algernon has a dog called Dorian*, the belief held by John might
be represented as the predicate statement

owns(Algernon, dog) ∧ name(dog, Dorian).

One can imagine a representation scheme in which such a belief is represented as
follows:

believes[John, owns(Algernon, dog) ∧ name(dog, Dorian)].

As you can see, the "predicate" *believes* is different from our usual predicates, in that
one of its arguments is an entire predicate logic sentence. We will not concern ourselves further with such schemes in this book.

14.3 Quantifiers

We have now established what is meant by a WFF, or *sentence*, of predicate logic.
The introduction to the discussion indicated that FOPL enables us to specify
logical statements that apply to all, or some of, the members of a set. Such statements are known as *quantified statements*. In this section, we examine the two
features of FOPL that make such statements possible.

14.3.1 The Existential Quantifier

There are many statements made in everyday life that seem quite trivial and yet are
logically quite complex. Consider the previous sentence, for example. It begins
"There are many statements …". This essentially means the same as "Some statements", or perhaps "One or more statements". In other words, what I am asserting
is that the proposition "made in everyday life that seem quite trivial and yet are
logically quite complex" holds for one or more statements.

Let us illustrate the immediately preceding discussion by using a more mundane
example. Algernon, our dog owner, actually owns several dogs. To represent this, we
need to say something like "Algernon owns several dogs". Put another way, we
could say "there are some dogs that Algernon owns". Put in a slightly more formal
way, we might say:

there are one or more objects, *x*, such that <u>*x is a dog and Algernon owns x*</u>.

You will appreciate that the underlined statement can be represented as a predicate
logic sentence as follows:

$$dog(x) \land owns(Algernon, x).$$

This does not represent exactly what the statement above says, as the first part of
the statement is not represented. In fact, FOPL provides a mechanism to represent
such statements. It is called the *existential quantifier*. It is written as a backward
facing letter "E", i.e. ∃. A FOPL sentence that directly represents our statement is:

$$(\exists x)(dog(x) \land owns(Algernon, x)).$$

Figure 14.2 shows how the above FOPL sentence represents the original English
sentence.

The above statement contains our first example of a *quantified variable*. This
is the variable *x*, which, since it is referred to by an *existential quantifier* (∃) is

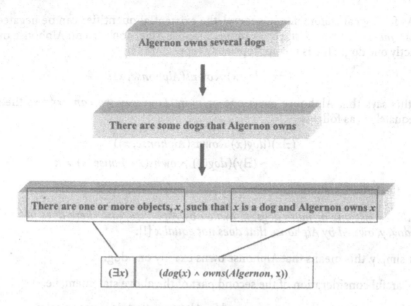

Figure. 14.2. How FOPL represents an English statement

known as an *existentially quantified variable*. The statement referred to by the quantifier, i.e.

$$(dog(x) \land owns(Algernon, x)),$$

is called the *scope* of the quantified variable. This is similar to the notion of the scope of variables in programming. For example, in the Pascal programming language, the scope of a variable defined within a procedure is that procedure (and any procedures and functions within that procedure, and so on).

Let us consider another example. Suppose we observe that Algernon owns dogs other than Dorian. We can express this fact in FOPL thus:

$$(\exists x)(dog(x) \land owns(Algernon, x) \land x \neq Dorian),$$

i.e. Algernon owns one or more dogs that are not Dorian.

As a final example, we consider a statement with two existentially quantified variables. We observe that on some sunny days, Algernon likes to drink a few beers. This can be expressed as

$$(\exists x)((day(x) \land sunny(x)) \rightarrow (\exists y)(beer(y) \land drinks(Algernon, y))).$$

Notice that the *scope* of *y* is only the statement after the →. However, as *y* does not occur in the statement to the left-hand side of the →, we can move the second quantifier to the beginning of the statement without affecting its meaning:

$$(\exists x)(\exists y)((day(x) \land sunny(x)) \rightarrow beer(y) \land drinks(Algernon, y))).$$

The part $(\exists x)(\exists y)$ can also be written $(\exists x, y)$. You should also note that the FOPL statement does not adequately represent the original in certain respects (see Exercise 14.2 at the end of this chapter).

As for logical statements in general, the existential quantifier can be negated to mean *there does not exist*. For example, suppose Algernon's friend Alphonse owns exactly one dog. This is not appropriately represented by

$$(\exists x)(dog(x) \wedge owns(Alphonse, x)),$$

as this says that Alphonse owns one or *more*. One way we *can* express the fact adequately is as follows:

$$(\exists x)(dog(x) \wedge owns(Alphonse, x))$$
$$\wedge \sim(\exists y)(dog(y) \wedge owns(Alphonse, y) \wedge x \neq y).$$

This final FOPL statement says:

> there exist one or more dogs, x, owned by Alphonse, and there does not exist a dog, y, owned by Alphonse that does not equal x [!].

Put simply, this means that Alphonse owns exactly one dog.

Careful consideration of the second part of the above statement, i.e.

> there does not exist a dog, y, owned by Alphonse that does not equal x

will reveal that this is another way of saying

> <u>for all</u> dogs, y, y is not owned by Alphonse or y = x.

This brings us neatly to the next FOPL facility we consider, known as the *universal quantifier*. This enables us to specify statements that hold over *all* objects in a domain.

14.3.2 The Universal Quantifier

Consider a familiar property that applies to positive integers, known as the *transitivity* of the "greater than" (>) relation. (We encountered transitivity in Chapter 12, where we saw that connectivity in a directed graph – in that case, a finite state recogniser (FSR) – is a transitive relation.) For any three numbers, x, y, and z, if $x > y$ and $y > z$, then $x > z$. This can be expressed in FOPL, by using what is known as the *universal quantifier*, which is written as an upside down letter "A", i.e. $\forall$. A suitable FOPL statement to do this is:

$$(\forall x)(\forall y)(\forall z)((x > y \wedge y > z) \rightarrow x > z).$$

As observed in relation to the existential quantifier, we could write the first part of the expression as $(\forall x, y, z)$. The scoping of variables also conforms to the same principles.

The exercises at the end of this chapter ask you to write FOPL statements to represent other well-known properties of the integers.

Given our discussion at the end of the preceding section, we can now rewrite the FOPL statement

$$(\exists x)(dog(x) \wedge owns(Alphonse, x))$$
$$\wedge \sim(\exists y)(dog(y) \wedge owns(Alphonse, y) \wedge x \neq y).$$

as

$$(\exists x)(dog(x) \wedge owns(Alphonse, x))$$
$$\wedge (\forall y)\sim(dog(y) \wedge owns(Alphonse, y) \wedge x \neq y).$$

As we shall shortly see, the rules of inference and equivalence for propositional logic given in Chapter 13 (Table 13.12) apply equally to FOPL (though FOPL has some additional rules – as you will see if you look ahead to Table 14.3). So the preceding statement could be rewritten as

$$(\exists x)(dog(x) \wedge owns(Alphonse, x))$$
$$\wedge (\forall y)((dog(y) \wedge owns(Alphonse, y)) \rightarrow x = y).$$

(You may like to convince yourself that this is so.) The above reads as

Alphonse owns a dog and all other dogs Alphonse owns are the same as that one.

This is essentially the same as saying "all dogs owned by Alphonse are the same dog", which in turn means "Alphonse owns exactly one dog".

FOPL expressions can feature as many universal and/or existential quantifiers as we need.

14.4 A "Blocks World" Example of FOPL Representation

Consider a simple "blocks world" problem. Imagine that a robot's task is the stacking of blocks, and the robot is equipped with a memory in which resides a FOPL representation. A rule that describes the movement of a block from the top of another block onto a clear block (a block with no other block on top of it) might be:

$$(\forall s)(\exists x, y, z)(clear(x, s) \wedge on(x, y, s) \wedge clear(z, s))$$
$$\rightarrow(\exists t)(t = s + 1 \wedge on(x, z, t) \wedge clear(y, t)).$$

This is an interesting example, as it illustrates a simple way in which temporal aspects may be coded in FOPL. We briefly discussed such issues earlier in this chapter. Each predicate has an additional argument (s and t are the variables used for this), which is meant to indicate a time value (simply represented as a non-negative integer – the robot is assumed to start at time 0). For example, Figure 14.3 illustrates a configuration of blocks at a given time (3), and the associated assignment of values to the arguments of the predicates that describe this. In this sense, all predicates featuring the same integer as their last argument represent a *state*, or situation. The robot would solve the problem by moving between successive states, at

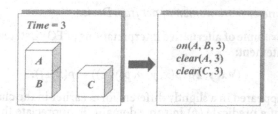

Figure 14.3. A blocks world configuration and its associated FOPL predicates

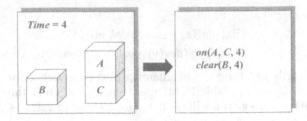

Figure 14.4. A successor state to that shown in Figure 14.3

each stage making only legal moves on blocks (for example, a block cannot be moved unless it is clear).

The robot's rule above thus specifies the moving of a clear block that is on top of a block in one state (at a given time) onto another clear block in the following state (i.e. at the given time plus 1). Figure 14.4 illustrates a suitable successor state to that shown in Figure 14.3, given the above rule. You are asked to develop the robot example further in the exercises.

We now turn to the semantics of FOPL.

14.5 The Semantics of FOPL: *Interpretation*

In the above discussion, hints have been made about how the semantics of FOPL are derived. Recall from above that FOPL statements are merely syntactic entities, and have no intrinsic meaning. The distinction between syntax and semantics was discussed in the context of formal languages in Chapter 2. As for any formal language (and clearly FOPL is such a thing), meaning is something that must be imposed from outside, as it were. For FOPL, the specification of meaning involves associating relationships, properties, objects, concepts, events and states from some world or domain with the FOPL entities predicates, functions and constants. Such a mapping is called an *interpretation*.

In general, any FOPL statement can have many possible interpretations. Consider an example FOPL statement from above:

$$(\exists x)(dog(x) \land owns(Algernon, x) \land x \neq Dorian).$$

This essentially means (according to our earlier informal interpretation):

Algernon owns one or more dogs apart from Dorian.

Now, suppose we attribute the meanings to our predicates and constants specified in Table 14.1. The statement now means:

Alfredo loves one or more women apart from Doris.

For a second example of alternative interpretations of FOPL statements, consider the following statement:

$$(\forall x, y, z)((p(x,y) \land p(y,z)) \rightarrow p(x,z)).$$

This statement appeared in a slightly different form earlier in the chapter. It defines the transitivity of a predicate (p) in some domain. To appreciate this, consider the domains and mappings shown in Table 14.2. In all the cases, the statement remains

Table 14.1. A possible interpretation of predicates and constants

Predicate/constant	Domain entity
dog	woman
owns	loves
Algernon	Alfredo
Dorian	Doris

Table 14.2. Some of the different possible interpretations of p in $(\forall x, y, z)((p(x,y) \wedge p(y,z)) \rightarrow p(x,z))$ that do not affect the truth of the statement

Interpretation of p	Domain
$=$	integers
$>$	integers
$<$	integers
$\geqslant$	integers
is descended from	people
is an ancestor of	people
is a family relation of	people
brighter than	colours
zoom in of	images
left pan of	images

true. They are all *valid interpretations* of the FOPL statement. You can probably think of many more mappings that would also be valid for the statement.

Next, we consider an appropriate set of "valid rules of inference", and how such rules can be applied to perform what is known as *classical FOPL reasoning*.

14.6 Problem Solving and Inference in FOPL

We now discover one method for using FOPL to represent domains and then solve problems in those domains. The method we use here is known as *classical reasoning*. It is very similar to the way we used propositional logic as a problem-solving tool in the previous chapter. We first consider the equivalencies and rules of inference for FOPL, and consider why we can be intuitively convinced that the rules are valid. We then turn to the problem-solving process itself.

14.6.1 Rules of Inference for FOPL

The rules of inference for FOPL are given in Table 14.3. As stated, rules 1–8 and 11–13 are exactly the same as those for propositional logic, and were discussed in Chapter 13. We therefore here discuss only those rules that refer explicitly to FOPL. Note that in rules 9 and 10 of Table 14.3, $p1(x)$ refers to an arbitrary FOPL expression in which the variable x appears, not necessarily to a single predicate.

Rule 9 was discussed earlier in the chapter. If there does not exist a single entity, x, for which the statement is true, then it is certainly true to say that for all entities, x, the statement is false. Consider Algernon's friend Alberto. He owns no dogs:

$$\sim(\exists x)(dog(x) \wedge owns(Alberto, x)).$$

According to rule 9, this can be represented:

$$(\forall x)\sim(dog(x) \wedge owns(Alberto, x)).$$

Table 14.3. Equivalence rules and rules of inference for FOPL

Rule no.	A	B	Comment
1	$\sim\sim p$	p	
2	$p \wedge q$	$q \wedge p$	
3	$p \vee q$	$q \vee p$	
4	$p \rightarrow q$	$\sim p \vee q$	
5	$\sim(p \wedge q)$	$\sim p \vee \sim q$	de Morgan's law
6	$\sim(p \vee q)$	$\sim p \wedge \sim q$	de Morgan's law
7	$p \wedge (q \vee r)$	$(p \wedge q) \vee (p \wedge r)$	Distribution of $\wedge$
8	$p \vee (q \wedge r)$	$(p \vee q) \wedge (p \vee r)$	Distribution of $\vee$
9	$\sim(\exists x)(p1(x))$	$(\forall x)\sim(p1(x))$	Also de Morgan's law
10	$\sim(\forall x)(p1(x))$	$(\exists x)\sim(p1(x))$	Also de Morgan's law

Note: In rules 9 and 10, $p1(x)$ is any sentence in which the quantified variable, x, appears. For example, $\sim(\exists x)(p(x) \vee q(x,y)) \leftrightarrow (\forall x)\sim(p(x) \vee q(x,y))$.

	Given	Assert	
11	$p \rightarrow q$, p	q	*modus ponens*
12	$p \rightarrow q$, $\sim q$	$\sim p$	*modus tollens*
13	$p \wedge q$	p (or q)	
14	$p1(X)$, where X is a constant, NOT a variable	$(\exists x)(p1(x))$, the constant X has been replaced by a variable	If $p1$ is true of an object, then it is true that there exists some object ...
15	$(\forall x)(p1(x))$	$p1(X)$ the variable x has been replaced by a constant	If $p1$ is true of all objects, then $p1$ is true of the constant X

Note: Rules 11–15 are *rules of inference*. p and q are any FOPL sentences. In rules 14 and 15, $p1(x)$ is any sentence in which the quantified variable, x, appears.

NB: Rules 1–8 and 11–13 are the rules for propositional logic from Table 13.12.

This becomes a little easier to verbalise if we apply rule 5 to get

$$(\forall x)(\sim dog(x) \vee \sim owns(Alberto, x)),$$

and then rule 4, giving

$$(\forall x)(dog(x) \rightarrow \sim owns(Alberto, x)).$$

The final expression can be verbalised as "anything that is a dog is not owned by Alberto".

Now consider rule 10 from Table 14.3. This rule says that a statement that is not true for everything must be false for something. Again, this should be intuitively clear. Though Alberto does not own any dogs, he owns some cats. Not all of his cats are black:

$$\sim(\forall x)((cat(x) \wedge owns(Alberto, x)) \rightarrow black(x)).$$

According to rule 10, this is the same as:

$$(\exists x)\sim((cat(x) \wedge owns(Alberto, x)) \rightarrow black(x)).$$

Once again, the application of certain rules, in sequence, can be used to manipulate the above statement until it becomes (see if you can work out which rules were applied to yield this):

$$(\exists x)(cat(x) \wedge owns(Alberto, x) \wedge \sim black(x)).$$

This can be stated as "there are one or more non-black cats owned by Alberto".

Next, we consider rule 14, from Table 14.3. Once again, $p1(x)$ refers to an arbitrary FOPL sentence in which the variable x appears. Rule 14 says that if something is true of a constant, X, then it is true to say that there exists a variable, x, for which that something is true. For example, if Alberto owns a cat called Tiddles, i.e.

$$cat(Tiddles) \wedge owns(Alberto, Tiddles),$$

then it is true to say

$$(\exists x)(cat(x) \wedge owns(Alberto, x)),$$

i.e. there exists a cat owned by Alberto. Note that rule 14 also tells us that

$$(\exists x)(cat(Tiddles) \wedge owns(x, Tiddles)),$$

i.e. there exists an owner of the cat, Tiddles, and even tells us

$$(\exists x, y)(cat(x) \wedge owns(y, x)),$$

i.e. there exists a cat that is owned by someone.

Finally, we consider rule 15 of Table 14.3. This rule tells us that if something is true for all things, x, then it is true for any given constant in our domain. This means we can replace any *universally quantified* variable with any constant from our domain, and then drop the $\forall$ (though only the one that refers to x). For example, consider our FOPL statement representing the transitivity of $>$ from earlier:

$$(\forall x)(\forall y)(\forall z)((x > y \wedge y > z) \rightarrow x > z).$$

Rule 15 tells us we can replace, say, z, with a constant. Since the domain is integers, let's replace z with a number, say 42. We now have

$$(\forall x)(\forall y)((x > y \wedge y > 42) \rightarrow x > 42),$$

i.e. for any two numbers, x and y, if x is greater than y and y is greater than 42, then x is greater than 42; which is clearly true.

However, rule 15 says we can replace *any* universally quantified variable by any constant, and the resulting statement will still be true. This leads to situations which may initially seem strange, but in fact are not when implication is interpreted correctly. To illustrate let us replace both x and y by 3, and z by 42. This leaves us with

$$(3 > 3 \wedge 3 > 42) \rightarrow 3 > 42.$$

The above looks rather peculiar, but nevertheless is still true: *if* 3 was greater than 3 and also greater than 42, then it would indeed be greater than 42. The fact is that 3 is neither greater than 3 nor greater than 42, so we cannot (by *modus ponens* – rule 11) infer that $3 > 42$.

Of course, in classical FOPL reasoning, we usually confine ourselves to manipulating the statements in ways that make sense in terms of our goal of reaching some kind of conclusion. We consider this next.

14.6.2 Solving Problems by Classical FOPL Reasoning

The above discussion should convince you that our rules of inference (Table 14.3) are logically sound. We now consider how to use the rules appropriately to solve problems in FOPL.

Table 14.4. Classical FOPL reasoning solves the lazy drunken students problem

Step (see text)	Rule applied (see Table 14.3)	FOPL
1		(a) $(\forall x)((lazy(x) \wedge drinks(x)) \rightarrow fail(x))$
		(b) $\sim fail(Fred)$
2		$(\exists x)(drinks(x) \rightarrow \sim lazy(x))$
3.1	15	$lazy(Fred) \wedge drinks(Fred) \rightarrow fail(Fred)$
3.2	12	$\sim(lazy(Fred) \wedge drinks(Fred))$
3.3	5	$\sim lazy(Fred) \vee \sim drinks(Fred)$
3.4	3	$\sim drinks(Fred) \vee \sim lazy(Fred)$
3.5	4	$drinks(Fred) \rightarrow \sim lazy(Fred)$
3.6	14	$(\exists x)(drinks(x) \rightarrow \sim lazy(x))$

To solve a problem using classical FOPL reasoning, we do the following:

Step 1 Represent the required knowledge in FOPL (the *axioms*)

Step 2 Represent the statement to be proved in FOPL

Step 3 See if the result of step 2 can be derived from the axioms by the repeated application of the rules in Table 14.3.

To illustrate this, let us consider the following problem:

Given that lazy students who drink beer fail their exams, and that Fred is a student who does not fail his exams, prove that there are one or more students who, if they drink beer, are not lazy.

We refer to this as the *lazy drunken students* problem. It will be revisited in the next chapter.

Table 14.4 shows the above problem-solving process being applied to the above problem. It can be seen that the statement to be proved is true. The exercises at the end of this chapter ask you to solve other problems of this nature.

14.7 The Nature of FOPL

The power of FOPL arises in the following way. We represent a domain in a consistent way by a set of FOPL statements, i.e. an appropriate interpretation is specified, and only true statements about that domain are represented by the FOPL. Then, *every FOPL statement derived through the application of valid rules of inference will be a true statement about that domain.*

Moreover, given a FOPL representation of a domain, there are no true statements about that domain that are a logical consequence of that initial representation that cannot be derived through FOPL. In this sense, we say that FOPL is *complete*.

However, one problem with FOPL is that, in terms used in Chapter 11, FOPL is *semidecidable*. It is possible to define algorithms that will decide if a true statement is indeed true, but that may not terminate at all if presented with a false statement as input. This may remind you of the discussion of solvability and the halting problem in Chapter 11, and in fact these issues are related. Indeed, a further associated issue is the decidability of the type 0 languages, also discussed in Chapter 11. Consider the application of a FOPL rule as if it were the application of a production in a grammar (in this discussion we return to the terminology of the earlier chapters of this book). You can see that FOPL has rules that both lengthen and shorten the current sentential form (unlike, say the type 1, 2 – ε *free* – and 3 grammars, in which sentential forms never shorten during a derivation). As was stated above, a valid FOPL representation cannot be used to derive false statements about a domain. This means that for false statements there is no sequence of rules of inference that will yield that statement. To be capable of dealing with this, the algorithm would need to be able to detect the difference between an arbitrarily long chain of inference that would lead to an appropriate conclusion and a non-terminating process that is going nowhere. In other words, in general, the algorithm would have to be capable of solving the halting problem, and as we have seen in earlier chapters, this is not possible.

However, it should be pointed out that there are many false statements that could be identified as such by a suitable algorithm. It is the general case that does not lend itself to a complete solution. In the final chapter of this book, the relationship between FOPL and Turing machines (TMs) will be considered in a little more detail.

One final feature of FOPL should be noted. In classical FOPL reasoning, statements are never deleted. The result of the application of a rule of inference is always an additional statement to the FOPL database. Thus, the database always grows in size when reasoning is carried out. This property is known as *monotonicity*. We say that FOPL reasoning is *monotonic*. There are such things as *non-monotonic* logics, but we are not interested in them here.

14.8 Conclusions

FOPL is clearly a very powerful tool for reasoning. Unfortunately, algorithms based on classical FOPL reasoning are likely to be extremely expensive in computational terms for very similar reasons to those relating to propositional logic discussed in the previous chapter. Indeed, for FOPL the problems are exacerbated, since, not only do we have a choice of rules to apply at each stage, we may also have a choice of values to assign to variables, and so on. Even in problems as simple as those considered above, numerous rules were applicable at almost every step of the proof. Most reasonably complex domains will require hundreds of axioms. One can imagine the need for extensive *backtracking* in algorithms that model classical reasoning techniques, to try previously unexplored potential paths to a solution. The problems associated with backtracking were discussed in the context of formal language recognition algorithms earlier in this book.

However, all is not lost. It turns out that there are techniques for representing and manipulating FOPL that result in more efficient reasoning systems. The next chapter is devoted to introducing one of these techniques, which is known as *resolution*.

14.9 Exercises

For exercises marked †, solutions, partial solutions, or hints to get you started appear in "Solutions to Selected Exercises" at the rear of the book.

14.1. Represent the rules for WFFs as an unambiguous context free grammar (CFG).

14.2. The statement

$$(\exists x)(\exists y)((day(x) \wedge sunny(x)) \rightarrow (beer(y) \wedge drinks(Algernon, y)).$$

Does not represent the fact that Algernon actually drinks the beer on the given sunny days. Amend the expression (introducing new predicates as required) to cater for this.

14.3. Use quantified FOPL statements to represent the following properties of integers:

(a) the "reflexivity" of "=" (i.e. that every number is equal to itself)

(b) the "symmetry" of "=" (i.e. that if $x = y$ then $y = z$)

(c) the "asymmetry" of "<" (or ">") (i.e. if x is less than y it is not true to say y is less than x)

(d) the infinity of the positive integers.

14.4. For the "blocks world" problem discussed earlier:

(a) Represent the "clear" relationship, i.e. if a block has nothing on top of it in a given state, then it is clear in that state.

(b) Given a FOPL representation of moving a block (based on that given earlier in the chapter) and representation from (a), represent a three-block initial configuration and use classical FOPL reasoning to derive a different configuration of blocks.

14.5[†]. Use classical FOPL reasoning techniques to solve the following problem:

All people who are politicians are liars or cheats. Any married cheat is having an affair. Algernon Cholmondsley-St John is not having an affair, and he is not a liar. Show that if Algernon is married, he is not a politician.

14.6. For the lazy drunken students problem, given that Brian drinks beer but does not fail his exams, prove that Brian is not lazy.

Notes

1. I pronounce this "fopple".

Chapter 15
Logic and Computation

15.1 Overview

In the last chapter we saw how to use *first order predicate logic* (FOPL) as a system for representation and reasoning. We noted that computational problems arise when attempting to model *classical* FOPL reasoning. This, the final chapter of the book, is mainly concerned with an introduction to a technique called *resolution*. This technique addresses some of the computational problems of FOPL.

We discuss:

- how to remove the quantifiers from FOPL statements, and then express the statements as a *conjunctive normal form database*
- the process called *unification*, which is fundamental to resolution
- how to use resolution to perform proof by *reductio ad absurdum* in conjunctive normal form databases
- why resolution is guaranteed to work
- how logic can be used in programming, by briefly considering the PROLOG programming language.

We complete the chapter, and the book, with a short discussion about the relationship between languages, machines, and logic.

15.2 A Computational Form of FOPL

In the preceding chapter, we discovered that FOPL is a powerful tool for reasoning and problem solving. We also discovered that it has severe computational limitations. We now see how FOPL can be manipulated so that it can be used by a computationally efficient process known as *resolution*. In order to illustrate the process, we revisit one of the problems that we solved using classical FOPL reasoning techniques in the previous chapter, i.e. the *lazy drunken students* problem:

> Given that lazy students who drink beer fail their exams, and that Fred is a student who does not fail his exams, prove that there are one or more students who, if they drink beer, are not lazy.

We represented the above problem statement as a series of FOPL statements. These are reproduced in Table 15.1. You will notice that we have also represented the statement to be proved in FOPL.

Table 15.1. The FOPL statements representing the *lazy drunken students* problem

English statement	FOPL form
Lazy students who drink beer fail their exams	$(\forall x)((lazy(x) \land drinks(x)) \rightarrow fail(x))$
Fred is a student who does not fail his exams	$\sim fail(Fred)$
Statement to be proved: There are one or more students who, if they drink beer, are not lazy	$(\exists x)(drinks(x) \rightarrow \sim lazy(x))$

The rest of this section is concerned with investigating the rules for converting any collection of FOPL statements into a set of clauses that we call a *conjunctive normal form database*. We will also use the term *clausal database*, or simply *database*, for convenience. The purpose of such a database will become apparent later in the chapter. The interesting thing about all of the transformations we apply to our FOPL statements is that nothing we do ever changes the truth value of the statements, only their form (in the parlance of earlier chapters, the *semantics* of the statements remains the same, only their *syntax* changes). You should keep that in mind as we proceed through the following stages; at each stage convince yourself that the meaning of the statements remains the same. Finally, you should appreciate that this is not an exhaustive treatment of this subject, and considers only simplified situations.

15.2.1 Getting Rid of ∀

The first thing we need to do is to remove the universal quantifier, $\forall$, from our statements. We can actually only remove it from a FOPL statement if it is at the beginning of the statement. If it is not, we have to take certain actions to move it to the front, before we are allowed to take it away. Once we have removed the $\forall$s from the front of a statement, we can simply assume that any variables in the statement that are not referred to by an $\exists$ are universally quantified.

In order to carry out many of the transformations in this section, we will be using the FOPL identities and rules of inference from Table 14.3, in the preceding chapter.

Let us take a simple example first. The statement:

$$(\forall k)(kangaroo(k) \rightarrow mammal(k))$$

might mean that all kangaroos are mammals. In this case we can simply remove the $\forall$, and assume it is there, i.e. the variable k in the statement is *universally quantified*.

It is important to remember that a universal quantifier cannot simply be removed if it is negated, i.e. has a $\sim$ in front of it, as in the following example.

$$\sim(\forall x)(car(x) \land wheels(x, 4))$$

This statement might mean "it is not the case that all cars have 4 wheels, which is clearly a true statement". Now suppose we removed the negated $\forall$ from the beginning of this statement. We would be left with:

$$car(x) \land wheels(x, 4).$$

As we are assuming that the *free* variables in all of our statements are universally quantified, we now have a statement that is clearly false. It means "every single object is a four wheeled car"! Neither can we simply negate the statement, having removed the negated ∀, as then we would be left with

$$\sim (car(x) \wedge wheels(x, 4))$$

which is

$$\sim car(x) \vee \sim wheels(x, 4) \quad \text{see rule 5 in Table 14.3.}$$

This is easier to express in English if we convert it to

$$car(x) \rightarrow \sim wheels(x, 4) \quad \text{see rule 4 in Table 14.3.}$$

Now, remembering that the variable is assumed to be universally quantified, this statement means "every car does not have four wheels". Still not right.

Looking at rule 10 of Table 14.3, we see that we can change our original statement into:

$$(\exists x) \sim (car(x) \wedge wheels(x) = 4).$$

The negated part of the above statement is one we encountered earlier, so the whole statement can be expressed:

$$(\exists x)(car(x) \rightarrow \sim wheels(x, 4)).$$

The preceding statement can be expressed: "There are one or more cars which do not have four wheels", which is clearly what we were saying in the first case. Unfortunately, we now have an existential quantifier, which we cannot just remove (for obvious reasons, given the above discussion). We find out how to treat these shortly, but first we consider one more example.

Consider the following statement:

$$(\forall x)(\exists y)(x = y \vee freq(x) < freq(y)).$$

To make things easier to follow, we will give the statement an informal interpretation. Suppose $freq(x)$ is a function that denotes the frequency of the note played when striking the piano key x. Suppose also that the domain is the set of keys on a standard piano. Then the statement can be said to mean:

For all keys, x, there exists a key, y, such that the frequency of key, x, is less than that of y (i.e. if x is not equal to y).

The statement therefore says that for every key on a piano except for one there is a key with a higher frequency. We can remove the ∀, as discussed above, leaving the statement:

$$(\exists y)(x = y \vee freq(x) < freq(y))$$

in which the variable x is assumed to be universally quantified. Once again, we cannot simply remove the ∃, as you will see if you imagine what the statement would mean without it.

15.2.2 Getting Rid of ∃

We now assume that no universal quantifiers appear in our statements, but that all variables that are not existentially quantified are universally quantified.

The problem now is to remove the ∃s from our statements. The discussion immediately above should convince you that it is not possible to simply remove them without radically changing the meaning of the statement. Consider the statement about our piano:

$$(\exists y)(x = y \lor freq(x) < freq(y)).$$

This is an interesting example, as, in the case of a piano, we know (or at least some of us do) the name of the key that represents the highest note. Let us denote that key by the constant $Rightmost\_key$. As y is the *existentially quantified* variable in our statement, we can simply replace each occurrence of y in the statement with the constant $Rightmost\_key$, and drop the ∃ altogether, giving us:

$$x = Rightmost\_key \lor freq(x) < freq(Rightmost\_key).$$

Once again bearing in mind that the variable x is universally quantified, we now have a statement that says "each key is the rightmost key or it is a key with a lower frequency note than the rightmost key". We have succeeded in getting rid of the ∃ by replacing the associated variables by a *constant*.

In the case of the piano, we were lucky in two respects. Firstly, we actually knew enough about the highest frequency key that we were able to name it. Secondly, we happened to know that the highest frequency key is unique. It is important to realise that neither of these considerations are represented in the statement itself, as the constants and predicates have no intrinsic meaning, as discussed in Chapter 14.

Now consider our car example from above:

$$(\exists x)(car(x) \rightarrow \sim wheels(x, 4)).$$

The problem in this case is how we arrive at a constant to represent the existentially quantified variable. The answer is a lot simpler than you may think; we simply make one up! Our statement, which is assumed to be true, asserts that there are one or more objects that have a given property. So what we do is make up a constant name to refer to one of those objects. We use a unique constant name, i.e. one that is not already being used to denote one of the objects in our domain, as, though we know there are objects in our domain for which the statement is true, we do not necessarily know precisely which objects these are. In this case, let us use $Car1$. Replacing the existentially quantified variable by this new constant gives us:

$$car(Car1) \rightarrow \sim wheels(Car1, 4).$$

This is still a true statement, as long as we agree that $Car1$ stands for one of the objects for which the statement is true. In any case, it certainly means that the statement cannot be used to refer to an object for which the statement is false, the use of the constant rules that out. We return to this point later on.

Unfortunately, our piano example oversimplifies things somewhat. Consider the following statement:

$$(\forall x, y)(\exists z)(aunt(x, y) \rightarrow (parent(z, y) \land sibling(z, x))).$$

This could mean "if x is the aunt of y, then there is a parent, z, of y who is a brother or sister to x". For any given collection of arguments that satisfy this rule, the parents will depend upon the two particular people who are associated by the *aunt* relation. In other words, we cannot put a constant into the statement in place of z, as a constant denotes only one object, and there are many z objects that could satisfy this rule. We would be saying that all the aunts in the world are sisters to the same person.

The above considerations apply whenever existentially quantified variables fall within the scope of the universal quantifier. In such cases, for each existentially quantified variable we make up a unique *function* name. We do not have to actually define the function, but merely *name* it. The function is put in the places in the statement where the respective existentially quantified variable appeared. Its arguments are the universally quantified variables. This is probably difficult to follow, so we will continue developing the example.

Let us call the new function F1. Replacing z in the statement, and removing the $\exists$, gives us (the newly introduced function calls have been underlined):

$$(\forall x, y)(aunt(x, y) \rightarrow (parent(\underline{F1(x, y)}, y) \wedge sibling(\underline{F1(x, y)}, x))).$$

This may seem rather strange, but essentially all it says is that in each case, the particular parent of y that is a sibling of x depends on (*is a function of*) the particular pair of people (x and y) that are related by the *aunt* relation.

Thus, the rule for removing existential quantifiers that are within the scope of $\forall$s is as follows:

For each existentially quantified variable, v_i, within the scope of universally quantified variables $x_1, x_2, ..., x_n$, introduce a unique function name, F_i. Replace every occurrence of v_i in the statement by $F_i(x_1, x_2, ..., x_n)$, then remove the $\exists$ that refers to v_i.

Note that the above rule makes it clear that we need a different function name for each existentially quantified variable, as in the following example. This time we use the domain of positive and negative integers:

$$(\forall x)(\exists y)(\exists z)(y = x + 1 \wedge z = x - 1).$$

Using F_y and F_z as the functions to replace y and z, respectively, and given that x is the sole universally quantified variable, the statement becomes:

$$(\forall x)(F_y(x) = x + 1 \wedge F_z(y) = x - 1).$$

The universal quantifier can then simply be removed, as we did above. If this all seems rather abstract, consider two possible definitions for the functions F_y and F_z. F_y could be a function that returns the result of adding 1 to its argument, while F_z could be a function that returns the result of subtracting 1 from its argument. The modified statement can then be expressed as "for every number, x, $x + 1$ is 1 more than x, and $x - 1$ is 1 less than x". However, as was stated previously, in most cases, we are not able to provide definitions for our made-up functions, nor do we need to.

We consider one final example, in the hope that this will make things absolutely clear. I suggest that you attempt to carry out the changes yourself before reading on. The statement is:

$$(\forall w, x, y, z)(\exists i, j)(w < x \wedge x < y \wedge y < z$$
$$\rightarrow i > w \wedge i < z \wedge j > w \wedge j < z \wedge i \neq j).$$

The transformed statement should be similar to this (your function names may be different):

$$w < x \wedge x < y \wedge y < z \to F_i(w, x, y, z) > w \wedge F_i(w, x, y, z)$$
$$< z \wedge F_j(w, x, y, z) > w \wedge F_j(w, x, y, z) < z \wedge F_i(w, x, y, z) \neq F_j(w, x, y, z).$$

The method for dealing with $\exists$, discussed above, is known as *Skolemisation*, after A.T. Skolem, the Norwegian mathematician who first formulated it in the 1920s. For this reason the made-up objects are often known as *Skolem* constants and *Skolem* functions.

Table 15.2 contains a more formal statement of the rules for removing quantifiers from FOPL statements, along with illustrative examples based on our statements above.

15.2.3 Conjunctive Normal Form Databases

At this point, we can assume that all of our statements have no quantifiers in them. The next stage is to transform the statements into conjunctive normal form clauses, to create our *database*, as mentioned earlier.

A formula in conjunctive normal form is of the form $C_1 \wedge C_2 \wedge \dots \wedge C_n, n \geqslant 0$ where each C_i is of the form $p_1 \vee p_2 \vee \dots \vee p_m, m \geqslant 0$. Each p_i is a predicate (possibly negated) applied to its arguments (in the parlance of Chapter 14, an *atom*). In other words, a conjunctive normal form formula is a sequence of "anded" together statements, each of which is a sequence of "ored" together predicates. When we create a conjunctive normal form database, however, we dispense with the $\wedge$ signs, and write out each C_i on a separate line. Each C_i is referred to as a *clause*. We often number the clauses, $C1 \dots Cn$, for convenience.

Let us illustrate the above by now converting our statements for the lazy drunken students problem (Table 15.1) into a conjunctive normal form database.

Table 15.2. The steps involved in removing quantifiers from FOPL statements (rules 9 and 10 are those in Table 14.3)

Step	Description	Examples
1	Move negations inwards using rules 9 and 10	$\sim(\exists x)(dog(x) \wedge cat(x))$ **becomes** $(\forall x)\sim(dog(x) \wedge cat(x))$
2	Remove leading $\exists$s, by replacing each existentially quantified variable by a distinct and unique Skolem constant	$(\exists x)(mammal(x) \wedge duck\_billed(x))$ **becomes** $mammal(X) \wedge duck\_billed(X)$ $(\exists x, y)(x < 10, y < 10, x/y < 5)$ **becomes** $Num1 < 10, Num2 < 10, Num1/Num2 < 5$
3	Remove any $\exists$s within the scope of $\forall$s, by replacing each existentially quantified variable by a distinct and unique Skolem function which has arguments $a_1, a_2, \dots, a_n$, where $a_1, \dots, a_n$ are the universally quantified variables	$(\forall x, y)(\exists z)(greater\_than(z, x + y)$ **becomes** $(\forall x, y)(greater\_than(Skol_1\_of(x, y), x + y)$ $(\forall x, y)(\exists u, v)(greater\_than(u - v, x + y))$ **becomes** $(\forall x, y)$ $(greater\_than(Skol_2\_of(x, y) - Skol_3\_of$ $(x, y), x + y))$
4	(Only $\forall$s now remain in quantified sentences) Remove all $\forall$s	$(\forall x, y)$ $(greater\_than(Skol_2\_of(x, y) -$ $Skol_3\_of(x, y), x + y))$ **becomes** $greater\_than(Skol_2\_of(x, y) - Skol_3\_of(x, y), x + y)$

First, the statement

$$(\forall x)((lazy(x) \wedge drinks(x)) \rightarrow fail(x)).$$

We can simply drop the $\forall$, as discussed earlier. Using Table 14.3, we know that $p \rightarrow q \leftrightarrow \sim p \vee q$. This gives us

$$\sim(lazy(x) \wedge drinks(x)) \vee fail(x).$$

Observing that the $\sim$ applies to the whole of the expression of the left-hand side of the $\vee$, we use one of de Morgan's laws, i.e. $\sim(p \wedge q) \leftrightarrow \sim p \vee \sim q$. This yields:

$$(\sim lazy(x) \vee \sim drinks(x)) \vee fail(x).$$

The brackets around the left-hand expression can be removed (see Exercise 13.5 of Chapter 13), giving us the final statement:

$$\sim lazy(x) \vee \sim drinks(x) \vee fail(x).$$

This last statement is one conjunctive normal form clause, as it contains only $\vee$s between the predicates (see the definition of conjunctive normal form above).

Now we turn to the next statement from Table 15.1, which is

$$\sim fail(Fred).$$

This is already in conjunctive normal form, and so constitutes our second clause.

We will not convert the statement to be proved into conjunctive normal form, yet, for reasons that will become clear in the next section. For the moment, we will transform some of our statements from above into conjunctive normal form clauses. The clauses for the lazy drunken students problem that we have derived so far are in Table 15.3.

Consider

$$aunt(x, y) \rightarrow (parent(F1(x, y), y) \wedge sibling(F1(x, y), x)).$$

This can be expressed as:

$$\sim aunt(x, y) \vee (parent(F1(x, y), y) \wedge sibling(F1(x, y), x)).$$

This (see Table 14.3) can be expressed as:

$$(\sim aunt(x, y) \vee parent(F1(x, y), y)) \wedge (\sim aunt(x, y) \vee sibling(F1(x, y), x)).$$

The above statement is in conjunctive normal form. It would be added to a database as two clauses:

1. $\sim aunt(x, y) \vee parent(F1(x, y), y)$
2. $\sim aunt(x, y) \vee sibling(F1(x, y), x)$

Similarly, our expression from earlier, i.e.

$$F_y(x) = x + 1 \wedge F_z(y) = x - 1,$$

Table 15.3. The clauses derived from the first two FOPL statements in Table 15.1

Original statement	Clauses
$(\forall x)((lazy(x) \wedge drinks(x)) \rightarrow fail(x))$ $\sim fail(Fred)$	C1: $\sim lazy(x) \vee \sim drinks(x) \vee fail(x)$ C2: $\sim fail(Fred)$

is already in conjunctive normal form, and comprises the two clauses:

1. $F_y(x) = x + 1$
2. $F_z(y) = x - 1$.

Finally, the statement from earlier,

$$x = Rightmost\_key \lor freq(x) < freq(Rightmost\_key),$$

constitutes a single conjunctive normal form clause.

It is an established result that there are algorithms that can convert any *quantifier-free* FOPL into conjunctive normal form. Here, our purpose in using conjunctive normal form is to show its application in a powerful proof method known as *resolution*. We consider this next.

15.3 Resolution

In this section we consider the computational technique for theorem proving known as *resolution*. Resolution operates on conjunctive normal form databases, as discussed above. First, we consider a process that is a key component of resolution, namely *unification*. Then, we consider the operation of resolution in an informal way. Finally, we consider the theoretical basis of the technique.

15.3.1 The Role of Unification in Resolution

Unification is a technique that involves making systematic substitutions in two predicates to make them textually identical. The reasons for doing this will become clear in the next section, since the operation of resolution requires the application of unification. Once again, our treatment of the subject is informal.

Some aspects of unification are easy to appreciate, if we consider our clausal databases, as described in the previous section, and recall that every variable in our clauses is universally quantified. This means that if we create a database from our FOPL *axioms* (for example, the first two statements in Table 15.1 are the axioms for the lazy drunken students problem), we assume that every single statement is true. If any of them were not true, then the whole database would essentially be false (you may like to consider why this is; we return to this point below), and would therefore be an unsuitable basis on which to prove anything.

As an illustrative example, let us consider the statements in Table 15.4. Statement S1 contains a universally quantified variable, x. The statement is thus assumed to be true for all objects in the domain. Let us assume that our only objects are the two denoted by the constants, *Kate* and *Wally*. The variable, x, in S1, can be replaced throughout by any of the constants in our domain without affecting the truth of the statement. S1 says "if x is a kangaroo, then x is a mammal". This statement as a whole is true for both Kate and Wally – *if* Wally is a kangaroo then he is a mammal. Similarly, with respect to S2 in Table 15.4, *if* Kate is a whale, then she is not a kangaroo. The same applies, of course, to the two clauses, C1 and C2 that were produced from S1 and S2, respectively.

It should be clear from the immediately preceding discussion that, in a conjunctive normal form database, we can replace a variable in a clause by any constant, as long as we carry out the replacement throughout that clause. The resulting amended clause will still represent a true statement. A little consideration will also reveal that the converse situation does not apply, i.e. we cannot replace a constant by a variable, since that would often lead to false statements. For example, if, in

Table 15.4. FOPL statements and CNF for the world of Kate and Wally

FOPL statement	Clauses
S1: $(\forall x)(kangaroo(x) \rightarrow mammal(x))$	C1: $\sim kangaroo(x) \lor mammal(x)$
S2: $(\forall y)(whale(y) \rightarrow \sim kangaroo(y))$	C2: $\sim whale(y) \lor \sim kangaroo(y)$
S3: $(\exists z)(mammal(z) \land \sim furry(z))$	C3: $mammal(M)$
	C4: $\sim furry(M)$
S4: $kangaroo(Kate)$	C5: $kangaroo(Kate)$
S5: $whale(Wally)$	C6: $whale(Wally)$
S6: $(\forall k)(\exists p)(kangaroo(k) \rightarrow (pouch(p) \land body\_part(k, p)))$	C7: $\sim kangaroo(k) \lor pouch(F1(k))$
	C8: $\sim kangaroo(k) \lor body\_part(k, F1(k))$
S7: $(\forall f)((mammal(f) \land \sim furry(f)) \rightarrow whale(f))$	C9: $\sim mammal(f) \lor furry(f) \lor whale(f)$
S8: $(\forall w)(whale(w) \rightarrow mammal(w))$	C10: $\sim whale(w) \lor mammal(w)$

clause C6 of Table 15.4, we replaced the constant *Wally* with a variable, say *v*, the resulting clause would say that every single object is a whale. We consider why this would represent severe problems in our database shortly.

A further consideration is that we cannot replace constants with other constants. Consider again clause C6 of Table 15.4. If we were to replace *Wally*, by *Kate*, we would have the statement *whale(Kate)*. You can appreciate what is wrong with this when you consider the relationship of C6 to C2. If we were to substitute *Kate* for *y* in the latter (which is a sound thing to do, as *y* is universally quantified). We would then have in our database three statements, i.e.

whale(Kate)

kangaroo(Kate)

~whale(Kate) ∨ ~kangaroo(kate),

that *cannot* co-exist if the database is consistent. As you will recall, the clauses in our database are "anded" together. So effectively, we have a database that says:

Kate is a whale and Kate is a kangaroo and Kate cannot be both a Whale and a Kangaroo.

It is clear from the preceding discussion that in any clause, we can replace a variable by a constant (but *not* vice versa), providing that we replace the variable by that constant throughout the clause. We could, if we wish, replace a variable by another variable, as long as the new variable is not already in the clause.

Finally, we consider *functions*. Clauses C7 and C8, in Table 15.4, were derived from statement S6, which features an existentially quantified variable in the scope of a universally quantified variable. C7 and C8 thus contain a *Skolem function*, F1. A function merely denotes a value, i.e. the result of applying that function to a particular argument or sequence of arguments. Thus, the function call merely denotes a constant in the domain, and can thus be replaced by any constant from the domain. In fact, the actual rules for replacing functions in clauses are a little more complex than are those for variables and constants, and we do not need to be too concerned with them here. For the sake of completeness, we include them in the rules for unification given later (Table 15.5).

We can use unification, as so far discussed, to draw inferences on the basis of our clauses. Consider clauses C2 and C6 from Table 15.4. C6 tells us that "Wally is a whale" is a fact (i.e. is true). Now let us replace *y*, throughout C2, by the constant *Wally*. C2 now says

$$\sim whale(Wally) \lor \sim kangaroo(Wally).$$

Table 15.5. The rules for unifying two predicates, $p1$ and $p2$, and their arguments (i.e. two *atoms*)

Rule no.	Description
1	The predicate names of $p1$ and $p2$ must be the same
2	$p1$ and $p2$ must have the same number of arguments
3	If the arguments of $p1$ are $(x_1, x_2, \ldots, x_n)$, $n > 0$, and the arguments of $p2$ are $(y_1, y_2, \ldots, y_n)$, each x_i must unify with the corresponding y_i, for $1 \leqslant i \leqslant n$
4	Constants unify only with the same constant or a variable
5	Functions unify with functions of the same name (if the arguments all unify) or with variables
6	Variables unify with constants, functions or variables
7	Substitutions must be carried out so that the least specific objects (variables) are replaced by more specific objects (functions and constants)
8	Any substitution of one entity for another must replace each occurrence of that entity throughout the arguments of both $p1$ and $p2$
9	If we cannot find a list of substitutions that would result in the two atoms being the same, they cannot be unified

For this statement to be true (and it *must* be true as it is an axiom, and the variable in it was universally quantified), either or both of the "ored" conditions must be true. However, we know that $\sim whale(Wally)$ is false, since $whale(Wally)$ is one of our facts. Therefore, $\sim kangaroo(Wally)$ must be true, or the whole statement would be false. We have proved that Wally is not a kangaroo.

One way of looking at what we did in the above example is that we took two clauses, $C2$ and $C6$, both of which contained the same predicate (in this case the predicate *whale*). For the moment, we ignore the fact that one of them was negated. In $C2$ the predicate was $whale(y)$ and in $C6$ the predicate was $whale(Wally)$. So we found a substitution, i.e. replacing y by $Wally$, that *made the two predicates and their arguments exactly the same*. Our substitution conformed to the rules outlined above, since we replaced a variable, y, by a constant, $Wally$. In other words, we *unified* the two predicates. The substitution was then made throughout $C2$, i.e. we replaced *all* occurrences of y with $Wally$. The significance of all this will become apparent below.

Unification is a procedure for determining a series of substitutions that will make two predicates and their arguments textually the same. Two predicates can be unified if they have the same name and the arguments all unify. The rules for unification can be found in Table 15.5. Table 15.6 gives some examples of unification in action. In the parlance of Chapter 11, unification is *totally decidable*, i.e. there are algorithms to determine whether or not two predicates and their arguments can be unified, and also derive the set of substitutions required to effect this. We now turn to resolution.

15.3.2 How to Do Resolution

What follows is an intuitive discussion of the process of resolution. Let us return to the example from the preceding discussion, where we established that *Wally* is not a kangaroo. The two clauses from Table 15.4 that were used, $C2$ and $C6$, were such that:

- one clause contained a predicate which was the negation of a predicate in the other and
- the two predicates in question could be unified.

Table 15.6. Examples of unification in action

Expressions	List of substitutions made	Unified form
$p(x, T)$ $p(y, T)$	x/y	$p(y, T)$
$p(f\text{-}of(y), T)$ $p(f\text{-}of(g\text{-}of(x)), x)$	$y/g\text{-}of(x),$ x/T	$p(f\text{-}of(g\text{-}of(T)), T)$
$p(x, T)$ $p(S, y)$	$x/S,$ y/T	$p(S, T)$
$p(f\text{-}of(y), y)$ $p(f\text{-}of(g\text{-}of(T)), x)$	$y/g\text{-}of(T),$ $x/g\text{-}of(T)$	$p(f\text{-}of(g\text{-}of(T)), g\text{-}of(T))$
$p(S, T)$ $p(x, x)$	–	Cannot be unified

In a sense, the two unifiable predicates "cancelled each other out", leaving us with the fact that we inferred, i.e. $\sim kangaroo(Wally)$. This fact was inferred by the application of sound rules of inference, which means we could add it to our database as a new clause.

Now consider $C3$ and $C9$ (again, from Table 15.4). This time we can unify $mammal(f)$ with $mammal(M)$ to produce $mammal(M)$, and then replace f by M throughout $C9$. One of the $mammal$ predicates is the negation of the other, so they cancel each other out, as in our previous example, leaving:

$$furry(M) \lor whale(M).$$

Now, in general it is the case that the two selected clauses that contain *unifiable* predicates may each consist of several predicates "ored" together, whereas in the examples immediately above, at least one of the chosen clauses contained only a single predicate. Consider what happens if we use $C2$ and $C9$ as the two clauses containing unifiable predicates (in this case, the *whale* predicate). To unify $whale(f)$ and $whale(y)$ we can replace f by y or y by f; it does not matter (since neither variable appears in the other clause). Let us replace f by y, which means we must do this throughout $C9$. As before, the unified predicates cancel each other out, as one is the negation of the other. However, this time we are left with:

$$\sim kangaroo(y) \lor \sim mammal(y) \lor furry(y).$$

As you can probably see, the first predicate comes from $C2$, the other two from $C9$. This new clause could also be added to our database.

The process of taking two clauses containing unifiable predicates p_1 and p_2, where once they are unified p_1 is the negation of p_2, and creating a third clause after removing the unified predicates is called *resolution*. We say we are *resolving* the two clauses together, and the newly created clause is called the *resolvent* of the two initial clauses. Figure 15.1 provides a schematic overview of this process.

In resolving two clauses, we create a third clause, which is then added to our database. The existing clauses in the database *do not change*, but rather the resolvent is an *additional* clause; it is important to remember this. In this respect, resolution on clausal databases is *monotonic*, as was discussed in the context of classical FOPL reasoning in Chapter 14.

The above discussion confirms the earlier assertion that resolution can be used to draw inferences. However, this is not how we usually use resolution. Consider again our database in Table 15.4. Let us now tell a lie: we introduce a statement

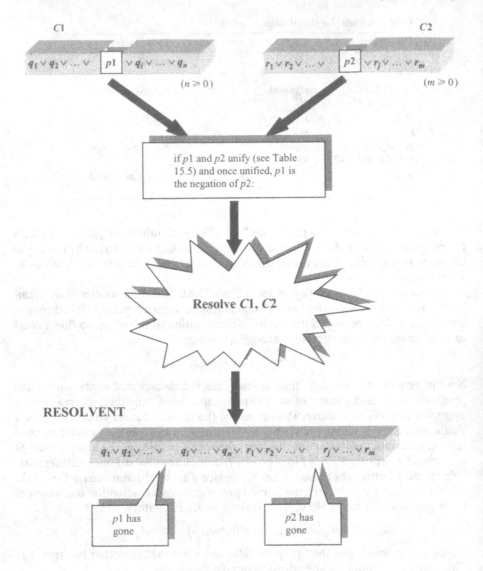

Figure 15.1. The process of *resolving* two clauses, *c1* and *c2*, and creating a *resolvent*

saying that Kate the kangaroo is bald. We enter this fact as clause $C11$ in our database (see Table 15.7). We will now carry out the sequence of resolutions specified in Table 15.8. Table 15.8 also shows the new clauses we must add to the database specified in Table 15.7. First, we resolve clauses $C11$ and $C9$, giving $C12$ (Table 15.8). Then, we resolve $C1$ and $C5$, giving $C13$, which we then resolve with $C12$, giving $C14$. $C2$ is then resolved with $C14$, resulting in $C15$. Now, what happens if we resolve $C5$ and $C15$? Before we consider this, let us be clear about the state of affairs that created this situation.

Starting with Table 15.7, we resolved pairs of clauses together, producing the resolvents as specified in Table 15.8, according to our rules for resolving clauses discussed above. However, we reached a situation wherein one clause ($C15$) was the

Table 15.7. Kate is now determined to be a bald kangaroo

FOPL statement	Clauses
S1: $(\forall x)(kangaroo(x) \rightarrow mammal(x))$	C1: $\sim kangaroo(x) \vee mammal(x)$
S2: $(\forall y)(whale(y) \rightarrow \sim kangaroo(y))$	C2: $\sim whale(y) \vee \sim kangaroo(y)$
S3: $(\exists z)(mammal(z) \wedge \sim furry(z))$	C3: $mammal(M)$
	C4: $\sim furry(M)$
S4: $kangaroo(Kate)$	C5: $kangaroo(Kate)$
S5: $whale(Wally)$	C6: $whale(Wally)$
S6: $(\forall k)(\exists p)(kangaroo(k) \rightarrow pouch(p) \wedge body\_part(k,p))$	C7: $\sim kangaroo(k) \vee pouch(F1(k))$
	C8: $\sim kangaroo(k) \vee body\_part(k, F1(k))$
S7: $(\forall f)((mammal(f) \wedge \sim furry(f)) \rightarrow whale(f))$	C9: $\sim mammal(f) \vee furry(f) \vee whale(f)$
S8: $(\forall w)(whale(w) \rightarrow mammal(w))$	C10: $\sim whale(w) \vee mammal(w)$
	C11: $\sim furry(Kate)$

negation of another ($C5$). Both clauses consist of only one predicate. This actually means that our database is *inconsistent*. Remember that *all of the clauses in conjunctive normal form database are "anded" together*. Here, then we have the following state of affairs:

$$C_1 \wedge C_2 \wedge \ldots \wedge C_n$$

where two of the clauses, say C_i and C_j, are such that

$$C_i = p \quad and \quad C_j = \sim p.$$

$p \wedge \sim p$ is *always false*. In a statement made up of an "anded" together series of clauses, if any pair of the clauses logically contradict each other, the *whole statement* is *false*.

Examination of the database reveals the problem. In this case, our representation is not sophisticated enough to deal with the fact that a kangaroo could be bald. Our database says that a non-furry mammal *must* be a whale, that a kangaroo is a mammal, and that Kate is a kangaroo. From this we can infer that Kate is a mammal, but this, along with our lie that Kate is not furry implies that Kate is a whale. But our rules tell us that an object cannot be both a kangaroo and a whale: hence the contradiction. However, our database was not inconsistent until we added clause $C11$ to it (Table 15.7). This implies that if we had added the negation of clause $C11$, i.e. *furry(Kate)*, instead of $C11$, our database would have remained consistent.

The immediately preceding discussion tells us something very important. For what we have really done is to prove that Kate is actually furry. Our "lie" represented by $C11$, is really the negation of a statement which is a *logical consequence* of our database. When added to the database and subjected to the processes of resolution combined with unification, our lie led to a contradiction. Thus, the *negation of our lie is true*.

Table 15.8. The sequence of resolutions carried out on the database in Table 15.7

Clauses resolved	Resolvent
C11, C9	C12: $\sim mammal(Kate) \vee whale(Kate)$
C1, C5	C13: $mammal(Kate)$
C12, C13	C14: $whale(Kate)$
C2, C14	C15: $\sim kangaroo(Kate)$
C5, C15	C16: ???

Resolution works as follows. We start with a consistent conjunctive normal form database, D. We take a FOPL statement, S, that we intend to prove. We negate S, i.e. $\sim S$, convert it into conjunctive normal form, and add it to our database. We repeatedly resolve pairs of clauses, adding the resolvent to D each time we carry out a resolution. If we reach a situation where we get two clauses in our database consisting of only one predicate, where one clause is the negation of the other (once the two clauses are unified), then S is a true statement. In such a case, we say that the resolvent of the two clauses is the *empty clause*. Figure 15.2 is a schematic overview of the process of resolution.

Resolution is thus a form of proof by *reductio ad absurdum*, a technique we have used in slightly different forms at various stages in this book. To prove something, we assume it is not true, and show that if we make this assumption we can, by using a logically sound chain of inference, derive a contradictory state of affairs. Thus, the negation of what we were trying to prove is false; the thing we are trying to prove is true.

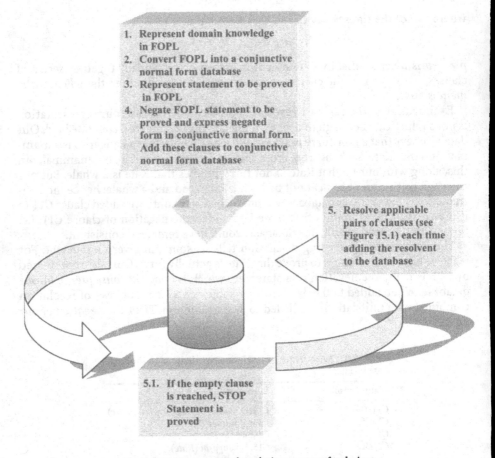

Figure 15.2. The process of resolution as a proof technique

Table 15.9. The database for the lazy drunken students problem, along with the conjunctive normal form of the negated statement to be proved (clauses $C3$ and $C4$)

Original Statement	Clauses
$(\forall x)((lazy(x) \wedge drinks(x)) \rightarrow fail(x))$	$C1: \sim lazy(x) \vee \sim drinks(x) \vee fail(x)$
$\sim fail(Fred)$	$C2: \sim fail(Fred)$
Negated statement to be proved:	
$\sim(\exists x)(drinks(x) \rightarrow \sim lazy(x))$	$C3: drinks(y)$
	$C4: lazy(z)$

To illustrate the process, we will now return to our *lazy drunken students* problem (Table 15.1). Remember that the statement we are trying to prove is:

$$(\exists x)(drinks(x) \rightarrow \sim lazy(x)).$$

As stated above, we must first negate this and express it as conjunctive normal form clauses. The negation is:

$$\sim(\exists x)(drinks(x) \rightarrow \sim lazy(x)).$$

This can be expressed as:

$$(\forall x) \sim (drinks(x) \rightarrow \sim lazy(x)).$$

We can simply remove the $\forall$:

$$\sim(drinks(x) \rightarrow \sim lazy(x)).$$

Since (see Table 14.3) $p \rightarrow q \leftrightarrow \sim p \vee q$, the above is equivalent to:

$$\sim(\sim drinks(x) \vee \sim lazy(x)).$$

One of de Morgan's laws (Table 14.3) tells us that $\sim(p \vee q) \leftrightarrow \sim p \wedge \sim q$, so we have

$$\sim\sim drinks(x) \wedge \sim\sim lazy(x),$$

which is:

$$drinks(x) \wedge lazy(x).$$

The final conjunctive normal form statement comprises two conjunctive normal form clauses. These are added to our database of Table 15.3, resulting in the database shown in Table 15.9. Notice that the variable names in the new clauses have been changed simply to avoid any confusion. Variables within any clause can be renamed as desired, as long as the renaming is done consistently throughout the clause, and the new name is not that of a variable that already features in the clause.

Table 15.10 shows a sequence of resolutions, and the resulting clauses, required to complete the proof. The negation of the statement to be proved, when added to our database, made our database inconsistent. Thus, the statement to be proved is true. There are, indeed, one or more students who, if they drink beer, are not lazy.

Table 15.10. The sequence of resolutions carried out on Table 15.9

Clauses resolved	Resolvent
$C3, C1$	$C5: \sim lazy(x) \vee fail(x)$
$C2, C5$	$C6: \sim lazy(Fred)$
$C4, C6$	$C7: <empty>$

It is important at this stage to emphasise two points about resolution, to avoid the development of misconceptions about the topic that I have observed to be sometimes held by students. As we have already noted, we resolve pairs of clauses until the "empty" clause is obtained (if it can be). Please take note of the following:

- It is <u>not</u> necessary to resolve all pairs of clauses together; only those necessary should be used.
- It is <u>not</u> essential to use all the clauses that represent the statement to be proved. In the lazy drunken students example, we did need to do this. However, in general, it may be that some of the clauses derived from our negated "statement to be proved" are redundant in terms of the proof process.

15.3.3 The Efficiency of Resolution

You may have noticed that resolution, as presented above, is not obviously any more efficient than the classical FOPL techniques of the previous chapter. Certainly, *monotonicity* (i.e. that each resolution of two clauses yields an additional clause for the database) is just as much of a problem with resolution as it is in classical reasoning. The same also applies if we carry out the process by simply attempting every possible resolution of applicable pairs of clauses at each stage. There are a number of strategies that can make resolution more efficient. Here, we will merely outline one of these here.

Since our initial database is consistent, it is the addition of the clauses derived from the statement we are trying to prove that result in the database becoming inconsistent. Therefore, it seems pertinent to ensure that such clauses feature as much as is possible in the procedure. A *control strategy* that reflects this assumption is known as the *set of support* strategy. At each stage, we attempt to use the added clauses (and the resolvents derived from these) as one or both of the two clauses to be resolved. Of course, we may not be able to do this at all stages. In such cases, we simply choose one of the other clauses.

A refinement to the set of support strategy also suggests that at each stage, we also try to ensure that one of our clauses consists of a single (possibly negated) predicate. This is known as the *unit preference* strategy. It is obvious why this is useful. Resolving two clauses in such cases always results in a resolvent with one less component predicate than the longer of the two clauses we are using.

As indicated earlier, when the above strategies cannot be applied, exhaustive (but systematic) sequences of resolutions are attempted. Thus, the strategies may help to make the process more efficient, but will not impede the ability of the resolution procedure to arrive at a proof, if one exists. In mathematical parlance, they do not affect the *completeness* of the resolution procedure.

We now turn to a brief discussion of the theoretical background of resolution. You may omit the section, if you are already convinced of the logical validity of the technique.

15.3.4 Why Resolution Works

In resolving two clauses C_i and C_j, we seek a predicate, p_i in C_i and a predicate p_j in C_j, such that p_i and p_j have the same predicate name, are unifiable, and where one is the negation of the other. We then create a new clause, C_R (which may be empty) by collecting together all predicates from the two clauses apart from p_i and p_j, and, if there are more than one such clauses, "oring" them all together. C_R is called the *resolvent* of C_i and C_j.

Resolution works on the principle that

$$((p_1 \vee p_2 \vee \ldots \vee p_n) \wedge (\sim p_n \vee p_{n+1} \vee \ldots \vee p_{n+m}))$$
$$\rightarrow p_1 \vee p_2 \vee \ldots \vee p_{n-1} \vee p_{n+1} \vee \ldots \vee p_{n+m},$$

where:

$$C_i = p_1 \vee p_2 \vee \ldots \vee p_n,$$
$$C_j = \sim p_n \vee p_{n+1} \vee \ldots \vee p_{n+m},$$
$$p_i = p_n,$$
$$p_j = \sim p_n, \quad \text{and}$$
$$C_R = p_1 \vee p_2 \vee \ldots \vee p_{n-1} \vee p_{n+1} \vee \ldots \vee p_{n+m}$$

If $C_i \wedge C_j$ is a true statement, then both sides of the $\wedge$ must be true. One of p_n, $\sim p_n$ must be false. The truth of the $\wedge$ thus depends on one or more of $p_1, p_2 \ldots p_{n-1}$ and/or one or more of $p_{n+1} \vee \ldots \vee p_{n+m}$. Put another way, the truth of $p_1 \vee p_2 \vee \ldots \vee p_{n-1} \vee p_{n+1} \vee \ldots \vee p_{n+m}$, is a *necessary condition* (see Chapter 13) for the truth of $C_i \wedge C_j$. To appreciate this, consider that if $p_1 \vee p_2 \vee \ldots \vee p_{n-1} \vee p_{n+1} \vee \ldots \vee p_{n+m}$ is false, then we know that every single p_i is false, which would reduce the left-hand side of the implication to $p_n \wedge \sim p_n$, which is, by definition, false.

Resolution, then, is simply another application of *modus ponens* (Table 14.3). The rule above tells us that, given $C_i \wedge C_j$, we can independently assert C_R. Thus, resolution is logically sound.

15.4 Logic in Action

The first popular *logic programming* system, the PROLOG (*PRO*gramming in *LOG*ic) language was developed in the early 1970s, and is still widely used today. Though PROLOG does not really use resolution as we have discussed above, it does use unification. A PROLOG program (also known as a PROLOG *database*) is written as a set of logical assertions (*facts* and *rules*). To run a PROLOG program, one enters a sequence of *atoms* (recall that an atom is a predicate along with its arguments), called a *query*. The atoms are assumed to be conjoined (i.e. "anded" together). PROLOG then tries to prove that the whole sequence of atoms is true, according to the facts and rules in the PROLOG program. One strange thing about PROLOG is that its only real outputs, in logical terms, are the word "yes" – the query is true – or "no". However, if there are variables in the query, and the query is true if the variables are each given appropriate values, PROLOG will return such an appropriate value for each of the variables in the query. What is more, PROLOG has a built-in *backtracking* mechanism, so you can repeatedly ask PROLOG to find an alternative way of proving the same query (and, once again, each time the query is proved the appropriate values for the variables are returned).

To illustrate the operation of PROLOG, we will consider a simple example, a variation on the lazy drunken students problem considered above. In PROLOG, the FOPL statement $(\forall x)((lazy(x) \wedge drinks(x)) \rightarrow fail(x))$ would be represented (note that in PROLOG, variables begin with a capital letter, while constants begin with a lower case letter):

```
fail_exams(X) :- lazy(X), drinks(X).                    (R1)
```

R1 should be read as if *x* drinks and *x* is lazy then *x* fails his/her exams. In PROLOG terms, the rule means that one way to prove that *x* fails his/her exams is to prove that *x* drinks and *x* is lazy (a comma between predicates is the PROLOG way of expressing ∧). PROLOG allows us to have as many rules with the same left-hand side as we like, for example:

```
lazy(Y) :- poor_attendance(Y).                          (R2)
lazy(Y) :- no_revision(Y).                              (R3)
lazy(Y) :- poor_coursework(Y).                          (R4)
```

There are thus three different ways to prove that someone is lazy (R2)–(R4). Now let us add some facts: .

```
no_revision(jim).                                       (F1)
drinks(jim).                                            (F2)
poor_coursework(jim).                                   (F3)
```

We could enter this query (the "?" is the PROLOG prompt):

```
?- fail_exam(jim).
```

To see if this is true, PROLOG would search for a fact that matched the query, or a rule with "fail exam" on its left-hand side. It searches downwards from the top of the database. In our case, (R1) is the only such rule. For PROLOG to apply a rule, the head of the rule must unify with the predicate in the query that is being considered. In this case, `fail_exam(X)` unifies with `fail_exam(jim)`, resulting in X being replaced by `jim` throughout the rule. PROLOG now has to try to prove `lazy(jim)` and then `drinks(jim)`. Note that, in a strictly logical sense, the order in which these two are proved is immaterial, but PROLOG always carries things out in this order. (R2) is tried first, but fails, as `poor_attendance(jim)` cannot be established. PROLOG does not give up when this happens, it just carries on to the next *lazy* rule (R3). This succeeds, since `no_revision(jim)` can be proved (it is fact (F1) in our database). Having proved the `lazy(jim)` part (R1), PROLOG now needs to try and prove `drinks(jim)`. This is (F2) in our database, so is proved. Thus the query assertion is true, and PROLOG answers "yes".

It can be seen from the above that what PROLOG actually does is to determine if the query represents a FOPL statement that is consistent with its database. That PROLOG has the built-in search capability indicated by the above example makes it a very powerful programming language. Of course, by Turing's thesis (Chapter 10), it is actually no more powerful than a Turing machine (TM) even if it is equipped with a potentially infinite memory. Hence, PROLOG is no more powerful than any other sufficiently powerful language. Nevertheless, PROLOG does demonstrate that logic can serve as the basis for computerised problem solving. It is worth learning to program in PROLOG partly because it seems to be based on a completely different *computational model* from most other languages. Yet, the wonder is that despite the seemingly dramatic differences between these computational models, we know that, by Turing's thesis, they are all fundamentally equivalent.

15.5 Languages, Machines, and Logic

In this book, we have explored the relationship between languages, computation, and logic. We have seen how formal logical systems are both linguistic and computational devices. We have seen how the distinction between syntax and semantics applies to languages and logic. We have considered the space and time complexity of language processing, computational, and logical reasoning systems.

In this chapter in particular, we have seen how logical reasoning can be expressed as a computational technique, and have briefly considered a programming language, PROLOG, that uses a form of FOPL as the basis of reasoning.

In Chapter 11, we discussed the unsolvability of the *halting problem*. This is essentially expressed as the inability of one sufficiently powerful computational system (i.e. a TM) to detect, in general, whether a computation being carried out by another such system will ever terminate. The way in which the result was established indicates that a system that is sufficiently powerful to carry out the *effective procedures* is also sufficiently powerful to incorporate *self-reference*. Moreover, the self-reference leads to *contradictions*, or absurdities.

In Chapter 11, we briefly mentioned the work of Kurt Gödel. Kurt Gödel was one of the most gifted logicians of all time. He was born early in the twentieth century in a part of what was then Austro-Hungary, and was later to become Czechoslovakia. However, he spent much of his life in the USA. In terms of our argument here, Gödel carried out his most significant work in the nineteen thirties, and it was a major influence on the work of Turing. We simplify things greatly here, in order to appreciate the main thrust of Gödel's result, now almost universally known as *Gödel's theorem*. Note the distinction between Turing's *thesis* (Chapter 10), i.e. an assertion that has yet to be disproved, and is almost certainly true, and Gödel's *theorem*, which was proved beyond doubt.

In the domain of FOPL, Gödel used coding schemes that are similar in spirit to those subsequently used by Turing to prove the unsolvability of the halting problem. He then showed that a consistent FOPL database could be used to generate a statement, P and its negation $\sim P$. Recall from the discussion earlier in this chapter that such a state of affairs was acceptable in the process of proof by resolution, since we *made* our database inconsistent by introducing the negation of what we assumed to be a true statement. However, for a supposedly consistent database to produce such a state of affairs means that the very system itself is *inconsistent*.

Once again, the problem is this. If the system is sufficiently powerful (and in Gödel's case, this actually means powerful enough to represent the basic axioms of *arithmetic*), it becomes *inconsistent*. If the system is not powerful enough, it is *incomplete* and cannot be used to prove some of the true statements in the domain.

It should be stated that the above is not meant to be a gloomy conclusion to this book. To me, it is part of the wonder of computational processes that such simple schemes can possess such power, and at the same time create such striking and insurmountable paradoxes. It is also just as exciting to see so many deceptively simple mechanical procedures that are practically effective only on the smallest sizes of input. Much of our daily work, in programming the digital computer, is devoted to finding ways around these obstacles. If this book has helped you to understand, even in part, from where such obstacles arise, then it will have achieved something.

We know from the world around us that theoretically limited as they may be, computers are highly useful in practical terms. Those of us who work with computers (which sometimes involves having to work against them!), can only benefit from an understanding of the intricate relationship between languages, machines and logic.

15.6 Exercises

For exercises marked †, solutions, partial solutions, or hints to get you started appear in "Solutions to Selected Exercises" at the rear of the book.

15.1[†]. Use resolution to solve the following problem (an exercise in classical reasoning from Chapter 14).

All people who are politicians are liars or cheats. Any married cheat is having an affair. Algernon Cholmondsley-St John is not having an affair, and he is not a liar. Show that if Algernon is married, he is not a politician.

15.2[†]. In Chapter 14, we considered the FOPL representation of the principle of *transitivity*:

$$(\forall x, y, z)((p(x, y) \land p(y, z)) \rightarrow p(x, z)).$$

Suppose *p* is interpreted as the arithmetic comparison operator $>$ (*greater than*). Given these facts:

$$24 > 3,$$
$$26 > 24,$$
$$3 > 1,$$

use resolution to prove that $26 > 1$.

15.3. Use the clauses in Table 15.4 to prove the following by resolution:

if Wilbur is a non-furry mammal, then he is a whale.

Then use classical reasoning, and the original FOPL statements (rather than the conjunctive normal form clauses) to prove the same statement.

15.4. Solve Exercise 14.6 of Chapter 14 using resolution.

Solutions to Selected Exercises

Chapter 2

2.1. (a) *Regular.* Every production has a lone non-terminal on its left-hand side, and the right-hand sides consist of either a single terminal, or a single terminal followed by a single non-terminal.

   (b) *Context free.* Every production has a lone non-terminal on its left-hand side. The right-hand sides consist of arbitrary mixtures of terminals and/or non-terminals one of which (*aAbb*) does not conform to the pattern for regular right-hand sides.

2.2. (a) $\{a^{2i}b^{3j}: i, j \geqslant 1\}$ i.e. *a*s followed by *b*s, any number of *a*s divisible by two, any number of *b*s divisible by three

   (b) $\{a^i a^j b^{2j}: i \geqslant 0, j \geqslant 1\}$ i.e. zero or more *a*s, followed by one or more *a*s followed by twice as many *b*s

   (c) { } i.e. the grammar generates no strings at all, as no derivations beginning with *S* produce a terminal string

   (d) $\{\varepsilon\}$ i.e. the only string generated is the empty string.

2.3. *xyz*, where $x \in (N \cup T)^*, y \in N$, and $z \in (N \cup T)^*$

   The above translates into: "a possibly empty string of terminals and/or non-terminals, followed by a single non-terminal, followed by another possibly empty string of terminals and/or non-terminals".

2.5. For an alphabet A, A^* is the set of all strings that can be taken from A including the empty string, ε. A regular grammar to generate, say $\{a, b\}^*$ is

$$S \rightarrow \varepsilon \mid aS \mid bS.$$

ε is derived directly from *S*. Alternatively, we can derive *a* or *b* followed by *a* or *b* or ε (this last case terminates the derivation), the *a* or *b* from the last stage being followed by a or b or ε (last case again terminates the derivation), and so on ...

Generally, for any alphabet, $\{a_1, a_2, ..., a_n\}$ the grammar

$$S \rightarrow \varepsilon \mid a_1 S \mid a_2 S \mid ... \mid a_n S$$

is regular and can be similarly argued to generate $\{a_1, a_2, ..., a_n\}^*$.

2.7. (b) The following fragment of the BNF definition for Pascal, taken from Jensen and Wirth (1975), actually defines all Pascal expressions, not only Boolean expressions.

<expression> ::= <simple expression> | <simple expression>
   <relational operator> <simple expression>
<simple expression> ::= <term> | <sign> <term> |
   <simple expression> <adding operator> <term>
<adding operator> ::= + | − | or
<term> ::= <factor> | <term> <multiplying operator> <factor>
<multiplying operator> ::= * | / | div | mod | and
<factor> ::= <variable> | <unsigned constant> | (<expression>) |
   <function designator> | <set> | not <factor>
<unsigned constant> ::= <unsigned number> | <string> |
   <constant identifier> | nil
<function designator> ::= <function identifier> |
   <function identifier> (<actual parameter>
                          {, <actual parameter>})
<function identifier> ::= <identifier>
<variable> ::= <identifier>
<set> ::= [<element list>]
<element list> ::= <element> {, <element>} | <empty>
<empty> ::=

Note the use of the empty string to enable an empty <set> to be specified (see Chapter 5).

2.10. Given a finite set of strings, each string x, from the set can be generated by regular grammar productions as follows:

$$x = x_1 x_2 ... x_n, \quad n \geq 1$$
$$S \rightarrow x_1 X_1$$
$$X_1 \rightarrow x_2 X_2$$
$$\vdots$$
$$X_{n-1} \rightarrow x_n.$$

For each string, x, we make sure the non-terminals X_i, are unique (to avoid any derivations getting "crossed").

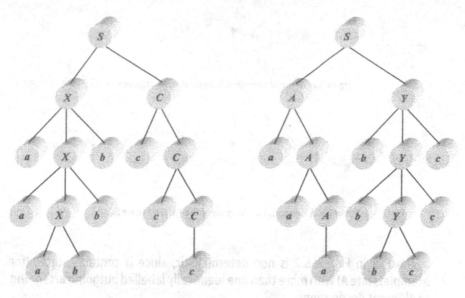

Figure S.1. Two derivation trees for the same sentence ($a^3b^3c^3$)

This applies to any finite set of strings, so any finite set of strings is a regular language.

Chapter 3

3.1. (b) $\{a^ib^jc^k: i, j, k \geqslant 1, i = j \text{ or } j = k\}$

    i.e. strings of *as* followed by *bs* followed by *cs*, where the number of *as* equals the number of *bs*, or the number of *bs* equals the number of *cs*, or both

  (c) Grammar *G* can be used to draw two different derivation trees for the sentence $a^3b^3c^3$, as shown in Figure S.1.

    The grammar is thus ambiguous.

3.2. (b) As for the Pascal example, the semantic implications should be discussed in terms of demonstrating that the same statement yields different results according to which derivation tree is chosen to represent its structure.

Chapter 4

4.1. (a) The FSR obtained directly from the productions of the grammar is shown in Figure S.2.

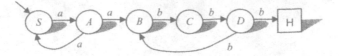

Figure S.2. A non-deterministic finite state recogniser

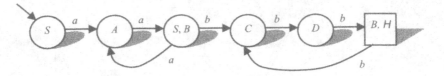

Figure S.3. A deterministic version of the FSR in Figure S.2

The FSR in Figure S.2 is non-deterministic, since it contains states (for example, state A) with more than one identically labelled outgoing arcs going to different destinations.

The deterministic version, derived using the subset method (null state removed) is in Figure S.3.

4.3. One possibility is to represent the FSR as a two-dimensional table (array), indexed according to (state, symbol) pairs. Table S.1 represents the FSR of Exercise 1(a) (Figure S.2).

In Table S.1, element (A, a), for example, represents the set of states $(\{S, B\})$ that can be directly reached from state A given the terminal a. Such a representation would be easy to create from the productions of a grammar that could be entered by the user, for example. The representation is also highly useful for creating the deterministic version. This version is also made more suitable if the language permits dynamic arrays (Pascal does not, but ADA, C, and Java are languages that do). The program also needs to keep details of which states are halt and start states.

An alternative scheme represents the FSR as a list of triples, each triple representing one arc in the machine. In languages such as Pascal or ADA, this

Table S.1. A tabular representation of the finite state recogniser in Figure S.2

States	Terminal symbols	
	a	b
S	A	–
A	S, B	–
B	–	C
C	–	D
D	–	B, H
H	–	–

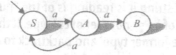

Figure S.4. Part of the finite state recogniser from Figure S.2

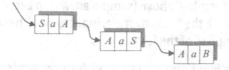

Figure S.5. The FSR fragment from Figure S.4 represented as a linked list of (state, arc symbol, next state) triples

scheme can be implemented in linked list form. For example, consider the part of the FSR from Exercise 4.1 (Figure S.2) shown in Figure S.4.

This can be represented as depicted in Figure S.5.

This representation is particularly useful when applying the reverse operation in the minimisation algorithm (the program simply exchanges the first and third elements in each triple). It is also a suitable representation for languages such as LISP or PROLOG, where the list of triples becomes a list of three element lists.

Since FSRs can be of arbitrary size, a true solution to the problem of defining an appropriate data structure would require dynamic data structures, even down to allowing an unlimited source of names. Although this is probably taking things to extremes, you should be aware when you are making such restrictions, and what their implications are.

Chapter 5

5.2. First of all, consider that a DPDR is not necessarily restricted to having one halt state. You design a machine, M, that enters a halt state after reading the first a, then remains in that state while reading any more as (pushing them on to the stack, to compare with the bs, if there are any). If there are no bs, M simply stops in that halt state, otherwise on reading the first b (and popping off an a) it makes a transition to another state where it can read only bs. The rest of M is an exact copy of M_3^d, of Chapter 5. If there were no bs, any as on the stack must remain there, even though M is accepting the string. Why can we not ensure that in this situation, M clears the as from its stack, but is still deterministic?

5.5. The reasons are similar to why arbitrary palindromic languages are not deterministic. When the machine reads the as and bs part of the string, it has no way of telling if the string it is reading is of the "number of *as* = number of *bs*", or the "number of *bs* = number of *cs*" type. It thus has to assume that the input string is of the former type, and backtrack to abandon this assumption if the string is not.

5.6. One possibility is to represent the PDR in a similar way to the list representation of the FSR described above (sample answer to Exercise 4.3, Chapter 4). In the case of the PDR, the "current state, input symbol, next state" triples would become "quintuples" of the form:

current state, input sym, pop sym, push string, new state

The program could read in a description of a PDR as a list (a file perhaps) of such "rules", along with details of which states were start and halt states.

The program would need an appropriate dynamic data structure (for example, linked list) to represent a stack. It may therefore be useful to design the stack and its operations first, as a separate exercise.

Having stored the rules, the program would then execute the algorithm in Table S.2.

As the PDR is deterministic, the program can assume that only one quintuple will be applicable at any stage, and can also halt its processing of invalid strings as soon as an applicable quintuple cannot be found. The non-deterministic machine is much more complex to model: I leave it to you to consider the details.

Table S.2. An algorithm to simulate the behaviour of a deterministic push down recogniser. The PDR is represented as quintuples

```
can-go := true
C := the start state of the PDR

while not(end-of-input) and can-go
    if there is a quintuple, Q, such that
        Q's current state = C, and
        Q's pop sym = the current symbol "on top" of the stack, and
        Q's read sym = the next symbol in the input
    then
        remove the top symbol from the stack
        set up ready to read the next symbol in the input
        push Q's push string onto the stack
        set C to be Q's next state
    else
        can-go := false
    endif
endwhile

if end-of-input and C is a halt state then
    return("yes")
else
    return("no")
endif
```

Chapter 6

6.3. Any FSR that has a loop on some path linking its start and halt states *in which there is one or more arcs not labelled with* ε recognises an infinite language.

6.4. In both cases, it is clear that the *v* part of the *uvw* form can consist only of *a*s or *b*s or *c*s. If this were not the case, we would end up with symbols out of their respective correct order. Then one simply argues that when the *v* is repeated the required numeric relationship between *a*s, *b*s, and *c*s is not maintained.

6.5. The language specified in this case is the set of all strings consisting of two copies of any string of *a*s and/or *b*s. To prove that it is not a CFL, it is useful to use the fact that we know we can find a *uvwxy* form for which $|vwx| \leqslant 2^n$, and *n* is the number of non-terminals in a Chomsky Normal Form grammar to generate our language. Let $k = 2^n$. There are many sentences in our language of the form $a^n b^n a^n b^n$, where $n > k$. Consider the *vwx* form as described immediately above. I leave it to you to complete the proof.

Chapter 7

7.1. This can be done by adding an arc labelled *x/x* (N) for each symbol, *x* in the alphabet of the particular machine (including the blank) from each of the halt states of the machine to a new halt state. The original halt states are then designated as non-halt states. The new machine reaches its single halt state leaving the tape/head configuration exactly as did the original machine.

Chapter 8

8.1. (The "or" FST). Assuming that the two input binary strings are of the same length, and are interleaved on the tape, a bitwise *or* FST is shown in Figure S.6.

8.2. An appropriate FST is depicted in Figure S.7.

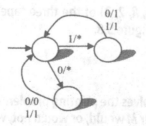

Figure S.6. A bitwise "or" finite state transducer. The two binary numbers are the same length, and interleaved when presented to the machine. The machine outputs "*" on the first digit of each pair

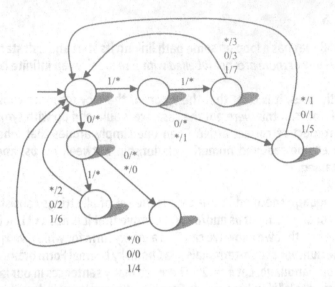

Figure S.7. A finite state transducer that converts a binary number into an octal number. The input number is presented to the machine in reverse, and terminated with "*". The answer is also output in reverse

Chapter 9

9.4. Functions are discussed in more detail in Chapter 11. A specification of a function describes what is computed and does not go into detail about how the computation is done. In this case, then, the TM computes the function:

$$f(y) = y \ div \ 2, \ y \geqslant 1.$$

9.5. (a) tape on entry to loop: $d1^{x+1}e$

tape on exit: $df^{x+1}e1^{x+1}$ i.e. the machine copies the $x + 1$ 1s between d and e to the right of e, replacing the original 1s by fs.

(b) $f(x) = 2x + 3$.

Chapter 10

10.3. The sextuple $(1, a, A, R, 2, 2)$ of the three tape machine M might be represented as shown in Figure S.8.

Chapter 11

11.1. Assume the TM, P, solves the printing problem by halting with output 1 or 0 according to whether M would, or would not, write the symbol s.

P would need to be able to solve the halting problem, or in cases where M was not going to halt P would not be able to write a 0, for "no".

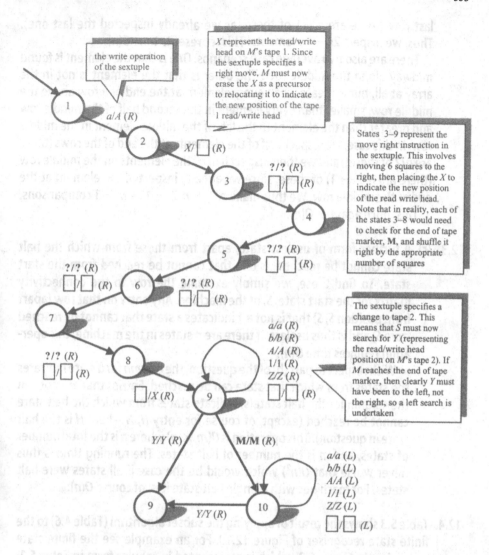

Figure S.8. A sketch of how a single tape TM, S, could model the sextuple (1, a, A, R, 2, 2) of a 3-tape machine, M, from Chapter 10. For how the 3 tapes are coded onto S's single tape, see Figures 10.21 and 10.22. This sequence of states applies when S is modelling state 1 of M and S's read write head is on the symbol representing the current symbol of M's tape 1

Chapter 12

12.1. There are two worst case scenarios. One is that the element we are looking
for is in position (n, 1), i.e. the leftmost element of the last row. The other is
that the element is not in the array at all, but is greater than every element
at the end of a row except the one at the end of the last row, and smaller
than every element on the last row. In both of these cases we inspect every
element at the end of a row (there are n of these), then every element on the

last row (there are $n-1$ of these, as we already inspected the last one). Thus, we inspect $2n-1$ elements. This represents time $O(n)$.

There are also two average case scenarios. One is that the element is found midway along the middle row. The other is that the element is not in the array at all, but is greater than every element at the end of a row above the middle row, smaller than every element in the second half of the middle row and greater than the element to the left of the middle element in the middle row. In this case, we inspect half of the elements at the end of the rows (there are $n/2$ of these), and we then inspect half of the elements on the middle row (there are $n/2-1$) of these, since we already inspected the element at the end of the middle row. We thus make $n/2 + n/2 - 1 = n - 1$ comparisons. This, once again, is $O(n)$.

12.3. (c) A further form of useless state, apart from those from which the halt state cannot be reached, is one that cannot be reached from the start state. To find these, we simply examine the row in the connectivity matrix for the start state, S, of the machine. Any entry on that row (apart from position S, S) that is not a 1 indicates a state that cannot be reached from S, and is thus useless. If there are n states in the machine, this operation requires time $O(n)$.

With respect to part (b) of the question, the *column* for a state indicates the states from which that state can be reached. Entries that are not 1 in the column for the halt state(s) indicate states from which the halt state cannot be reached (except, of course for entry H, H where H is the halt state in question). This operation is $O(m \times n)$ where n is the total number of states, and m is the number of halt states. The running time is thus never worse that $O(n^2)$ which would be the case if all states were halt states. For machines with a single halt state it is, of course $O(n)$.

12.4. Table S.3 shows the result of applying the subset algorithm (Table 4.6) to the finite state recogniser of Figure 12.10. For an example see the finite state recogniser in Figure S.2, which is represented in tabular form in Table S.1.

Table S.3. The deterministic version of the finite state recogniser from Figure 12.10, as produced by the subset algorithm of Chapter 4 (Table 4.6). There are eight states, which is the maximum number of states that can be created by the algorithm from a 3-state machine

States	Terminal symbols					
	a	b	c	d	e	f
1 (start state)	1_2	1_2_3	N	1_3	3	2
2	N	N	2_3	N	N	N
3 (halt state)	N	N	N	N	N	N
1_2	1_2	1_2_3	2_3	1_3	3	2
1_3 (halt state)	1_2	1_2_3	N	1_3	3	2
2_3 (halt state)	N	N	2_3	N	N	N
1_2_3 (halt state)	1_2	1_2_3	2_3	1_3	3	2
N (null state)	N	N	N	N	N	N

Here, the tabular form is used in preference to a diagram, as the machine has a rather complex structure when rendered pictorially.

Chapter 13

13.1. The *NOR* gate representation of the *and* operator is depicted in Figure S.9 (cf. de Morgan's law for the intersection of two sets, in Figure 6.7). The truth table in Table S.4 verifies the representation. Note the similarity between this representation and the *NAND* gate representation of the *or* operator given in Figure 13.5.

13.2. (a) Figure S.10 shows the engineer's circuit built up entirely from *NAND* gates. Note that we have taken advantage of one of de Morgan's laws (Table 13.12), to express $C + \sim D$ as $\sim(\sim C . D)$, i.e. $(\sim C$ NAND $D)$. This involves an intermediate stage of expressing $C + \sim D$ as $\sim\sim(C + \sim D)$ – rule 1 of Table 13.12.

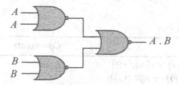

Figure S.9. The *NOR* gate representation of the Boolean and (.) operator

Table S.4. Verification by truth table that the *NOR* gate representation, $\sim(\sim(A + A) + \sim(B + B))$ is equivalent to $A . B$

A	B	(P) $A + A$	(R) $\sim P$	(Q) $B + B$	(S) $\sim Q$	(T) $R + S$	(Result 1) $\sim T$	(Result 2) $A . B$
0	0	0	1	0	1	1	0	0
0	1	0	1	1	0	1	0	0
1	0	1	0	0	1	1	0	0
1	1	1	0	1	0	0	1	1

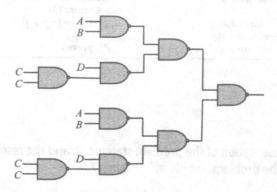

Figure S.10. The engineer's circuit of Chapter 13, i.e. $\sim(A . B) . (C + \sim D)$, using only *NAND* gates

Table S.5. Verification by truth table that $p \wedge p$ is equivalent to p

p	$p \wedge p$
0	0
1	1

13.5. (c) The truth table showing the equivalence of $(p \wedge p)$ with p is given in Table S.5.

The significance of this is that whenever we have a proposition of this form, for example $(s \rightarrow t) \wedge (s \rightarrow t)$, we can simply remove one of the equal parts, in this case giving $(s \rightarrow t)$.

Chapter 14

14.5. The representation of the problem statement, and the reasoning required to solve the problem, are given in Table S.6.

Table S.6. Classical reasoning in FOPL

FOPL	Comments
A1. $(\forall x) (politician(x) \rightarrow (liar(x) \vee cheat(x)))$	Axioms
A2. $(\forall y)((married(y) \wedge cheat(y)) \rightarrow affair(y))$	
A3. $\sim affair(Alg)$	
A4. $\sim liar(Alg)$	
P: $married(Alg) \rightarrow \sim politician(Alg)$	Statement to be proved
$politician(Alg) \rightarrow (liar(Alg) \vee cheat(Alg))$	From A1
$\sim politician(Alg) \vee (liar(Alg) \vee cheat(Alg))$	$(p \rightarrow q) \leftrightarrow (\sim p \vee q)$
$\sim politician(Alg) \vee cheat(Alg)$	(X) With A4: $p \vee 0 \leftrightarrow p$ (see Exercise 5(e), Chapter 13)
$politician(Alg) \rightarrow cheat(Alg)$	$(p \rightarrow q) \leftrightarrow (\sim p \vee q)$
$(married(Alg) \wedge cheat(Alg)) \rightarrow affair(Alg)$	From A2
$\sim(married(Alg) \wedge cheat(Alg))$	With A3: *modus tollens*
$\sim married(Alg) \vee \sim cheat(Alg)$	de Morgan's law
$cheat(Alg) \rightarrow \sim married(Alg)$	(Y) $(p \rightarrow q) \leftrightarrow (\sim p \vee q)$
$(politician(Alg) \rightarrow cheat(Alg)) \wedge$ $(cheat(Alg) \rightarrow \sim married(Alg))$	$X \wedge Y$
$politician(Alg) \rightarrow \sim married(Alg)$	Transitivity of implication (see Exercise 5(a), Chapter 13)
$\sim politician(Alg) \vee \sim married(Alg)$	$(p \rightarrow q) \leftrightarrow (\sim p \vee q)$
$\sim married(Alg) \vee \sim politician(Alg)$	
$married(Alg) \rightarrow \sim politician(Alg)$	P is proved

Chapter 15

15.1. The representation of the problem statement, and the resolutions required to solve the problem, are given in Table S.7.

Table S.7. Proof by resolution. This is the same problem as Exercise 14.5 of Chapter 14

Proof	Comments
A1. $(\forall x)(politician(x) \rightarrow (liar(x) \lor cheat(x)))$	Axioms
A2. $(\forall y)(married(y) \land cheat(y)) \rightarrow affair(y))$	
A3. $\sim affair(Alg)$	
A4. $\sim liar(Alg)$	
P: $married(Alg) \rightarrow \sim politician(Alg)$	Statement to be proved
C1. $\sim politician(x) \lor liar(x) \lor cheat(x)$	From A1
C2. $\sim married(y) \lor \sim cheat(y) \lor affair(y)$	From A2
C3. $\sim affair(Alg)$	A3
C4. $\sim liar(Alg)$	A4
C5. $married(Alg)$	From $\sim P$ (negated statement to be proved)
C6. $politician(Alg)$	
C7. $\sim politician(x) \lor liar(x) \lor \sim married(x) \lor affair (x)$	C1 and C2 resolved
C8. $liar(Alg) \lor \sim married(Alg) \lor affair(Alg)$	C6 and C7 resolved
C9. $liar(Alg) \lor affair(Alg)$	C8 and C5 resolved
C10. $affair(Alg)$	C9 and C4 resolved
C11. <empty>	C10 and C3 resolved. P is proved

15.2. The representation of the problem statement, and the resolutions required to solve the problem, are given in Table S.8.

Table S.8. Using the transitivity of the *greater than* relation to prove that $26 > 1$ by resolution

Proof	Comments
A1. $(\forall x, y, z)(p(x, y) \land p(y, z)) \rightarrow p(x, z))$	Axioms
A2. $p(24, 3)$	$24 > 3$
A3. $p(26, 24)$	$26 > 24$
A4. $p(3, 1)$	$3 > 1$
P. $p(26, 1)$	Statement to be proved
C1. $\sim p(x, y) \lor \sim p(y, z) \lor p(x, z)$	From A1
C2. $p(24, 3)$	A2
C3. $p(26, 24)$	A3
C4. $p(3, 1)$	A4
C5. $\sim p(26, 1)$	$\sim P$ (negated statement to be proved)
C6. $\sim p(24, z) \lor p(26, z)$	C1 and C3 resolved
C7. $p(26, 3)$	C6 and C2 resolved
C8. $\sim p(3, z) \lor p(26, z)$	C7 and C1 resolved
C9. $p(26, 1)$	C8 and C4 resolved
C10. <empty>	C9 and C5 resolved. P is proved

Further Reading

The following are some suggested titles for further reading. Notes accompany most of the items. Some of the titles refer to articles that describe practical applications of concepts from this book.

The numbers in parentheses in the notes refer to chapters in this book.

Church A. (1936) An Unsolvable Problem of Elementary Number Theory. American Journal of Mathematics, 58, 345–363.

Church and Turing were contemporaneously addressing the same problems by different, but equivalent means. Hence, in books such as Harel's we find references to the "Church–Turing thesis", rather than "Turing's thesis". (9–11).

Cohen D. I. A. (1996) Introduction to Computer Theory. John Wiley, New York. 2nd edition.

Covers some additional material such as regular expressions, Moore and Mealy machines (in this book our FSTs are Mealy machines) (8). Discusses relationship between multi-stack PDRs (5) and TMs.

Floyd R.W. and Beigel R. (1994) The Language of Machines: An Introduction to Computability and Formal Languages. W.H. Freeman, New York.

Includes a discussion of regular expressions (4), as used in the UNIX™ utility "egrep". Good example of formal treatment of minimisation of FSRs (using equivalence classes)(4).

Harel D. (1992) Algorithmics: The Spirit of Computing. Addison-Wesley, Reading, MA. 2nd edition.

Study of algorithms and their properties, such as complexity, big O running time (12) and decidability (11). Discusses application of finite state machines to modelling simple systems (8). Focuses on 'counter programs': simple programs in a hypothetical programming language.

Harrison M.A. (1978) Introduction to Formal Language Theory. Addison-Wesley, Reading, MA.

Many formal proofs and theorems. Contains much on closure properties of languages (6).

Hopcroft J.E. and Ullman J.D. (1979) Introduction to Automata Theory, Languages and Computation. Addison-Wesley, Reading, MA.

Discusses linear bounded TMs for context sensitive languages (11).

Jensen K. and Wirth N. (1975) Pascal User Manual and Report. Springer-Verlag, New York.

Contains the BNF and Syntax chart descriptions of the Pascal syntax (2). Also contains notes referring to the ambiguity in the "if" statement (3).

Kain R.Y. (1972) Automata Theory: Machines and Languages. McGraw-Hill, New York.

Formal treatment. Develops Turing machines before going on to the other abstract machines. Discusses non-standard PDRs (5) applied to context sensitive languages.

Minsky M.L. (1967) Computation: Finite and Infinite Machines. Prentice Hall, Englewood Cliffs, NJ.

A classic text, devoted to an investigation into effective procedures (11). Very detailed on most aspects of computer science. Of particular relevance is description of Shannon's 2-state TM result (12), and reference to unsolvable problems (11). The proof we use in this book to show that FSTs cannot perform arbitrary multiplication (8) is based on Minsky's.

Murdocca M. (2000) Principles of Computer Architecture. Addison Wesley, Reading, MA.

Computer architecture books usually provide useful material on Boolean logic and its application in digital logic circuits (13). This book also has sections on reduction of logical circuits.

Kelley D. (1998) Automata and Formal Languages: An Introduction. Prentice Hall, London.

Covers most of the introductory material on regular (4) and context free (5) languages, also has chapters on Turing machine language processing (7), decidability (11) and computational complexity (12).

Post E. (1936) Finite Combinatory Processes – Formulation 1. Journal of Symbolic Logic, 1, 103–105.
  Post formulated a simple abstract string manipulation machine at the same time as did Turing (9).
  Cohen (see above) devotes a chapter to these "Post" machines.

Rayward-Smith V.J. (1983) A First Course in Formal Language Theory. Blackwell, Oxford, UK.
  The notation and terminology for formal languages we use in this book is based on Rayward-Smith.
  Very formal treatment of regular languages (plus regular expressions), FSRs (4), and context free languages and PDRs (5). Includes Greibach normal form (as does Floyd and Beigel) an alternative CFG
  manipulation process to Chomsky Normal Form (5). Much material on top-down and bottom-up
  parsing (3), LL and LR grammars (5), but treatment very formal.

Rich E. and Knight K. (1991) Artificial Intelligence. McGraw-Hill, New York.
  Artificial intelligence makes much use of representations such as grammars and abstract machines.
  In particular, machines called recursive transition networks and augmented transition networks
  (equivalent to TMs) are used in natural language processing. AI books are usually good for learning
  about first order predicate logic and resolution (14–15), since AI practitioners are interested in using
  FOPL to solve real world reasoning problems.

Tanenbaum A.S. (1998) Computer Networks. Prentice-Hall, London. 3rd edition.
  Discusses FSTs (8) for modelling protocol machines (sender or receiver systems in computer networks).

Turing A. (1936) On Computable Numbers with an Application to the Entscheidungs problem.
  Proceedings of the London Mathematical Society, 42, 230–265.
  The paper in which Turing introduces his abstract machine, in terms of computable numbers rather
  than computable functions. Also includes his notion of a universal machine (10). A paper of remark-
  able contemporary applicability, considering that Turing was considering the human as computer,
  and not machines.

Winston P.H. (1992) Artificial Intelligence. Addison-Wesley, Reading, MA (see Rich), 3rd edition.

Wood D. (1987) Theory of Computation. John Wiley, Chichester, UK.
  Describes several extensions to PDRs (5). Introduction to proof methods, including the pigeonhole
  principle (also mentioned by Harel) on which both the repeat state theorem (6, 8) and the *uvwxy*
  theorem (6) are based.

Index